ADVANCES IN CHEMICAL PHYSICS

VOLUME XCII

Advances in
CHEMICAL PHYSICS

Edited by

I. PRIGOGINE

University of Brussels
Brussels, Belgium
and
University of Texas
Austin, Texas

and

STUART A. RICE

Department of Chemistry
and
The James Franck Institute
The University of Chicago
Chicago, Illinois

VOLUME XCII

AN INTERSCIENCE® PUBLICATION
JOHN WILEY & SONS, INC.
NEW YORK • CHICHESTER • BRISBANE • TORONTO • SINGAPORE

CONTRIBUTORS TO VOLUME XCII

PHIL ATTARD, Department of Physics, Faculty of Science, Australian National University, Canberra, Australia

R. HILFER, Institute of Physics, University of Oslo, Oslo, Norway, and Institut für Physik, Universität Mainz, Mainz, Germany

MARC T. M. KOPER, Department of Electrochemistry, Debye Institute, Utrecht University, Utrecht, The Netherlands

PIER LUIGI LUISI, Institut für Polymere, Zürich, Switzerland

INTRODUCTION

Few of us can any longer keep up with the flood of scientific literature, even in specialized subfields. Any attempt to do more and be broadly educated with respect to a large domain of science has the appearance of tilting at windmills. Yet the synthesis of ideas drawn from different subjects into new, powerful, general concepts is as valuable as ever, and the desire to remain educated persists in all scientists. This series, *Advances in Chemical Physics*, is devoted to helping the reader obtain general information about a wide variety of topics in chemical physics, a field that we interpret very broadly. Our intent is to have experts present comprehensive analyses of subjects of interest and to encourage the expression of individual points of view. We hope that this approach to the presentation of an overview of a subject will both stimulate new research and serve as a personalized learning text for beginners in a field.

I. Prigogine
Stuart A. Rice

CONTENTS

ELECTROLYTES AND THE ELECTRIC DOUBLE LAYER

PHIL ATTARD

Department of Physics, Faculty of Science, Australian National University, Canberra, ACT, 0200, Australia

CONTENTS

Advances in Chemical Physics, Volume XCII, Edited by I. Prigogine and Stuart A. Rice.
ISBN 0-471-14320-0 © 1996 John Wiley & Sons, Inc.

1. INTRODUCTION

A. Background

Particles in polar solvents usually develop surface charges, due either to dissociation of chemical groups on the surface, or to chemical binding or physical adsorption of ions from the electrolyte. This surface charge is balanced by an equal and opposite net charge of ions in the electrolyte. These counterions are those that have either dissociated from the surface, or they are the expartners of the adsorbed ions, since the electrolyte was originally electroneutral. Although definitely attracted to the oppositely charged surface, due to entropy these counterions remain dispersed and mobile in the solvent in the vicinity of the surface. This spatial separation of charge is termed the electric double layer.

The electric double layer affects the properties of individual particles, and also the interactions between them. Usually, two similarly charged particles in a solvent will experience a mutual repulsion, as one might expect, but the repulsion is not given simply by Coulomb's law because of the presence of the counterions in the double layer. At large separations there is no interaction because the "bare" charge on the particles is screened by the neutralizing counterions that surround them; at small separations their diffuse double layers begin to overlap and the ions in them must rearrange, which gives rise to a force. This electric double-layer repulsion is opposed by the van der Waals attraction that exists between all particles, and it is the precise balance between these two that is important. The dominance of one or the other, for example, primarily determines whether colloidal particles remain dispersed and mobile in the solvent, or whether they coagulate and float or precipitate out of solution. Similarly, the properties of individual particles or aggregates can depend on the nature of the surrounding double layer. For instance, membranes such as those forming biological cells, or those stabilizing bubbles or foams, are often comprised of amphiphilic molecules that have a solvophilic head and a solvophobic tail. These molecules typically spontaneously assemble to form a bilayer film with the tails in the interior away from the solvent, and the heads wet by the solvent. For polar solvents such as water, the head groups often have a net charge that is balanced by an electric double layer. It is the properties of the latter that determines whether the membrane forms in the first place, and if so, the geometry and curvature of the resultant structure.

These two examples—the stability of colloidal dispersions and the formation of bilayer membranes—are indicative of the widespread occurrence of the electric double layer. The approach and fusion of biological cells (conception, infection, and recognition), the tertiary conformation of proteins, and the transport of ions and other molecules across cell membranes, are all fundamental life processes involving the electric double layer at surfaces in an aqueous environment. The swelling of clays affects the drainage of soils in agriculture, the stability of building foundations, and the consistency of drilling muds. This swelling is directly determined by the double layer repulsion between the clay platelets. Technological applications include the adhesion of glues, inks, and paints, the formation and cohesion of ceramics, the propagation of cracks in metals and glasses, and the lubrication of surfaces in proximal motion. The ability to predict these phenomena, and to control and to use these systems, relies upon a quantitative understanding of the electric double layer.

Several experimental techniques have been developed that enable the double layer to be characterized in various systems. An essential requirement is that the method be surface specific, since one has to discriminate between the properties of the double layer and those of the bulk electrolyte; for most properties the former are swamped by the latter. Voltage and capacitance measurements give information about the double layer in the vicinity of the electrode and its dependence on the electrolyte. Electrophoresis measures the mobility of particles in an applied electric field, and this is related to the double layer potential near the surface. Electrolysis and electrodeposition experiments quantify ion transport in the double layer. Various spectroscopic, X-ray, and optical techniques can be surface specific, probing the state of the surface and the adjacent electrolyte. Neutron and X-ray scattering measures the correlations between submicroscopic particles, which is related to the interaction free energy due to the overlap of double layers between their surfaces. The interaction free energy can also be measured directly between macroscopic surfaces using force balance techniques, including the atomic force microscope, or by applying osmotic stress. Less directly it can be inferred from colloid sedimentation or flotation rates.

The task of theory is to provide a framework within which these measurements can be quantitatively interpreted, and to rationalize double layer phenomena more generally. Clearly if one is to describe so many diverse systems one must simplify the problem to its essential ingredients. One hopes first to deduce principles that hold in general, and hence to address exceptional or specific behavior by including more complicated effects as the need arises. The abstract representation of the electric double layer fundamentally consists of uniformly charged surfaces, either planar or with constant curvature, immersed in an electrolyte comprised of simple ions in a dielectric continuum. More sophisticated models of the particles include the effects of discrete surface charges (fixed or mobile), nonuniformly curved, flexible, or rough surfaces, and image charges due to the dielectric boundaries necessarily present. In dense suspensions the double layer interactions between the particles are not simply pairwise additive, as is assumed in the usual treatments. For the electrolyte, ion size and shape may become involved, as well as their polarizability. It may also become necessary to go beyond the continuum picture and to include the solvent as a distinct molecular species. These possibilities notwithstanding, the essence of the double layer consists of charged surfaces and mobile ions. It will be seen that this already provides a challenge for theory, and that a rich and varied behavior is encompassed by the minimal model.

The theoretical techniques that are applied to the electric double layer are those of classical statistical mechanics, which describes the submicroscopic behavior of atoms and molecules. The traditional approach is the Poisson–Boltzmann theory, which is a mean-field method that takes the density of ions in the diffuse layer near the charged surface to be proportional to the Boltzmann factor of the average electrostatic potential; the latter is related to the ion density by Poisson's equation. With the advent of computers, more sophisticated numerical treatments have prevailed. These are primarily Monte Carlo and molecular dynamics simulations, whose exactitude is limited only by the computer time and size, and integral equation theories, which though fundamentally approximate are very efficient. Whereas the simulations follow the motion of individual ions in the double layer, integral equations are based on mathematical relationships between the ion density profiles and the ion pair correlation functions, either those of the bulk electrolyte or those in the double layer. Density functional techniques are variational procedures that minimize the double layer free energy with respect to the ion density profiles; the free energy is approximated by an integral involving the bulk direct correlation function. The Ornstein–Zernike equation is the fundamental integral equation of statistical mechanics. This equation provides a relation between the direct and total correlation functions. It is solved by iteration, invoking some closure relation such as the hypernetted chain approximation. The pair correlation functions that appear in it are either those between ions in the presence of the particle (inhomogeneous integral equations), or they are between the particle and the ions (singlet integral equations). In the latter case the ion density profile of the double layer corresponds to the particle-ion distribution function function. These rather sophisticated numerical treatments have revealed not only quantitative deficiencies in the classic Poisson–Boltzmann approximation for the double layer, but even qualitatively different behavior, particularly at higher ionic strengths and/or higher surface charges.

As an example of a quantitative deficiency, the Poisson–Boltzmann theory can be fit to the numerical data for the potential drop across the double layer provided that one uses an *effective* surface charge rather than the actual surface charge. A rather surprising qualitative difference was the prediction, obtained by several different methods, that the force between two similarly charged surfaces can be *attractive*. There are two distinct regimes for this nonclassical behavior. At small separations the force changes from repulsive to attractive as the surface charge and electrolyte concentration is increased. This is the van der Waals regime, and the attractions are due to the electrostatic correlations between the ions confined between the surfaces; the effect is most pronounced for high ion couplings, which for aqueous electrolytes at room temperature means multivalent ions. Second, there is the large separation regime, where asymptotic analysis reveals an intimate connection between the interaction due to overlapping double layers, the double layer density profiles at an isolated surface, and the ion–ion correlation functions in the bulk electrolyte. At low concentrations the asymptotic interaction is monotone repulsive, as predicted by the Poisson–Boltzmann approximation; it has the same functional form as in the classical theory but with effective parameters, (e.g., decay length and surface charge), which can be expressed as integrals of bulk and particle correlation functions. At higher electrolyte couplings, the bulk radial distribution function becomes oscillatory, and so does the asymptotic force between two similarly charged surfaces; it is alternately attractive and repulsive with the same period and decay length of the bulk ion–ion correlation functions. These asymptotic and other analyses have rationalized the discrepancies between the classic Poisson–Boltzmann prediction and the recent rather sophisticated numerical treatments. These analyses have made it possible to interpret a number of novel and unusual experimental measurements. It is these newer theoretical approaches and analyses that are the subject of this chapter.

B. Review of Reviews

The original book of Verwey and Overbeek [1] remains the classic text on the electric double layer as the fundamental basis of colloid stability, and on its description by the Poisson–Boltzmann approximation (Gouy–Chapman theory). Two recent reviews in this series [2, 3] have focused on the isolated planar double layer. Carnie and Torrie [2] give a comprehensive coverage of theories for the primitive model electrolyte, including the modified Poisson–Boltzmann theory, the singlet Ornstein–Zernike approach, and comparison with simulations. Blum [3] extended the coverage to the inhomogeneous Ornstein–Zernike equation.

More specific reviews include that of Lozada–Cassou [4], who gives singlet hypernetted chain results for the primitive model electric double layer, and that of Patey and Torrie [5] who cover discrete solvent effects at a similar level of approximation. Reviews of the bulk Coulomb fluid include the mathematical treatments of sum rules by Martin [6], and of the soluble two-dimensional plasma by Alastuey [7]. The structure and dynamics of ionic liquids have been reviewed by Parrinello and Tosi [8].

C. Scope

This chapter concentrates on the theory of the equilibrium electric double layer, using the simplest model for the charged surfaces and the primitive model for the electrolyte. Regrettably, solvent effects and the application to experiment are discussed only superficially. The modern integral equation methods of statistical mechanics are emphasized, and the related analytic techniques. The double layer interactions between particles receive the most attention, although one theme concerns the relationship of these with the double layer of an isolated particle, and with the properties of the bulk electrolyte. This chapter is composed of several sections: a formally exact analysis of bulk electrolytes (Section II), analysis and asymptotics of the double layer (Section III), modern computational methods (Section IV) and numerical results (Section V), and elaborations of the basic model (Section VI). Sections II and III are somewhat mathematical, whereas in the remainder a more discursive approach is taken. This chapter concludes with a discussion of the relevance and application of these modern theories to experiment.

The primitive model electrolyte is defined in Section II, and the behavior of the ion correlation functions in the homogeneous electrolyte is analyzed in the remainder of Section II. The fact that these decay exponentially is established in Section II.B, along with the electroneutrality and moment conditions. In Section II.C it is shown that the classic Debye screening length has to be changed in order for the modified Debye–Hückel theory for finite sized ions to satisfy the Stillinger–Lovett second moment condition. This foreshadows the formally exact asymptotic analysis of Section II.D, where the actual screening length is determined by equations involving the short-range part of the correlation functions. Here it is shown that the Yukawa form for the ion pair correlation function, as given by Debye–Hückel theory, is formally exact asymptotically but with effective parameters. As the electrolyte coupling is increased three asymptotic regimes are identified: monotonic, charge oscillations, and density oscillations.

The connection with the electric double layer is made in Section III where the intimate relation between the properties of the bulk electrolyte and those of the double layer is described. For example, it is shown that the decay length of the ion profiles in the diffuse layer next to the charged surface is the same as the screening length of the bulk electrolyte (i.e., not the Debye length). Spherical solutes are treated in Section III.A and an isolated planar wall in Section III.B, where it is shown that asymptotically the double layer has linear Poisson–Boltzmann form but with an effective surface charge. The algebraic correlations between ions along a wall are also derived. In Section III.C the properties of interacting planar electric double layers are determined, and again the asymptote has a linear Poisson–Boltzmann form that involves the effective surface charge for the isolated surface. Section IV concludes with a summary of methods to treat the double layer in various geometries, including cylinders, pores, and dumbbells; the latter regards two interacting particles as a single solute.

The discussion of computational methods in Section IV focuses mainly on singlet integral equations, since arguably their combination of accuracy and computational efficiency makes feasible a reliable analysis of experimental data. The singlet hypernetted chain approximation is treated in Section IV.A, and methods for including bridge diagrams are discussed. Expressions for the solvation free energy are derived for this approximation, and this provides an alternative pathway to the contact theorem for the pressure. Inhomogeneous integral equations are mentioned in Section IV.B, a brief description of density functional methods can be found in Section IV.C, and simulations are covered in Section IV.D.

Section V provides some numerical results and comparisons for the planar double layer. The effective surface charge, which converts a fit made with the linear Poisson–Boltzmann approximation to the actual surface charge, is calculated in Section V.A, and the hypernetted chain approximation is compared with the analytic effective Poisson–Boltzmann approximation. In Section V.B the interaction pressure due to two walls is calculated in the wall–wall and dumbbell hypernetted chain approximations and tested against simulation and inhomogeneous integral equation results.

Section VI summarizes extensions of the basic model of the double layer that include different physical effects. Concentrated dispersions, discrete, variable, and rough surface charges, dielectric images, and molecular solvents, are all briefly discussed. This chapter concludes with a discussion of the relevance and implications for experiment. The focus is on the use of the simplest model of the double layer, and on the utility of the Poisson–Boltzmann theory in the light of the nonclassical behavior revealed by more advanced theories.

II. CORRELATION FUNCTIONS OF THE BULK ELECTROLYTE

The electric double layer is an inhomogeneous fluid in the sense that the ion densities vary in space in the vicinity of the charged particle or wall. This is in contrast to the bulk electrolyte where the ion densities are uniform and constant. Similarly, in the bulk the ion pair correlation functions are isotropic and depend only on the separation of the ions, whereas in the double layer they are not only anisotropic but also inhomogeneous. They depend on from three (for a spherical macroion, or for one or two planar walls, which have the highest symmetry) to six coordinates of the two ions.

Despite these qualitative differences, it turns out that the techniques used to analyze the electric double layer are virtually the same as those used for a bulk electrolyte, since what is common and essential to both is the long-ranged Coulomb potential. Moreover, it will be seen that the bulk electrolyte determines in a quantitative fashion many of the properties of the electric double layer, in particular the asymptotic decay of the density profiles and the interaction between overlapping double layers. Hence, this section is concerned with the analysis of the correlation functions of bulk electrolytes, and it is in the later sections that the methods and results are applied to the electric double layer. Section II.B the moment conditions are given, and it is proven that the pair correlation functions in the bulk electrolyte decay exponentially, Section II.C treats Debye–Hückel theory with a modification for the Debye screening length; and Section II.D is concerned with a formal analysis of the asymptotic behavior of the pair correlation functions, again in the bulk.

A. The Primitive Model

The minimal model of the double layer (uniformly curved and charged surfaces, and ions in a continuum solvent) first requires the specification of the electrolyte. In this chapter classical statistical mechanics will be used. In general, quantum mechanics affects small particles, such as photons and electrons, but the dynamics of atoms, molecules, and ions are governed by Newton's equations of motion. Thus classical statistical mechanics in general suffices to describe the behavior of most fluids; argon, which is one of the smaller atoms, in the liquid state has a quantum correction of the order of 1% for the internal energy and of the order of 10% for the pressure [9]. In classical statistical mechanics, the kinetic energy or momentum integrals are independent Gaussians, and hence their contribution to the free energy is trivial; it is the potential energy, which depends on the configurations of the particles, whose contribution is highly nontrivial.

An electrolyte, in general, consists of ions and solvent molecules. In most of this chapter the primitive model will be used. This model ignores the solvent molecules explicitly, and subsumes their effect into a continuum dielectric constant, ε. Accordingly, the ions interact via Coulomb's (Coul) law *in media*

$$u_{\alpha\gamma}^{\text{Coul}}(r) = \frac{q_\alpha q_\gamma}{\varepsilon r} \tag{2.1}$$

Here q_α is the charge on ions of species α, $\varepsilon = 4\pi\varepsilon_0\varepsilon_r$ is the total permittivity (ε_0 is the permittivity of free space, and ε_r is the relative dielectric constant of the medium), and r is the separation between the ions. This Coulomb potential is strictly an interaction free energy (cf. the temperature dependence of the dielectric constant), and corresponds to first the Born–Oppenheimer procedure for averaging out the electron contribution, and second the McMillan–Mayer representation, where the solvent coordinates have been integrated out of the problem. The "dielectric screening" of the ionic interaction by the solvent is valid in the limit of large separations; incorporating the solvent contributions via the dielectric constant is exact asymptotically. At intermediate separations one begins to see departures from Coulomb's law *in media*. For example, there is a repulsive cavity term due to exclusion of polarizable solvent by the ion that decays as r^{-4} [10]. There are also many-body contributions due to ionic polarizability that decay as r^{-6}, and hence the assumption of pairwise additivity is an approximation. In addition, at still closer separations one expects oscillatory behavior as the molecular size of the solvent becomes important [5], as well as solvent-induced many-body interactions. In the primitive model all of these effects are ignored, and the exact asymptote that is Coulomb's law *in media* is applied at all separations.

In addition to their charge, the primitive model recognizes that ions have size, which prevents them from overlapping. This excluded volume effect is incorporated into the model by adding to Coulomb's law a short-range repulsion, the most common choice being the hard-sphere (hs) potential,

$$u_{\alpha\gamma}^{hs}(r) = \begin{cases} \infty & r < d_{\alpha\gamma} \\ 0 & r > d_{\alpha\gamma} \end{cases} \tag{2.2}$$

where additive hard-sphere diameters are used, $d_{\alpha\gamma} = (d_\alpha + d_\gamma)/2$, d_α being the hard-sphere diameter of ions of type α. Again the hard-sphere repulsion is an approximation that roughly models the Pauli exclusion of electrons, which is what in reality prevents molecular overlap. It is true of course that $u(r) \to \infty$, $r \to 0$, but in fact the repulsion is softer than that used in the hard-sphere model (sometimes a r^{-12} form is used), and again there are three-body contributions [11]. The ion diameter that is used in the primitive model includes approximately the first solvation shell, since it is larger than the bare ion diameter of ionic crystals. A particularly simple version of the model is the so-called restricted primitive model, which is a binary symmetric electrolyte with all ions being the same size and the two species having equal but opposite charge.

B. Electroneutrality, Moments, and Screening

This section is concerned with establishing electroneutrality and the exponential screening of the correlation functions in the electrolyte. Most would take these facts to be self-evident, and, while Debye shielding has been rigorously established at vanishing concentration [12–16], there does not appear to be a more general proof of exponential behavior. Of course, one person's rigor is another's mortis, and hopefully the analysis below will convince rather than convict.

For a multicomponent fluid, the Ornstein–Zernike equation is [17]

$$h_{\alpha\gamma}(r) = c_{\alpha\gamma}(r) + \sum_{\lambda} \rho_{\lambda} \int h_{\alpha\lambda}(s) c_{\lambda\gamma}(|\mathbf{r} - \mathbf{s}|) \, d\mathbf{s} \qquad (2.3)$$

where h and c are the total and the direct correlation functions, ρ is the number density, and the Greek subscripts index the species. The total correlation function is directly related to the more familiar radial distribution function, $h_{\alpha\gamma}(r) = g_{\alpha\gamma}(r) - 1$; $g_{\alpha\gamma}(r)$ is proportional to the probability of finding ions of type α and γ at a separation r. This can be written in matrix form

$$\underline{\underline{H}}(r) = \underline{\underline{C}}(r) + \int \underline{\underline{H}}(s)\underline{\underline{C}}(|\mathbf{r} - \mathbf{s}|) \, d\mathbf{s} \qquad (2.4)$$

which factors in Fourier space

$$\underline{\underline{\hat{H}}}(k) = \underline{\underline{\hat{C}}}(k) + \underline{\underline{\hat{H}}}(k)\underline{\underline{\hat{C}}}(k) \qquad (2.5)$$

The symmetric matrices have components

$$\{\underline{\underline{H}}(r)\}_{\alpha\gamma} = \rho_{\alpha}^{1/2}\rho_{\gamma}^{1/2}h_{\alpha\gamma}(r) \qquad (2.6)$$

and

$$\{\underline{\underline{C}}(r)\}_{\alpha\gamma} = \rho_{\alpha}^{1/2}\rho_{\gamma}^{1/2}c_{\alpha\gamma}(r) \qquad (2.7)$$

As mentioned above, the long-range part of the pair potential in the primitive model electrolyte follows Coulomb's law *in media*, $u_{\alpha\gamma}^{\text{Coul}}(r) = q_{\alpha}q_{\gamma}/\varepsilon r$. Thus, one defines the dyadic matrix [18]

$$\underline{\underline{Q}} = \frac{4\pi\beta}{\varepsilon} \underline{q}\,\underline{q}^{T} \qquad (2.8)$$

where $\beta = 1/k_{B}T$ is the inverse of the thermal energy, (k_{B} is Boltzmann's constant and T is the absolute temperature), and where $\underline{q}^{T}$ is the row vector corresponding to the transpose of the column vector $\underline{q}$, which has components

$$\{\underline{q}\}_{\alpha} = \rho_{\alpha}^{1/2}q_{\alpha} \qquad (2.9)$$

The matrix $\underline{\underline{Q}}$ has a number of convenient properties. Its trace is related to the Debye length,

$$\text{Tr}\{\underline{\underline{Q}}\} = \frac{4\pi\beta}{\varepsilon} \underline{q}^T \underline{q} = \frac{4\pi\beta}{\varepsilon} \sum_\alpha \rho_\alpha q_\alpha^2 = \kappa_D^2 \qquad (2.10)$$

it is essentially idempotent,

$$\underline{\underline{Q}}^{n+1} = \kappa_D^{2n} \underline{\underline{Q}} \qquad (2.11)$$

and, although $\underline{\underline{Q}}$ itself is singular, $\text{Det}\{\underline{\underline{Q}}\} = 0$, one has

$$(\underline{\underline{I}} + \alpha\underline{\underline{Q}})^{-1} = \underline{\underline{I}} - \frac{\alpha}{1 + \alpha\kappa_D^2} \underline{\underline{Q}} \qquad (2.12)$$

1. Moment Conditions

All of the analysis below is based upon the fundamental assumption that the total correlation function is integrable,

$$\int h_{\alpha\gamma}(r) \, d\mathbf{r} \equiv \hat{h}_{\alpha\gamma}(0) < \infty \qquad (2.13)$$

where the circumflex denotes the three-dimensional radial Fourier transform. With the exception of the spinodal, it is axiomatic that this always holds. It will be shown below that the direct correlation function goes as $c_{\alpha\gamma}(r) \sim -\beta u_{\alpha\gamma}^{\text{Coul}}(r)$, $r \to \infty$. Hence, one defines

$$\chi_{\alpha\gamma}(r) \equiv c_{\alpha\gamma}(r) + \beta u_{\alpha\gamma}^{\text{Coul}}(r) \qquad (2.14)$$

where for the present $\chi_{\alpha\gamma}(r)$ is assumed to be of shorter range than the Coulomb potential. (Later it will be shown to be exponentially decaying.)

The Fourier transform of the Ornstein–Zernike equation, Eq. (2.5), may be written

$$\underline{\underline{\hat{H}}}(k) = \underline{\underline{\hat{\chi}}}(k) - \underline{\underline{Q}}k^{-2} + \underline{\underline{\hat{H}}}(k)\underline{\underline{\hat{\chi}}}(k) - \underline{\underline{\hat{H}}}(k)\underline{\underline{Q}}k^{-2} \qquad (2.15)$$

where the k^{-2} term is the transform of the Coulomb potential. Now in the limit $k \to 0$, $k^2\hat{\chi}(k) \to 0$, since $\chi(r)$ decays faster than the Coulomb potential. Hence, multiplying both sides by k^2 and taking the limit, in view of the integrability of the total correlation function, one must have

$$0 = -\underline{\underline{Q}} - \underline{\underline{\hat{H}}}(0)\underline{\underline{Q}} \qquad (2.16)$$

Explicitly, this is the electroneutrality condition

$$q_\alpha = -\sum_\gamma \rho_\gamma q_\gamma \int h_{\alpha\gamma}(r)\, d\mathbf{r} \qquad (2.17)$$

which expresses the fact that each ion is surrounded by a cloud of ions bearing a net equal and opposite charge. Note that although $\underline{\underline{H}}$ and $\underline{\underline{C}}$ commute, the three matrices $\underline{\underline{H}}$, $\underline{\underline{\chi}}$, and $\underline{\underline{Q}}$ in general do not commute, except at $k = 0$.

The electroneutrality condition is also called the zeroth moment condition, and it is here appropriate to define the moments. In general, the transform of an integrable function possesses a small-k Taylor expansion; in the case of radial functions only even powers of k appear. One has

$$\hat{\underline{\underline{H}}}(k) \sim \underline{\underline{H}}^{(0)} + \underline{\underline{H}}^{(2)}k^2 + \underline{\underline{H}}^{(4)}k^4 + \cdots \qquad k \to 0 \qquad (2.18)$$

and similarly for $\hat{\underline{\underline{\chi}}}(k)$. This expression may be obtained by expanding the integrand of the Fourier transform, and the moments are defined as

$$\underline{\underline{H}}^{(2n)} = \frac{4\pi(-1)^n}{(2n+1)!} \int_0^\infty \underline{\underline{H}}(r) r^{2n+2}\, dr \qquad (2.19)$$

In general, only a finite number of moments exist; if $h(r) \sim r^{-\eta}$, $r \to \infty$, then the moment integral is divergent for $2n + 2 - \eta > -1$. All moments exist for an exponentially short-ranged function, and the task is to show that this is indeed the case for the correlation functions of the electrolyte.

The electroneutrality condition provides constraints on the sums of the zeroth moments of the total correlation functions; for an m-component electrolyte in the most general case there are $m(m+1)/2$ total correlation functions and m constraints. For the case of a binary electrolyte, the $\hat{h}_{\alpha\gamma}(0)$ can in fact be expressed in terms of the isothermal compressibility [8, 19–22],

$$\rho_+^{-1} + \hat{h}_{++}(0) = \rho_-^{-1} + \hat{h}_{--}(0) = \hat{h}_{+-}(0) = \frac{k_B T}{\rho}\left(\frac{\partial \rho}{\partial p}\right)_T \qquad (2.20)$$

where $\rho = \rho_+ + \rho_-$ and p is the pressure. This result holds only for the binary electrolyte; more generally the individual $\hat{h}_{\alpha\gamma}(0)$ are given as linear combinations of the molecular chemical potential derivatives of the molecular densities, since the individual ion chemical potentials are undetermined. The chemical potential derivative of ion distribution functions of a binary electrolyte have been given [23]. Suttorp and van Wonderen [24, 25] give results for an ionic mixture in the presence of a neutralizing background (multicomponent jellium). According to Kirkwood–Buff theory [26], the isothermal compressibility can also be expressed in terms of the direct correlation function

$$\beta \chi_T^{-1} \equiv \rho \left(\frac{\partial \beta p}{\partial \rho} \right)_T = \underline{\rho}^T [\underline{\underline{I}} - \underline{\underline{\hat{C}}}(0)] \underline{\rho} \tag{2.21}$$

where $\{\underline{\rho}\}_\alpha = \rho_\alpha^{1/2}$. Here the direct correlation function may be replaced by its short-ranged part because $\underline{\rho}^T \underline{q} = 0$. In the case of the binary electrolyte one has

$$\beta \chi_T^{-1} = \rho - \rho_+^2 \hat{\chi}_{++}(0) - 2\rho_+ \rho_- \hat{\chi}_{+-}(0) - \rho_-^2 \hat{\chi}_{--}(0) \tag{2.22}$$

and it does not appear possible to express the individual $\hat{\chi}_{\alpha\gamma}(0)$ in terms of measurable thermodynamic parameters.

The zeroth moment conditions arose from the equality of the coefficients of k^{-2} in the small-k Taylor expansion of the Fourier transform of the Ornstein–Zernike equation. It depended only on the long-range nature of the Coulomb potential, the assumption that the total correlation function is integrable, and the assumption that the direct correlation function decayed no slower than the Coulomb potential. The second moment condition uses the coefficients of k^0, and it is necessary to establish the integrability of $\chi(r)$, (i.e., the existence of its zeroth moment). At this stage this will simply be assumed; in Section II.B.2 a much stronger result will emerge, namely, that $\chi(r)$ is exponentially short ranged.

Using the (assumed) integrability of $\chi(r)$ and equating the coefficients of k^0 in the small k Taylor expansion of the Ornstein–Zernike equation one obtains

$$\underline{\underline{H}}^{(0)} = \underline{\underline{\chi}}^{(0)} + \underline{\underline{H}}^{(0)} \underline{\underline{\chi}}^{(0)} - \underline{\underline{H}}^{(2)} \underline{\underline{Q}} \tag{2.23}$$

Premultiplying by $\underline{\underline{Q}}$ gives

$$\underline{\underline{Q}}\underline{\underline{H}}^{(0)} = \underline{\underline{Q}}\underline{\underline{\chi}}^{(0)} + \underline{\underline{Q}}\underline{\underline{H}}^{(0)} \underline{\underline{\chi}}^{(0)} - \underline{\underline{Q}}\underline{\underline{H}}^{(2)} \underline{\underline{Q}} \tag{2.24}$$

or, using Eq. (2.16),

$$-\underline{\underline{Q}} = -\underline{\underline{Q}}\underline{\underline{H}}^{(2)} \underline{\underline{Q}} \tag{2.25}$$

Explicitly, this is,

$$1 = \frac{-4\pi\beta}{6\varepsilon} \sum_{\gamma\lambda} q_\gamma q_\lambda \rho_\gamma \rho_\lambda \int h_{\gamma\lambda}(r) r^2 \, d\mathbf{r} \qquad (2.26)$$

This is the second moment condition, which was first given by Stillinger and Lovett [27, 28]. The current derivation that equates Fourier coefficients is due to Mitchell et al. [29].

Outhwaite [30] showed that the second moment condition can be written as a condition on the zeroth moment of the mean electrostatic potential about an ion. The latter is also called the fluctuation potential and is

$$\psi_\alpha(r) = \frac{q_\alpha}{\varepsilon r} + \sum_\gamma \rho_\gamma \int h_{\alpha\gamma}(s) \frac{q_\gamma}{\varepsilon |\mathbf{r} - \mathbf{s}|} \, d\mathbf{s} \qquad (2.27)$$

in terms of which the Ornstein–Zernike equation may be rewritten

$$\underline{\underline{H}}(r) = -\beta \underline{\psi}(r) \underline{q}^T + \underline{\underline{\chi}}(r) + \int \underline{\underline{H}}(s) \underline{\underline{\chi}}(|\mathbf{r} - \mathbf{s}|) \, d\mathbf{s} \qquad (2.28)$$

where $\{\underline{\psi}(r)\}_\alpha = \rho_\alpha^{1/2} \psi_\alpha(r)$. Premultiplying both sides by $\underline{Q}$, integrating over $\mathbf{r}$, and using the electroneutrality condition, one obtains

$$1 = \beta \underline{q}^T \hat{\underline{\psi}}(0) \qquad (2.29)$$

or

$$1 = \beta \sum_\alpha \rho_\alpha q_\alpha \int \psi_\alpha(r) \, d\mathbf{r} \qquad (2.30)$$

This may be shown to be equivalent to the second moment condition, Eq. (2.26), by using Poisson's equation (integrate twice by parts, and sum appropriate multiples of the two integrands to give the Laplacian in spherical coordinates [30]). This form of the condition, as a constraint on the total ion fluctuation potential, is in some ways more fundamental than the form involving the second moments of the total correlation functions; the simple form of the latter is peculiar to a uniform electrolyte, whereas the former holds in the electric double layer for arbitrary charge inhomogeneities [31].

For an m-component electrolyte there are m zeroth moment conditions but only one second moment condition; the latter provides a constraint on the total sum of the second moments of the ion pair correlation functions but does not determine the individual second moments. Like the electroneutrality condition, summing the second moments with weights equal to the ionic charges results in the cancelation of the terms involving the terms involving the $\hat{\chi}_{\alpha\gamma}(0)$, which accounts for the universal nature of the result; it is independent of any short-range interactions between the ions. Simply adding the correlation functions (i.e., without the ionic charge weights) results in various sum rules and fluctuation formulas that involve thermodynamic derivatives; these have been explicitly analysed for a multicomponent ionic mixture in the presence of a neutralizing background [24, 25].

The fourth moment of the total correlation function is important because it is related to the decay length of the electrolyte. In contrast to the zeroth and second moments, all the sum rules for the fourth moment are of a nonuniversal character. This can be illustrated by equating the coefficients of k^2 in the Ornstein–Zernike equation,

$$\underline{\underline{H}}^{(2)} = \underline{\underline{\chi}}^{(2)} + \underline{\underline{H}}^{(2)}\underline{\underline{\chi}}^{(0)} + \underline{\underline{H}}^{(0)}\underline{\underline{\chi}}^{(2)} - \underline{\underline{H}}^{(4)}\underline{\underline{Q}} \qquad (2.31)$$

The second moment of the shortened direct correlation function (whose existence at this stage is assumed), can be eliminated by premultiplying by $\underline{\underline{Q}}$ and using the zeroth moment condition,

$$\underline{\underline{Q}}\underline{\underline{H}}^{(2)} = \underline{\underline{Q}}\underline{\underline{H}}^{(2)}\underline{\underline{\chi}}^{(0)} - \underline{\underline{Q}}\underline{\underline{H}}^{(4)}\underline{\underline{Q}} \qquad (2.32)$$

Finally, the second moment equation can be used to relate the fourth moment of the total correlation function to the two zeroth moments,

$$\underline{\underline{Q}}\underline{\underline{H}}^{(4)}\underline{\underline{Q}} = (\underline{\underline{H}}^{(0)} - \underline{\underline{\chi}}^{(0)} - \underline{\underline{\chi}}^{(0)}\underline{\underline{H}}^{(0)})(\underline{\underline{I}} - \underline{\underline{\chi}}^{(0)}) \qquad (2.33)$$

Hence, summing the fourth moments with ionic charge weights gives a result dependent on the non-Coulombic interactions between the ions, and which in general cannot be simply expressed as thermodynamic properties of the electrolyte. In the case of the one component plasma, where fluctuations in charge are equivalent to fluctuations in number density, the fourth moment of the total correlation function is proportional to the isothermal compressibility, which to the same order equals the screening length of the plasma [32–35]; results for the multicomponent jellium have also been established [24, 25]. For the restricted primitive model (symmetric binary electrolyte), Eq. 2.33 reduces to

$$\kappa_D^4(h_{++}^{(4)} - h_{+-}^{(4)}) = \chi_{++}^{(0)} - \chi_{+-}^{(0)} - 1/(\rho_+ + \rho_-) \qquad (2.34)$$

As mentioned above, although the sum of the $\chi^{(0)}$ equals the isothermal compressibility, their difference has no physical interpretation. Hence, it is not possible to give the fourth moment of the total correlation function in terms of a measurable thermodynamic property. Within a linearized hydrodynamic approximation, retaining only terms to order k^4, the fourth moment of the charge–charge correlation function of the restricted primitive model has been identified with the decay length [8, 20, 21].

2. Exponential Screening

The above moment conditions relied upon the assumptions that the direct correlation function went like the negative of the Coulomb potential, and that the remainder was integrable. These assumptions will now be justified by establishing a stronger condition, namely, that either the short-range part of the direct correlation function decays as the square of the total correlation function, or it decays exponentially. This will be sufficient to show that the total correlation function decays exponentially. Rigorous mathematical bounds have established the Debye–Hückel theory as the zero coupling limit of the electrolyte, including the exponential Debye shielding of the total correlation function [12–16]. These works and the present analysis seek to establish exponential behavior from deeper principles, which is in contrast to most other studies where the exponential behavior is assumed from the outset (the clustering hypothesis, e.g., [6, 18, 36, 37]).

The argument is based upon the exact closure to the Ornstein–Zernike equation [17],

$$h_{\alpha\gamma}(r) = -1 + \exp[-\beta u_{\alpha\gamma}^{\mathrm{Coul}}(r) - c_{\alpha\gamma}(r) + h_{\alpha\gamma}(r) + d_{\alpha\gamma}(r)]$$

$$= -1 + \exp[h_{\alpha\gamma}(r) - \chi_{\alpha\gamma}(r) + d_{\alpha\gamma}(r)] \qquad (2.35)$$

where $d(r)$ is the bridge function. Now, since the left-hand side decays to zero for large r, [$h(r)$ is integrable by fiat], then so must the exponent, and in the asymptotic limit one linearizes the exponential to obtain

$$h_{\alpha\gamma}(r) \sim -\chi_{\alpha\gamma}(r) + h_{\alpha\gamma}(r) + d_{\alpha\gamma}(r) \qquad r \to \infty \qquad (2.36)$$

where the neglected terms decay as the square of the pair correlation functions.

If the total correlation function is exponentially decaying, which is to be proved, then it can be shown that both the direct correlation function and the bridge function are also exponentially decaying (see below). At this stage the contrary will be assumed, namely, that the total correlation function decays as in integrable power law, $h(r) \sim r^{-\eta}$, $\eta > 3$.

The bridge function consists of diagrams comprised of h bonds, and there are no nodal points between the root points [17]. This means that at least two h bonds must bridge between the root points, and the individual diagrams of $d(r)$ decay at least as fast as the square of $h(r)$ [38, 39]. This result is certainly true if $h(r)$ decays as an integrable power law because then the convolution integrals are dominated by regions with the field point close to one or other of the root points [because $(r/2)^{-2\eta} \ll r^{-\eta}$]. To apply this result to the bridge function itself, which contains an infinite sum of such binodal diagrams, it is sufficient to observe that the range of the individual binodal diagrams does not change as the number of field points increases, and that the trinodal and higher order diagrams decay faster than this. (This is in contrast to, e.g., the Gaussian fluid, where one can show for the series diagrams that the range of each diagram is proportional to the square root of the number of nodal points; the individual diagrams decay as Gaussians, but the infinite sum of them is renormalized to give exponential decay.) One concludes that the bridge function goes as

$$d_{\alpha\gamma}(r) \sim \mathcal{O} h_{\alpha\gamma}(r)^2 \qquad r \to \infty \qquad (2.37)$$

at least when $h(r) \sim r^{-\eta}$, $\eta > 3$. Strictly speaking, what appears on the right-hand side should be the most long ranged of the various $h(r)$. It will be shown later that in fact the total correlation functions between pairs of ions all have the same range; for the present the analysis may be interpreted as applying to the longest ranged of the correlation functions.

There then remains from the linearization of the closure only the short-range part of the direct correlation function [the $h_{\alpha\gamma}(r)$ being subtracted from both sides]. Evidently, this can decay no slower than the square of the total correlation function (because there is nothing left to cancel any more slowly decaying parts). One concludes that

$$\chi_{\alpha\gamma}(r) \sim \mathcal{O} h_{\alpha\gamma}(r)^2 \qquad r \to \infty \qquad (2.38)$$

which is sufficient to ensure that the zeroth and second moments of $\chi(r)$ exist, as was assumed above in the proof of the second and fourth moment conditions. For fluids that interact with integrable power law potentials, the coefficient of this term is just the second density derivative of the chemical potential [39]. This also completes the proof of the earlier assumption that

$$c_{\alpha\gamma}(r) \sim -\beta u_{\alpha\gamma}^{\text{Coul}}(r) \qquad r \to \infty \qquad (2.39)$$

This last result is rather well known and well used; the present derivation represents a detailed attempt at a proof (see also [39]). It is only in exceptional circumstances that the result is known not to hold. In fluids near the spinodal line or the critical point it may possibly be dominated by the square of the total correlation function [40]. In the case of what may be called infinitely short-ranged fluids (e.g., the hard-sphere fluid or the Gaussian fluid), both the total and the direct correlation functions exhibit exponential decay, which dominates this direct contribution from the pair potential.

The above shows that if $h(r)$ decays as an integrable power law, then $\chi(r) \sim \mathcal{O}h(r)^2$, $r \to \infty$. It is now shown that if $h(r)$ is exponentially decaying, then all of the moments of $\chi(r)$ exist, and hence it is also exponentially decaying. The proof is by induction. Assume that all of the $\underline{\underline{\chi}}^{(n)}$ exist for $n \leq 2m - 2$, where $m \geq 1$ is an integer. Since $h(r)$ is exponential (by present assumption) all of the $\underline{\underline{H}}^{(\ell)}$ exist. Now equate the coefficients of k^{2m} in the small-k Taylor expansion of the Ornstein–Zernike equation,

$$\underline{\underline{H}}^{(2m)} = \underline{\underline{\chi}}^{(2m)} + [\underline{\underline{H}}^{(2m)}\underline{\underline{\chi}}^{(0)} + \underline{\underline{H}}^{(2m-2)}\underline{\underline{\chi}}^{(2)} + \cdots + \underline{\underline{H}}^{(0)}\underline{\underline{\chi}}^{(2m)}] + \underline{\underline{H}}^{(2m+2)}\underline{\underline{Q}}$$

$$(2.40)$$

which gives $\underline{\underline{\chi}}^{(2m)}$ as the sum of a finite number of moments, which by assumption are themselves finite. Hence, if all the moments $\leq 2m - 2$ exist, then $|\underline{\underline{\chi}}^{(2m)}| < \infty$. [Note that the compressibility must be positive, $\underline{I} + \underline{\hat{H}}(0) > 0$.] The zeroth moment exists (by inspection), and hence by induction all moments of $\chi(r)$ exist. Thus if $h(r)$ decays exponentially, then so does $\chi(r)$. Hence, either all moments of $\chi(r)$ exist, or, if $h(r)$ decays as an integrable power law, then $\chi(r)$ has at least as many moments as $h(r)$ (since it decays as the square of the latter).

The existence of $\hat{h}(0)$ and the fact that $\chi(r)$ is either exponentially decaying or of no longer range than $h(r)$ are all that are required to establish the exponential decay of the total correlation function. The proof is again by induction on the moments. Assume that all moments, $\underline{\underline{H}}^{(n)}$, exist for $n \leq 2m$, where $m \geq 1$ is an integer. Since $\chi(r)$ is either exponential or at least as short ranged as $h(r)$, all of the $\underline{\underline{\chi}}^{(n)}$, $n \leq 2m$ also exist. Equation (2.40) gives $\underline{\underline{H}}^{(2m+2)}$ as the sum of a finite number of moments, which by assumption are themselves finite. Hence, if all the moments $\leq 2m$ exist, then $|\underline{\underline{H}}^{(2m+2)}| < \infty$. Since the zeroth moment exists by fundamental assumption, then, by induction, all moments of the total correlation function exist. This proves that the total correlation function must be at least exponentially decaying.

From the above the short-range part of the direct correlation function is also at least exponentially decaying, and, from Eq. (2.36), so is the bridge function. For the case of a simple fluid when $h(r)$ decays as an integrable power law, both $\chi(r)$ and $d(r)$ decay as its square (see above). For the present electrolyte with an exponentially decaying $h(r)$, $\chi(r) - d(r) \sim h(r)^2/2$, $r \to \infty$, but the two functions do not necessarily individually decay as the square of the total correlation function; Kjellander and Mitchell [37] assumed that $h(r)$ decayed exponentially and showed that at high electrolyte concentrations the decay lengths of $\chi(r)$ and of $d(r)$ may be larger than $\kappa^{-1}/2$ (but not as large as κ^{-1}), where κ^{-1} is the decay length of $h(r)$. In what follows, it will be assumed that $\chi(r)$ and $d(r)$ have a decay length strictly less than that of $h(r)$. The mathematical justification of this requires some detailed diagrammatic analysis, but the basic idea is that $\chi(r)$ is the multiply connected subset of diagrams of $h(r)$, and hence cannot be of longer range. This is perhaps clearest for the bridge function where there are always two or more h bonds connected in parallel between the root points, and hence $d(r)$ is more short-ranged than $h(r)$, and consequently so is $\chi(r)$.

This completes the proof that the total correlation function is exponentially decaying in a bulk electrolyte, where "exponential" covers both monotonic and damped sinusoidal behavior. Obviously, undamped sinusoidal behavior is precluded in a disordered fluid, and there is no evidence that the total correlation function decays faster than exponential. The fact that the total correlation function for an electrolyte is more short-ranged than the pair potential is in marked contrast to fluids with integrable power law potentials; for these the total correlation function decays at the same rate as the pair potential, with a coefficient proportional to the isothermal compressibility [38, 39, 41]. (Hence, the total correlation function is of exactly the same range as the full direct correlation function.) An analogous result holds in the case of the conditionally convergent dipolar fluid [42, 43]. For the case of fluids with infinitely short-ranged pair potentials (e.g., the Gaussian fluid or the hard-sphere fluid), the total correlation function is of longer range than the pair potential.

The preceding discussion considered only the effect of the long-range Coulomb potential, since one does not expect the short-range interactions between the ions to qualitatively change the behavior deduced above. (These additional interactions cannot be too ill-behaved because of the fundamental assumption of a finite compressibility.) Such short-ranged potentials are present in any realistic model of an electrolyte, and include the soft or hard core repulsion that prevents ion overlap, and also perhaps an r^{-6} dispersion attraction. One's expectation is that the charge–charge correlations will remain exponentially screened, and that the density–density correlations will decay in proportion to any power law potentials that are present [21, 41]. (In the context of the present proof, only a finite number of moments exist for these power law potentials.) For the case of an infinitely short-ranged potential such as the hard sphere, one expects there to be exponential decay of the density–density correlation function; at low density that decay length will be shorter than the electrostatic decay length, but at high packing it should be longer, and both types of correlations will be oscillatory. (The individual ion–ion correlation functions will all decay with the same decay length, namely, the longer one of the two.)

In addition to modifying the interactions between the ions, one can envisage adding solvent as a specific molecular species. The civilized model electrolyte includes a multipolar solvent, and, since the multipoles can be represented as a sum of discrete charges each of which is screened, then the solvent multipolar interactions themselves are screened. Hence, the ion–ion, ion–solvent, and solvent–solvent correlation functions should all decay exponentially [44–46]. However, if the ions themselves carry multipole moments, only the charge–charge correlations are exponentially screened; the nonspherical projections of the ion pair correlation functions (such as the dipole–dipole) decay as power laws [6]. In summary then, irrespective of the specific short-range interactions between ions, or the presence of solvent or other additives, in an electrolyte the long-range Coulomb potential can be expected to cause the charge–charge correlations to decay exponentially.

3. n Ions

The result for the pair correlation function is now extended to the many-body correlation functions and it is shown that these are also exponentially screened. This finding implies that the charge and multipole moments of an ion cluster are compensated by the surrounding electrolyte [47, 48]. There is nothing unexpected in these results; an electrolyte that screens the charge on a single ion will obviously screen the individual charges on a cluster, and at large distances either the net charge nor the multipole moments are visible.

Define $h_{\alpha^n}^{(n)}(\mathbf{r}^n)$ to be the n-body total correlation function for ions of type $\alpha_1, \alpha_2, \ldots, \alpha_n$ at $\mathbf{r}_1, \mathbf{r}_2, \ldots, \mathbf{r}_n$. Consider its behavior as $\mathbf{r}_n \to \infty$ (by this is meant $r_i/r_n \to 0$, $1 \leq i < n$). The function $h^{(n)}$ comprises the connected diagrams with n root points, and it may be split into those diagrams in which the root point at $\mathbf{r}_n$ hangs from a node (nodal diagrams) and those in which it does not (binodal and higher diagrams). By topological reduction, the bond between the root point and the node of the nodal diagrams is the pair total correlation function. Those nodal diagrams in which the node is also a root point go as $h(r)$. Denote by $\mathbf{r}_0$ the position of the field point node in the remaining nodal diagrams. In the asymptotic limit the dominant contribution to the integral occurs when the nodal field point is around the remaining ions at the origin, $r_0/r_n \to 0$. This is because the nodal point must be connected to the $n - 1$ remaining ions by at least binodal points due to the topological reduction, and stretching them is like stretching the bridge function. Hence, any diagrams with $r_0/r_n = \mathcal{O}(1)$ are of order $h(r)^2$ [more precisely, they have the same range as the pair bridge function, which as argued above is shorter than that of $h(r)$], whereas the diagrams with $r_0/r_n \to 0$ go like $h(r)$. In the multinodal diagrams h bonds also link the multinodal points to the root points. One concludes that the binodal and higher diagrams decay at least as fast as $h(r)^2$ as $\mathbf{r}_n \to \infty$. Hence, asymptotically the many-ion total correlation function is dominated by the mononodal diagrams, which decay as the pair total correlation function, $h_{\alpha^n}^{(n)}(\mathbf{r}^n) \sim \mathcal{O}h(r_n)$, $r_n \to \infty$, which from the above is exponentially decaying.

The many-ion total correlation function $h^{(n+1)}$ is the connected subset of the many-ion distribution function, $g_{\alpha^{n+1}}^{(n+1)}(\mathbf{r}^{n+1})$, which represents the probability of finding the $n + 1$ ions in that particular configuration in the electrolyte. One may single out the ion α_{n+1}, and the function $\rho_{\alpha_{n+1}} g^{(n+1)}/g^{(n)}$ may be interpreted as the density profile of ions of type α_{n+1} in the presence of an n ion. This consists of a uniform bulk electrolyte contribution, which arises from the diagrams in which the ion α_{n+1} is disconnected from the rest, and a distance-dependent contribution, which includes $h_{\alpha^{n+1}}^{(n+1)}(\mathbf{r}^{n+1})$, that consists of diagrams in which α_{n+1} is connected to one or more of the remaining n ions. By the same arguments as above, this latter contribution is exponentially decaying. Now, in general, the n ion has a net charge and multipole moments, which give rise to power law interaction potentials. If these charges or moments were not screened (i.e., neutralized or compensated) by the surrounding electrolyte, then the density would decay to its bulk value as a power law. Since the rate of decay is exponential, one concludes that the charge and the multipole moments of an ion cluster are perfectly screened by the surrounding electrolyte.

Martin [6] reviewed sum rules in electrolytes, using a somewhat more mathematical approach than here. These sum rules show that if the n-ion distribution function decays faster than $r^{-\ell}$, then the n ion plus the surrounding electrolyte has no multipole moments of order ℓ or less [6, 47, 48]. If the correlations are exponentially screened, as was shown above, then this gives rise to an infinite number of sum rules on the n-ion distribution function. The (ℓ, m) sum rule on the $(n + 1)$ ion distribution function is [6]

$$\sum_{\alpha_{n+1}} q_{\alpha_{n+1}} \int \left[\rho_{\alpha_{n+1}}(\mathbf{r}_{n+1}) \frac{g_{\alpha^{n+1}}^{(n+1)}(\mathbf{r}^{n+1})}{g_{\underline{\alpha}^n}^{(n)}(\mathbf{r}^n)} \right] |\mathbf{r}_{n+1}|^\ell Y_{\ell m}(\hat{\mathbf{r}}_{n+1}) \, d\mathbf{r}_{n+1}$$

$$= -Q_{\ell m}^{\text{ext}} - \sum_{j=1}^{n} q_{\alpha_j} |\mathbf{r}_j|^\ell Y_{\ell m}(\hat{\mathbf{r}}_j) \quad (2.41)$$

The bracketed part of the integrand represents the density of ions of type α_{n+1} in the presence of an n ion and an external field. ($Q^{\text{ext}} = 0$ corresponds to the bulk electrolyte with uniform density, $\rho_\alpha(\mathbf{r}) = \rho_\alpha$.) Since $Y_{\ell m}$ is the spherical harmonic, the left side represents the (ℓ, m) multipole moment of this inhomogeneous electrolyte, and the right side is the negative of the multipole moment of the n ion and of the external charge distribution. The case $\ell = m = 0$ yields the electroneutrality condition.

For the homogeneous electrolyte, $Q^{\text{ext}} = 0$, the $n = 0$ case gives the bulk electroneutrality condition,

$$\sum_{\alpha=1}^{m} q_\alpha \rho_\alpha \int_V d\mathbf{r} = V \sum_{\alpha=1}^{m} q_\alpha \rho_\alpha = 0 \quad (2.42)$$

which restricts the concentrations of the species in the electrolyte. The $n = 1$ case yields the zeroth moment condition given above

$$\sum_{\alpha=1}^{m} q_\alpha \rho_\alpha \int g_{\alpha\gamma}(r_{\alpha\gamma}) \, d\mathbf{r}_\alpha = \sum_{\alpha=1}^{m} q_\alpha \rho_\alpha \int h_{\alpha\gamma}(r_{\alpha\gamma}) \, d\mathbf{r}_\alpha = -q_\gamma \quad (2.43)$$

since $h_{\alpha\gamma}(r) = g_{\alpha\gamma}(r) - 1$.

For the electric double layer for $n = 0$, $\ell = m = 0$ one has

$$\sum_{\alpha=1}^{m} q_\alpha \int_V \rho_\alpha(\mathbf{r}) \, d\mathbf{r} = -Q^{\text{ext}} \quad (2.44)$$

and, for $n = 1$, $\ell = m = 0$,

$$\sum_{\alpha=1}^{m} q_\alpha \int \rho_\alpha(\mathbf{r}) g_{\alpha\gamma}(\mathbf{r}_\alpha, \mathbf{r}_\gamma) \, d\mathbf{r}_\alpha = -Q^{\text{ext}} - q_\gamma \quad (2.45a)$$

or, in terms of the inhomogeneous total correlation function,

$$h_{\alpha\gamma}(\mathbf{r}_\alpha, \mathbf{r}_\gamma) = g_{\alpha\gamma}(\mathbf{r}_\alpha, \mathbf{r}_\gamma) - 1,$$

$$\sum_{\alpha=1}^{m} q_\alpha \int \rho_\alpha(\mathbf{r}) h_{\alpha\gamma}(\mathbf{r}_\alpha, \mathbf{r}_\gamma)\, d\mathbf{r}_\alpha = -q_\gamma \qquad (2.45b)$$

This shows the local screening of each ion in the double layer.

C. Debye–Hückel Theory

The exponential screening of ion correlations in electrolytes is well known. And nearly as widespread is the use of the Debye length as the decay length, both in the electrolyte and in the double layer. In this section, a simple demonstration is given illustrating that even at the level of Debye–Hückel theory the screening length of the electrolyte is different from the Debye length.

The linearized Debye–Hückel theory, originally derived on the basis of the Poisson–Boltzmann approximation [49], may also be obtained by neglecting the short-ranged part of the direct correlation function,

$$\hat{\underline{\chi}}(k) = 0 \qquad (2.46)$$

Note that this is an approximation even for point ions. It is expected to become an increasingly better approximation as the coupling between the ions is reduced, for example, in monovalent electrolyte at low concentrations. The Ornstein–Zernike equation (2.5) becomes

$$\hat{\underline{H}}(k) = -(\underline{I} + \underline{Q}k^{-2})^{-1}\underline{Q}k^{-2}$$

$$= -\left(\underline{I} - \frac{1}{k^2 + \kappa_D^2}\underline{Q}\right)\underline{Q}k^{-2}$$

$$= \frac{-1}{k^2 + \kappa_D^2}\underline{Q} \qquad (2.47)$$

where Eqs. (2.11) and (2.12) have been used. This has inverse transform

$$\underline{H}(r) = \frac{-e^{-\kappa_D r}}{4\pi r}\underline{Q} \qquad (2.48)$$

or, in component form,

$$h_{\alpha\gamma}(r) = \frac{-\beta q_\alpha q_\gamma}{\varepsilon r} e^{-\kappa_D r} \qquad (2.49)$$

Hence, the Debye–Hückel theory predicts that the ion correlation functions are exponentially decaying, with the decay length being the Debye length, and with the amplitude proportional to the product of the charges on the ions.

The Debye–Hückel approximation only satisfies the electroneutrality condition if this exponential form holds for all of r, not just asymptotically. For ions all with a hard core of diameter d, $h_{\alpha\gamma}(r) = -1$, $r < d$. This means that the electroneutrality condition is no longer satisfied by applying Eq. (2.49) beyond the core. Retaining the Yukawa form, Eq. (2.49), but scaling the prefactor to satisfy electroneutrality, one obtains the modified linearized Debye–Hückel approximation,

$$
h_{\alpha\gamma}(r) = \begin{cases} -1 & r < d \\ \dfrac{-\beta q_\alpha q_\gamma e^{\kappa_D d}}{\varepsilon[1 + \kappa_D d]} \dfrac{e^{-\kappa_D r}}{r} & r > d \end{cases} \tag{2.50}
$$

A problem with this linearized theory is that if allows the co-ion radial distribution function to become negative. For this reason the Yukawa form is often applied instead to the potential of mean force,

$$
w_{\alpha\gamma}(r) = \frac{q_\alpha q_\gamma}{\varepsilon} \frac{e^{-\kappa_D r}}{r} \qquad r > d \tag{2.51}
$$

When exponentiated this gives the radial distribution function, $g_{\alpha\gamma}(r) = \exp - \beta w_{\alpha\gamma}(r)$. The justification for this lies in the traditional derivation of Debye–Hückel theory, which sets the potential of mean force equal to the charge times the mean electrostatic potential, and solves Poisson's equation for the latter by linearizing the radial distribution function. When the nonlinear terms are negligible, the linear and the nonlinear total correlation functions will be approximately equal and the theory could be said to be internally consistent. But, when the nonlinear terms make a contribution, it is arguably better to use the exponential of the potential of mean force even though this is not fully consistent with the linearization that was assumed to solve Poisson's equation. This nonlinear version has the same asymptotic form as the modified Debye–Hückel approximation, but experience shows that overall it tends to be a better approximation, in part because it always remains physical at small separations.

For the case when α is a macroion, the Debye–Hückel result is called the linear Poisson–Boltzmann approximation for an isolated spherical macroion, and $\rho_\gamma g_{\alpha\gamma}(r)$ represents the density profile for ions of type γ in the double layer about the macroion. It is not possible to obtain an analytic solution for the nonlinear Poisson–Boltzmann approximation in spherical geometry. If both α and γ are taken to be macroions, then the Debye–Hückel $w_{\alpha\gamma}(r)$ represents an approximation for their interaction. This, however, is not the same as solving the linear Poisson–Boltzmann equation for two interacting macroions, since the solution of the latter involves two-center Bessel function expansion [50–52]. The "linear" in linear Poisson–Boltzmann refers only to the local proportionality between the electrostatic potential and the density; the potential itself is clearly a nonlinear function of distance. The presence of a second macroion changes the boundary conditions such that the potential between two macroions is not strictly the pairwise sum of the potentials due to the isolated macroions (except as an approximation when one can ignore the presence of the second sphere in the boundary condition, possibly at asymptotic separations). This is an example of how the level of approximation depends on whether one regards both solutes as fixed and the source of a single external field, or whether one treats the second solute the same as the electrolyte and allows it to respond to the field due to the first solute. A similar situation has already arisen in the discussion of n ions, and will occur again in the treatment of interacting planar walls. Numerical solutions of the nonlinear Poisson–Boltzmann equation for two spheres have been given [53–55].

1. Self-Consistent Screening Length

The other problem with the (linear) modified Debye–Hückel result is that it does not obey the second moment condition. This was noted by Stillinger and Lovett [28], who pointed out that for finite-sized ions the modified Debye–Hückel result could only satisfy their condition by allowing for an effective screening length rather than the Debye length. That is,

$$
h_{\alpha\gamma}(r) = \begin{cases} -1 & r < d \\ \dfrac{-\beta q_\alpha q_\gamma e^{\kappa d}}{\varepsilon[1 + \kappa d]} \dfrac{\kappa^2}{\kappa_D^2} \dfrac{e^{-\kappa r}}{r} & r > d \end{cases} \tag{2.52}
$$

This satisfies the electroneutrality condition, and κ is determined by the second moment condition [18],

$$
1 = \frac{-4\pi\beta}{6\varepsilon} \sum_{\alpha\gamma} q_\alpha \rho_\alpha q_\gamma \rho_\gamma \int_d^\infty \frac{-\beta q_\alpha q_\gamma e^{\kappa d}}{\varepsilon[1 + \kappa d]} \frac{\kappa^2}{\kappa_D^2} \frac{e^{-\kappa r}}{r} 4\pi r^4 \, dr
$$

$$
= \frac{\kappa_D^2}{\kappa^2} \frac{1 + \kappa d + (\kappa d)^2/2 + (\kappa d)^3/6}{1 + \kappa d} \tag{2.53}
$$

The assumption of purely exponential profiles is only expected to be valid for small κd, and one expects this result to correct the Debye–Hückel theory in at least this regime. This expression only yields sensible results for $\kappa_D d \leq \sqrt{6}$, which is the upper bound established by Stillinger and Lovett [27] for monotonic ion correlations, $g_{++}(r) \leq g_{+-}(r)$, $r > d$, in the restricted primitive model. At higher concentrations than this the correlations *must* be oscillatory, although in practice the oscillations occur before the bound is reached.

For an m-component mixture there are in general $m(m + 1)/2$ independent total correlation functions. In Section II.D it will be shown that in the asymptotic regime the matrix of these is dyadic and that they have Yukawa form, as in the linear Debye–Hückel approximation. The present approximation is to apply the asymptote everywhere beyond the core, which means that one has only $m + 1$ unknowns (essentially the m effective charges and the decay length). By happy coincidence there are m electroneutrality conditions and one second moment condition, and hence in this approximation it is always possible to determine the decay length of a mixture of ions of arbitrary valence and diameter. (The equations given explicitly above only apply to ions of equal size.) It is also possible to implement a nonlinear version of this approximation, where it is the potential of mean force that is taken as the dyadic Yukawa beyond the core. One must now solve the $m + 1$ equations numerically, and one finds that this exponential approximation is more accurate for multiply charged ions than the analytic linear version given above.

Figure 1 shows the actual screening length of aqueous restricted primitive model electrolytes relative to the Debye length, and compares the self-consistent Debye–Hückel result given here with the results of the more sophisticated hypernetted chain approximation. Equation (2.53) relates the two-dimensionless parameters, κd and $\kappa_D d$, and hypernetted chain results confirm that the latter is an appropriate characteristic for monovalent electrolytes in this regime [18, 56]. In the case of divalents at nonvanishing concentrations the results do depend on the diameter of the ions [18, 56], and $\kappa_D d$ is less appropriate. From the figure it may be seen that the actual screening length at low concentrations is initially greater than the Debye length for divalents ($\kappa < \kappa_D$), whereas for monovalents it is mainly less than the Debye length ($\kappa > \kappa_D$). For divalents the linear self-consistent Debye–Hückel approximation, Eq. (2.53), is inaccurate, but the exponential version lies closer to the hypernetted chain results. Initially, the decay length decreases with increasing electrolyte concentration (greater screening), but there comes a point where the hypernetted chain results predict that the decay length increases with increasing concentration. For monovalents this occurs at about $\kappa_D d = 1.3$,

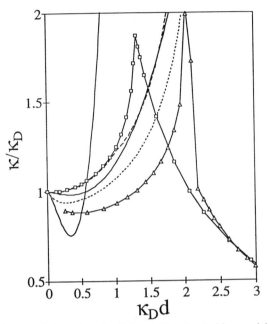

Figure 1. The actual decay length of the restricted primitive model in terms of the Debye length. The symbols are hypernetted chain calculations (squares are monovalent, $d = 4$ Å, triangles are divalent, $d = 4$ Å, data from [18]), the dashed line is the self-consistent linear Debye–Hückel result, Eq. (2.53), and the dotted line is the exponentiated version for divalents. The solid curves are the low-density expansion [37], Eq. (2.55), with the divalent having the deeper minimum. Here and in the remaining figures the temperature is $T = 300$ K and the relative permittivity is $\varepsilon_r = 78.5$, unless otherwise noted.

which is the transition from monotonic exponential decay to damped sinusoidal behavior, where the counterion and co-ion correlation functions oscillate out of phase. This will be discussed in detail in Section II.D, but beyond this there is another transition to a core-dominated regime where the ion correlation functions oscillate in phase. This can be seen in Fig. 1 at high concentrations where the divalent and monovalent results coincide.

There have been a number of results for the screening length of Coulomb fluids [8, 18, 20, 22, 36, 37, 56–59]. Of particular relevance to the present section are those based upon the linearized Poisson–Boltzmann equation [60, 61], and the analytic solution of the mean spherical approximation [62–64]. For the restricted primitive model the latter gives $\kappa d = -1 + \sqrt{(1 + 2\kappa_D d)} = \kappa_D d[1 - \kappa_D d/2 + \cdots]$. On the other hand, the expression given by Stillinger and Kirkwood [65], based upon a charge moment expansion at low concentrations goes like $\kappa d = \kappa_D d[1 + (\kappa_D d)^2/4 + \cdots]$, which to the exhibited order agrees with Eq. (2.53). Stell and Lebowitz [36] use a high-temperature–low-density expansion and obtain corrections to the Debye length in terms of the correlation functions of a reference hard-sphere fluid. For a symmetric electrolyte they give

$$\frac{\kappa^2}{\kappa_D^2} = 1 - \kappa_D^2 \int_0^\infty h^{(0)}(r) r \, dr + \cdots$$

$$= 1 + \kappa_D^2 d^2/2 + \cdots \qquad (2.54)$$

which again agrees with Eq. (2.53). Mitchell and co-workers [37, 56–59] also derived low-density expansions for the screening length; for a symmetric electrolyte they give [37]

$$\frac{\kappa}{\kappa_D} = 1 + \frac{\beta^2 q^4 \kappa_D^2}{12\varepsilon^2} \ln \frac{\beta e^2 \kappa_D}{\varepsilon} + \mathcal{O}(\kappa_D^2) \qquad (2.55)$$

In contrast to Eq. (2.53) and to the expansion of Stell and Lebowitz [36], this leading term in the expansion is independent of the ion diameter and approaches unity from below. In Fig. 1 it is compared to the hypernetted chain data, which is expected to be accurate in the regime shown. For the monovalent electrolyte the expansion is qualitatively correct up to $\kappa_D d \approx 1.5$, although not quite as accurate as the self-consistent Debye–Hückel result, Eq. (2.53). For the divalent electrolyte the low-density expansion is obviously inapplicable at physical electrolyte concentrations. Nevertheless, it is in qualitative accordance with the hypernetted chain results in predicting that at lower concentrations the screening length of the divalent electrolyte is greater than the Debye length (i.e., the ratio κ/κ_D is less than unity). Mitchell and co-workers [37, 57, 58] find that the primary departure from the Debye length at vanishing concentrations is larger for an asymmetric electrolyte than for a symmetric electrolyte. In particular, they find that κ approaches κ_D linearly from above for an asymmetric electrolyte [58]. Stell and Lebowitz [36], also find an effect due to asymmetry (their leading order term also vanishes for a symmetric electrolyte); however, they argue for the one component plasma that the screening length should be greater than the Debye length. Whether asymmetry causes a large shift in the screening length (compared to the symmetric electrolyte), remains to be checked at nonvanishing concentrations by accurate numerical calculations.

D. Asymptotic Analysis

This section analyzes the asymptotic behavior of the ion–ion correlation functions for the bulk electrolyte. One of the things to be determined is the transition from monotonic oscillator decay of the correlation functions, first discussed for electrolytes by Kirkwood [65–67]; in general fluids this transition is called the Fisher–Widom line [68]. In electrolytes, three different asymptotic regimes may be identified, and these are shown rather clearly in Fig. 2. At low concentrations (more precisely low-ion coupling) there is monotonic decay of the total correlation function; the counterion density about an ion always exceeds the co-ion density. This is the exponential behavior of Debye–Hückel theory. At higher concen-

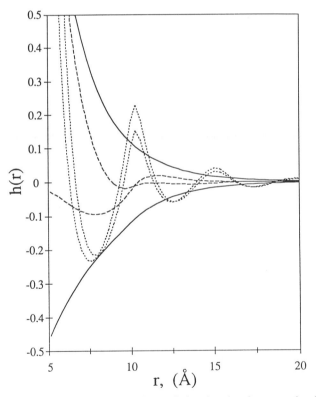

Figure 2. The hypernetted chain total correlation function for a restricted primitive model monovalent electrolyte with $d = 5\,\text{Å}$. Results are for the monotonic electrostatic regime ($0.5\,M$, solid curves), the oscillatory electrostatic regime ($2\,M$, dashed curves), and the oscillatory core regime ($5\,M$, dotted curves). The counterion curve is greater than the corresponding co-ion curve near contact. Redrawn from [18].

trations one finds oscillatory behavior. In the first place the oscillations are in the charge density (electrostatic-dominated oscillatory decay), and the counterion and co-ion densities oscillate out of phase. At the highest couplings it is the number density that oscillates (core-dominated oscillatory decay), and here the counterion and co-ion densities oscillate in phase. Note the difference in period of these two oscillatory regimes: In the core-dominated regime the period is close to the ionic diameter, whereas in the electrostatic regime it is much larger. Roughly speaking, the charge oscillations correspond to alternating shells of positive and negative charge, and hence the period in this case is at least $2d$, and can be considerably larger than this close to the monotonic-oscillatory transition. (It is likely that electrolytes with more than two components have more than two oscillatory asymptotic phases, depending on their symmetry and the linear combinations that may be formed.)

In the core-dominated regime for the restricted primitive model of Fig. 2, the density–density correlation function has a larger decay length than the charge–charge correlation function, and vice versa in the electrostatic dominated regime. It has to be understood that there is a unique decay length common to all the individual ion–ion pair correlation functions (in this example the larger of the density–density and the charge–charge); it is only particular linear combinations of the individual functions that can decay with different rates. This point, which generally holds in mixtures, has received some recent attention [69–72], although it has probably always been well known; for example, the literature is extensive on *the* decay length of electrolytes. As is discussed in detail below, the asymptotic decay rate is determined by the location in Fourier space of the poles of the pair correlation functions, and the matrix of the latter is, by the Ornstein–Zernike equation, an inverse matrix times the matrix of the direct correlation functions. The poles correspond to the singularities (i.e., the vanishing of the determinant) of this inverse matrix. Hence, these singularities are common to all of the pair correlation functions.

The asymptotic behavior of electrolytes has been studied in the context of the departure from the Debye length [8, 20, 22, 36, 56–64], and has received more detailed attention recently [18, 37, 59, 73]. The analysis of this section follows that of [18] rather closely. The results of Section II.B are used, namely, that the ion correlation functions decay exponentially because the direct correlation function goes like the Coulomb potential plus a short-ranged function. By manipulating the Ornstein–Zernike equation in Fourier space, expressions are obtained for the amplitude and decay length of the correlation functions. The asymptote is formally identical to the linear Debye–Hückel total correlation function, but with effective parameters.

1. Monotonic Asymptotic Decay

The Ornstein–Zernike equation (2.15) may be solved for the total correlation function,

$$\hat{\underline{H}}(k) = (\underline{I} - \hat{\underline{\chi}}(k) + \underline{Q}k^{-2})^{-1}(\hat{\underline{\chi}}(k) - \underline{Q}k^{-2})$$

$$= \left(\underline{I} + (\underline{I} - \hat{\underline{\chi}}(k))^{-1}\underline{q}\,\underline{q}^T\,\frac{4\pi\beta}{\varepsilon k^2}\right)^{-1}(\underline{I} - \hat{\underline{\chi}}(k))^{-1}\left(\hat{\underline{\chi}}(k) - \underline{q}\,\underline{q}^T\,\frac{4\pi\beta}{\varepsilon k^2}\right)$$

$$= \left(\underline{I} + \tilde{\underline{q}}(k)\underline{q}^T\,\frac{4\pi\beta}{\varepsilon k^2}\right)^{-1}(\underline{I} - \hat{\underline{\chi}}(k))^{-1}\left(\hat{\underline{\chi}}(k) - \underline{q}\,\underline{q}^T\,\frac{4\pi\beta}{\varepsilon k^2}\right) \qquad (2.56)$$

Here an effective charge function has been defined,

$$\tilde{\underline{q}}(k) = (\underline{I} - \hat{\underline{\chi}}(k))^{-1}\underline{q} \qquad (2.57a)$$

which may be equivalently written

$$\tilde{\underline{q}}(k) = \underline{q} + \hat{\underline{\chi}}(k)\tilde{\underline{q}}(k) \qquad (2.57b)$$

Now the first inverse is readily evaluated [cf. Eq. (2.12)]

$$\left(\underline{I} + \tilde{\underline{q}}(k)\underline{q}^T\,\frac{4\pi\beta}{\varepsilon k^2}\right)^{-1} = \underline{I} - \frac{4\pi\beta/\varepsilon}{k^2 + \Lambda(k)^2}\,\tilde{\underline{q}}(k)\underline{q}^T \qquad (2.58)$$

where a function has been defined that will become the screening length [cf. Eq. (2.10)],

$$\Lambda(k)^2 = \frac{4\pi\beta}{\varepsilon}\,\tilde{\underline{q}}^T(k)\underline{q} \qquad (2.59)$$

Now, if κ exists such that $\Lambda(i\kappa) = \kappa$, then the total correlation function will have a pole at $k = i\kappa$, which determines its asymptotic behavior. [This assumes that $\chi(r)$ is more short ranged than $h(r)$, and hence that any singular behavior of $\hat{\chi}(k)$ occurs further from the origin than this κ.] This pole corresponds to the vanishing of the denominator, which is a scalar multiplying the remaining matrices. Hence, each element of the matrix of total correlation functions has a pole located at $\Lambda(i\kappa)$, and hence each will have the same asymptotic decay. Neglecting the regular part, one has

$$\underline{\underline{\hat{H}}}(k) \sim \frac{-4\pi\beta/\varepsilon}{k^2 + \Lambda(k)^2} \, \underline{\tilde{q}}(k)\underline{q}^T(\underline{\underline{I}} - \underline{\underline{\hat{\chi}}}(k))^{-1}\left(\underline{\underline{\hat{\chi}}}(k) - \underline{q}\,\underline{q}^T \frac{4\pi\beta}{\varepsilon k^2}\right)$$

$$= \frac{-4\pi\beta/\varepsilon}{k^2 + \Lambda(k)^2} \, \underline{\tilde{q}}(k)\underline{\tilde{q}}^T(k)\left(\underline{\underline{\hat{\chi}}}(k) - \underline{q}\,\underline{q}^T \frac{4\pi\beta}{\varepsilon k^2}\right)$$

$$= \frac{-4\pi\beta/\varepsilon}{k^2 + \Lambda(k)^2} \, \underline{\tilde{q}}(k)(\underline{\tilde{q}}^T(k)\underline{\underline{\hat{\chi}}}(k) - \underline{q}^T\Lambda(k)^2/k^2)$$

$$\sim \frac{-4\pi\beta/\varepsilon}{k^2 + \Lambda(k)^2} \, \underline{\tilde{q}}\,\underline{\tilde{q}}^T , \qquad k \to i\kappa \qquad (2.60)$$

Here the second equality follows from Eq. (2.57a) and the final equality from Eq. (2.57b). Also, $\underline{\tilde{q}} \equiv \underline{\tilde{q}}(i\kappa)$. This is in similar form to the Debye–Hückel result, Eq. (2.47), and in order to cast it in identical form, that is to exhibit the residue explicitly, one needs the Taylor expansion of the denominator about $k = i\kappa$,

$$k^2 + \Lambda(k)^2 \sim (k - i\kappa)\left(2i\kappa + \frac{4\pi\beta}{\varepsilon}\underline{q}^T\underline{\tilde{q}}'\right) + \cdots$$

$$\sim (k^2 + \kappa^2)\left(1 + \frac{4\pi\beta}{2i\kappa\varepsilon}\underline{q}^T\underline{\tilde{q}}'\right) + \mathcal{O}(k - i\kappa)^2 \qquad (2.61)$$

Here

$$\underline{\tilde{q}}' = \frac{\partial\underline{\tilde{q}}(k)}{\partial k}\bigg|_{k=i\kappa}$$

$$= \frac{\partial\underline{\underline{\hat{\chi}}}(k)}{\partial k}\bigg|_{k=i\kappa}\underline{\tilde{q}} + \underline{\underline{\hat{\chi}}}(i\kappa)\frac{\partial\underline{\tilde{q}}(k)}{\partial k}\bigg|_{k=i\kappa}$$

$$= (\underline{\underline{I}} - \underline{\underline{\hat{\chi}}}(i\kappa))^{-1}\frac{\partial\underline{\underline{\hat{\chi}}}(k)}{\partial k}\bigg|_{k=i\kappa}\underline{\tilde{q}} \qquad (2.62)$$

Hence, one defines

$$\nu \equiv 1 + \frac{4\pi\beta}{2i\kappa\varepsilon}\underline{q}^T\underline{\tilde{q}}'$$

$$= 1 + \frac{4\pi\beta}{2i\kappa\varepsilon}\underline{q}^T(\underline{\underline{I}} - \underline{\underline{\hat{\chi}}}(i\kappa))^{-1}\underline{\underline{\hat{\chi}}}'(i\kappa)\underline{\tilde{q}}$$

$$= 1 + \frac{4\pi\beta}{2i\kappa\varepsilon}\underline{\tilde{q}}^T\underline{\underline{\hat{\chi}}}'(i\kappa)\underline{\tilde{q}} \qquad (2.63)$$

One now has

$$\hat{\underline{H}}(k) \sim \frac{-4\pi\beta}{\varepsilon\nu} \frac{\tilde{\underline{q}}\,\tilde{\underline{q}}^T}{k^2 + \kappa^2} \qquad k \to i\kappa \qquad (2.64)$$

with inverse

$$h_{\alpha\gamma}(r) \sim \frac{-\beta\tilde{q}_\alpha\tilde{q}_\gamma}{\varepsilon\nu} \frac{e^{-\kappa r}}{r} \qquad r \to \infty \qquad \mathrm{Im}\{\kappa\} = 0 \qquad (2.65)$$

Note that if $\chi = 0$, Eq. (2.57a) implies that $\tilde{q} = q$, Eq. (2.59) yields $\kappa = \kappa_D$, and $\bar{\bar{\mathrm{E}}}$q. (2.63) shows that $\nu = 1$. In other words, the exact asymptote, Eq. (2.65), reduces to the Debye–Hückel result, Eq. (2.49).

Recall that $\tilde{q}$ is given by Eq. (2.57a) evaluated at $k = i\kappa$. With this definition the actual screening length, Eq. (2.59), assumes a form similar to the Debye length [18, 36, 37, 59]

$$\kappa^2 = \frac{4\pi\beta}{\varepsilon}\,\tilde{\underline{q}}^T\underline{q} = \frac{4\pi\beta}{\varepsilon} \sum_\alpha \rho_\alpha \tilde{q}_\alpha q_\alpha \qquad (2.66)$$

The quantity $\varepsilon\nu$, which appears in the denominator of the asymptote, can be related to the nonlocal electrostatic susceptibility of the electrolyte [37]; in this connection this dielectric response function was approximated by Lovett and Stillinger [61] and used to find the asymptotic decay length of the electrolyte. Stell and Lebowitz [36] formally obtained the same asymptote and screening length, except they set $\nu = 1$. To leading order in βq^2, they give an approximation for the effective charges in terms of properties of the reference system (in this case it would be a hard-sphere fluid),

$$\tilde{q}_\alpha \approx q_\alpha + \sum_\gamma q_\gamma \rho_\gamma \int h^0_{\alpha\gamma}(r)\,d\mathbf{r} = \sum_\gamma q_\gamma \frac{\partial\rho_\alpha}{\partial\beta\mu^0_\gamma} \qquad (2.67)$$

where the superscript denotes the reference system. This first correction to the bare charge vanishes in the symmetric electrolyte and in this case higher order terms are required [36].

The effective charges represent the response of the ions to the mean electrostatic potential [37, 59]. This can be seen from the Fourier transform of the Ornstein–Zernike equation written in terms of the latter, Eq. (2.28),

$$\underline{\hat{H}}(k) = -\beta \underline{\hat{\psi}}(k)\underline{q}^T + \underline{\hat{\chi}}(k) + \underline{\hat{H}}(k)\underline{\hat{\chi}}(k)$$

$$= -\beta \underline{\hat{\psi}}(k)\underline{q}^T(\underline{I} - \underline{\hat{\chi}}(k))^{-1} + \underline{\hat{\chi}}(k)(\underline{I} - \underline{\hat{\chi}}(k))^{-1}$$

$$\sim -\beta \underline{\hat{\psi}}(k)\underline{\tilde{q}}^T \qquad k \to i\kappa \qquad (2.68)$$

The last line follows since $\chi(r)$ is more short-ranged than $h(r)$, and hence only the first term contributes to the asymptote. In other words, the mean electrostatic potential about an ion has the same range as the total correlation function, and one has

$$h_{\alpha\gamma}(r) \sim -\beta \tilde{q}_\alpha \psi_\gamma(r) \qquad r \to \infty \qquad (2.69)$$

which shows that it is the effective charge on the ion that gives its response to the mean electrostatic potential. Now from symmetry $h_{\alpha\gamma}(r) = h_{\gamma\alpha}(r)$, and one concludes that $\psi_\gamma(r) \propto \tilde{q}_\gamma$. The precise relationship follows from Eq. (2.65),

$$\psi_\gamma(r) \sim \tilde{q}_\gamma \frac{e^{-\kappa r}}{\varepsilon \nu r} \qquad r \to \infty \qquad (2.70)$$

In the asymptotic regime the potential of mean force has linear Poisson–Boltzmann form, $w_{\alpha\gamma}(r) \sim \tilde{q}_\alpha \psi_\gamma(r) \sim \tilde{q}_\alpha \tilde{q}_\gamma \psi(r)$, which is significant because the full symmetry of the pair correlation functions is preserved. This is not the case for generalizations of Debye–Hückel theory based upon the nonlinear Poisson–Boltzmann approximation. Here the mean electrostatic potential about each ion is no longer strictly proportional to the response of that ion (i.e., the charge), and consequently the pair correlation functions are not symmetric, which is a long-standing criticism of that approximation [74–76].

For an m-component electrolyte, in the most general case there are $m(m + 1)/2$ pair total correlation functions, but the dyadic nature of the asymptote given here indicates that in this regime only m of them are independent. A similar situation occurs for fluid mixtures that interact with integrable power law potentials. In general, one has

$$h_{\alpha\gamma}(r) \sim -\beta \sum_{\lambda\delta} \frac{\partial \ln \rho_\alpha}{\partial \beta \mu_\lambda} u_{\lambda\delta}(r) \frac{\partial \ln \rho_\delta}{\partial \beta \mu_\gamma} \qquad r \to \infty \qquad (2.71)$$

For the case of a dyadic potential, $u_{\alpha\gamma}(r) \sim a_\alpha a_\gamma r^{-n}$, $r \to \infty$, this reduces to

$$h_{\alpha\gamma}(r) \sim -\beta \tilde{a}_\alpha \tilde{a}_\gamma r^{-n} \qquad r \to \infty \qquad n > 3 \qquad (2.72)$$

with

$$\tilde{a}_\alpha = \sum_\lambda q_\lambda \frac{\partial \ln \rho_\alpha}{\partial \beta \mu_\lambda} = \sum_\lambda q_\lambda \frac{\partial \ln \rho_\lambda}{\partial \beta \mu_\alpha} \qquad (2.73)$$

In the asymptotic regime there are only m-independent quantities, which are linear combinations of the chemical potential derivative of the densities. The assumed dyadic nature of the pair potential is not as specialized as it may at first appear; in the case of a Lennard–Jones mixture the coefficient of the r^{-6} term represents the product of the polarizabilities of the individual atoms.

2. Oscillatory Asymptotic Decay

One needs to allow for the possibility that κ is complex, which corresponds to oscillatory solutions. In this case one takes twice the real part of the asymptote given above, as may be seen as follows. By definition $h_{\alpha\gamma}(r)$ is a real, even function of r, and hence its Fourier transform is even, $\hat{h}_{\alpha\gamma}(-k) = \hat{h}_{\alpha\gamma}(k)$, and any series expansion in k has real coefficients, $\hat{h}_{\alpha\gamma}(k) = \hat{h}_{\alpha\gamma}(\bar{k})$, where the overbar denotes the complex conjugate. In other words,

$$\hat{\underline{\underline{H}}}(-k) = \hat{\underline{\underline{H}}}(k) \qquad (2.74a)$$

and

$$\overline{\hat{\underline{\underline{H}}}(k)} = \hat{\underline{\underline{H}}}(\bar{k}) \qquad (2.74b)$$

These two results imply that there are four poles located at $\pm i\kappa$, and $\pm i\bar{\kappa}$, and the singular part may be written

$$\hat{\underline{\underline{H}}}^s(k) = \frac{\underline{\underline{A}}}{k - i\kappa} + \frac{\underline{\underline{A}}'}{k + i\kappa} + \frac{\underline{\underline{B}}}{k - i\bar{\kappa}} + \frac{\underline{\underline{B}}'}{k + i\bar{\kappa}}$$

$$= \frac{\underline{\underline{A}}}{k^2 + \kappa^2} + \frac{\underline{\underline{B}}}{k^2 + \bar{\kappa}^2}$$

$$= \frac{\underline{\underline{A}}}{k^2 + \kappa^2} + \frac{\overline{\underline{\underline{A}}}}{k^2 + \bar{\kappa}^2} \qquad (2.75)$$

since Eq. (2.74a) implies that $A = -A'$ and $B = -B'$, and Eq. (2.74b) implies that $B = \bar{A}$. Hence, when the conventional Fourier inverse is evaluated by closing the contour in the upper half plane (choosing κ to be in the first quadrant), one picks up the residue at $k = +i\kappa$, which is $A/(2i\kappa)$, and also its complex conjugate from the residue from the pole at $k = -i\bar{\kappa}$, which is $\bar{A}/(2i\bar{\kappa})$. The sum of these two is twice the real part of either one, and one has

$$h_{\alpha\gamma}(r) \sim 2\,\mathrm{Re}\left\{\frac{-\beta\tilde{q}_\alpha\tilde{q}_\gamma}{\varepsilon\nu}\frac{e^{-\kappa r}}{r}\right\} \qquad r \to \infty \qquad \mathrm{Im}\{\kappa\} \neq 0 \qquad (2.76)$$

The abrupt disappearance of the factor of 2 as the poles coalesce on the imaginary axis suggests nonanalyticity in the amplitude of the correlation functions at the transition from monotonic to oscillatory decay. In fact, the amplitude becomes infinite, as will now be shown.

Let the pole just move off the imaginary k axis, $\kappa = \kappa_r + i\kappa_i$, $\kappa_i \to 0$. Expanding Eq. (2.59) one obtains

$$\kappa_r + i\kappa_i = \left(\frac{4\pi\beta}{\varepsilon}\underline{q}^T\underline{\tilde{q}}(i\kappa_r - \kappa_i)\right)^{1/2}$$

$$\sim \left(\frac{4\pi\beta}{\varepsilon}\underline{q}^T\underline{\tilde{q}}(i\kappa_r)\right)^{1/2}\left[1 - \frac{\kappa_i}{2}\frac{\underline{q}^T\underline{\tilde{q}}'(i\kappa_r)}{\underline{q}^T\underline{\tilde{q}}(i\kappa_r)} + \mathcal{O}(\kappa_i^2)\right]$$

$$= \kappa_r - \frac{\kappa_i}{2\kappa_r}\frac{4\pi\beta}{\varepsilon}\underline{q}^T\underline{\tilde{q}}' \qquad\qquad (2.77)$$

By equating the coefficients of κ_i one obtains

$$\frac{4\pi\beta}{2\kappa_r\varepsilon}\underline{q}^T\underline{\tilde{q}}' = -i \qquad \kappa_i \to 0 \qquad (2.78)$$

or from the first equality of Eq. (2.63),

$$\nu \to 0 \qquad \kappa_i \to 0 \qquad (2.79)$$

That is, the amplitude of the total correlation function becomes infinite at the oscillatory to monotonic transition.

One has to be careful in interpreting this result. The vanishing of ν means that the next term in the Taylor expansion (2.61) is nonnegligible. Hence, the denominator corresponds to a double pole, and the residue comes from the linear term in the numerator, which ensures that the asymptotic behavior remains exponential. For infinitesimal but nonzero ν, the present formulas give the strict asymptote, but the regime of applicability moves to ever larger separations. For fixed r, as $\nu \to 0$ the $h(r)$ have a large contribution from the (finite) second term. (This limit has been discussed in the context of a dominant and a subdominant pole coalescing on, and then parting off, the imaginary axis [37, 73].) For fixed nonzero ν, as $r \to \infty$ the $h(r)$ are given by the strict asymptote. So even though the amplitude diverges, so do the relevant separations, and consequently thermodynamic properties such as the internal energy remain finite at the monotonic–oscillatory transition.

3. Core Domination

The analysis of Section II.D.2 is formally exact, and the Fourier transforms of the total correlation functions are guaranteed to have a pole at $k = i\kappa$, for κ satisfying Eqs. (2.57a) and (2.59). Nevertheless, this pole does not necessarily determine the asymptotic behavior of the correlations, because there could be another pole, $k = i\xi$, with $\mathrm{Re}\{\xi\} < \mathrm{Re}\{\kappa\}$, even assuming that κ represents the solution of the preceding equations with smallest real part. Such a qualitatively different pole would correspond to the matrix $\underline{I} - \hat{\underline{\chi}}(i\xi)$ being singular.

The implicit reason for splitting the matrix $\underline{I} - \hat{\underline{C}}(k)$ in the fashion of the preceding section was the assumption that the matrix $\underline{Q}$ was the most important in the asymptotic regime. This would be the case when electrostatics determined the asymptotic behavior, hence the title of the section. However, at high densities one might expect the short-range interactions to become important asymptotically, and this regime might be termed the "core-dominated asymptote," which is the subject of this section.

This regime will not be treated in full generality, but instead the following restriction will be observed. Denote the amplitude of the total correlation functions in the asymptotic regime by $a_{\alpha\gamma}$. Then, because electrostatic effects are of shorter range, locally the asymptotic charge density about an ion must vanish, and one has

$$\sum_\alpha q_\alpha \rho_\alpha a_{\alpha\gamma} = 0 \qquad\qquad (2.80)$$

It is emphasized that this is an exact result that holds in the core-dominated asymptotic regime. One way of satisfying this equation is if

$$a_{\alpha\gamma} = a \qquad \text{all species} \qquad\qquad (2.81)$$

What follows is predicated on this restriction. Two cases can be mentioned where this equation will hold exactly with no approximation. First, there is the general binary electrolyte, in which case this is the only possible solution. Second, there is a multicomponent electrolyte with the short-range interactions between the ions being identical; since it is the latter that determine the asymptote, then the total correlation functions between all the species must be asymptotically equal.

The condition Eq. (2.81) means that the short-range part of the direct correlation function goes like

$$\hat{\underline{\chi}}(k) \sim \underline{\rho}\underline{\rho}^T x(k) \qquad k \to i\xi \tag{2.82}$$

where $\{\rho\}_\alpha = \rho_\alpha^{1/2}$. (This follows from the Ornstein–Zernike equation solved for the direct correlation function near the pole.) Note that $\underline{\rho}^T\underline{q} = 0$, and hence near the pole

$$(\underline{I} - \hat{\underline{C}}(k))^{-1} \sim (\underline{I} - \underline{\rho}\underline{\rho}^T x(k) + \underline{q}\underline{q}^T 4\pi\beta/\varepsilon k^2)^{-1}$$

$$= \underline{I} + \frac{\underline{\rho}\underline{\rho}^T x(k)}{1 - \underline{\rho}^T\underline{\rho}x(k)} - \frac{\underline{q}\underline{q}^T 4\pi\beta/\varepsilon k^2}{1 + \underline{q}^T\underline{q}4\pi\beta/\varepsilon k^2} \tag{2.83}$$

Now it is the middle term that has the pole at $k = i\xi$, and neglecting the remaining regular parts one obtains

$$\hat{\underline{H}}(k) \sim \frac{\underline{\rho}\underline{\rho}^T x(k)}{1 - \underline{\rho}^T\underline{\rho}x(k)} (\underline{\rho}\underline{\rho}^T x(k) - \underline{q}\underline{q}^T 4\pi\beta/\varepsilon k^2)$$

$$= \frac{\underline{\rho}^T\underline{\rho}x(k)^2}{1 - \underline{\rho}^T\underline{\rho}x(k)} \underline{\rho}\underline{\rho}^T$$

$$\sim \frac{-2i\xi x(i\xi)}{\underline{\rho}^T\underline{\rho}x'(i\xi)(k^2 + \xi^2)} \underline{\rho}\underline{\rho}^T \qquad k \to i\xi \tag{2.84}$$

Here a Taylor expansion of the denominator has been used, together with the fact that ξ satisfies

$$\underline{\rho}^T\underline{\rho}x(i\xi) = \rho x(i\xi) = 1 \tag{2.85}$$

where the total number density is $\rho = \Sigma_\alpha \rho_\alpha = \underline{\rho}^T\underline{\rho}$. One concludes that

$$h_{\alpha\gamma}(r) \sim 2\,\text{Re}'\left\{\frac{-2i\xi}{4\pi\rho^2 x'(i\xi)} \frac{e^{-\xi r}}{r}\right\} \qquad r \to \infty \tag{2.86}$$

The notation Re$'\{z\}$ means to take the real part of z if z is complex, and take half of z if z is real. (In this regime ξ will always be complex because the analysis only makes sense when the correlation functions are oscillatory.) In contrast to Section II.D.2, this analysis holds for uncharged particles, subject to the restriction (2.81). In the core dominated regime the ion–ion total correlation functions decay as damped sinusoids, with a prefactor that corresponds to the unique amplitude that they have in common. The charge on an ion is very rapidly neutralized by the surrounding counterions, and the correlations that persist are due to packing of the ions in the dense electrolyte or molten salt. One anticipates that the period of the oscillations will correspond to the size of the ions, wheres in the case of charge oscillations the period was greater than about twice the ionic diameter.

III. ANALYSIS OF THE DOUBLE LAYER

There are two approaches to analyzing the electric double layer: one can regard the charged macroparticles as solutes in the electrolyte and take the infinite dilution limit of this species of the multicomponent mixture (singlet method), or one can regard these particles as fixed and as the source of an external field causing a nonuniform distribution of ions in their vicinity (inhomogeneous method). These are but alternative viewpoints of the same phenomena, and both can be used to obtain formally exact results. In many ways the distinction between them is not so much conceptual as practical and historical, and the relative numerical merits of the two approaches will be discussed in more detail under approximate methods (Section IV), and under results (Section V). For asymptotic analysis, which forms a large part of this section, the single approach is the most useful, and here it is used to formulate the electric double layer in various geometries.

The charged particles that are the source of the electric double layer are taken here to be solutes at infinite dilution. Section III.A treats spherical macroions, and establishes their charge neutralization by the excess of counterions in the double layer, the exponential decay of the latter with the decay length of the bulk electrolyte, and the fact that the asymptotic interaction between two macroions scales with the square of the effective charge of an isolated macroion. Isolated and interacting planar solutes are treated in Sections III.B and C, both with a view to establishing the wall–ion and the wall–wall Ornstein–Zernike equations, and to determining the asymptotic behavior of the planar double layer. The Derjaguin approximation is derived in Section III.C.2, and its domain of applicability delineated. The section concludes with the application of the singlet approach to other solute geometries, specifically cylinders, pores, and dumbbells.

A. Spherical Solutes

1. Double Layer Moments

The analysis of Section III.B is here extended to the double layer about a charged solute, following [18]. For the multicomponent electrolyte, one of the components is taken to be a spherical solute, species $\alpha = 0$ with charge q_0. One is interested in the double layer about this solute, which is treated in isolation by taking the infinite dilution limit, $\rho_0 = 0$. One then focuses on the solute–ion contributions to the multicomponent Ornstein–Zernike equation (1.3) by defining a vector of solute–ion total correlation functions

$$\{\underline{H}(r)\}_\gamma = \rho_\gamma^{1/2} h_{0\gamma}(r) \qquad \gamma > 0 \tag{3.1}$$

and similarly for the direct correlation functions. The density of ions in the double layer about the solute is $\rho_\gamma(r) = \rho_\gamma[h_{0\gamma}(r) + 1]$. Since $\rho_0 = 0$, there is no solute–solute contribution to the convolution integral, and the solute–ion Ornstein–Zernike equation becomes

$$\underline{H}(r) = \underline{C}(r) + \int \underline{\underline{C}}(s)\underline{H}(|\mathbf{r} - \mathbf{s}|)\, ds \tag{3.2}$$

with Fourier transform

$$\hat{\underline{H}}(k) = \hat{\underline{C}}(k) + \hat{\underline{\underline{C}}}(k)\hat{\underline{H}}(k)$$

$$= \hat{\underline{\chi}}(k) - \frac{4\pi\beta q_0}{\varepsilon k^2}\underline{q} + \hat{\underline{\underline{\chi}}}(k)\hat{\underline{H}}(k) - \underline{\underline{Q}}\hat{\underline{H}}(k)k^{-2} \tag{3.3}$$

The proof by induction that the ion–solute correlation functions are exponentially decaying is almost unchanged from the ion–ion case treated in Section II.B.2. Since the exact solute–ion closure is identical to the ion–ion closure, it can be concluded that the short-range part of solute–ion direct correlation function, $\chi(r)$, is either exponentially decaying or more short ranged than $\underline{H}(r)$. Then by equating the various powers of k in the Fourier transform of the Ornstein–Zernike equation, and by using the established fact that the necessary moments of $\underline{\chi}(r)$ exist, it is found that both solute–ion correlation functions are exponentially short ranged. That is to say that the ion densities in the double layer around a charged solute decay to their bulk electrolyte concentrations at an exponential rate. The function $\chi_{0\gamma}(r)$ is also exponentially decaying, and one has

$\chi_{0\gamma}(r) - d_{0\gamma}(r) \sim h_{0\gamma}(r)^2/2$, $r \to \infty$. For the asymptotic analysis below, one assumes that the sort-range part of solute–ion direct correlation function is more short ranged than the solute–ion total correlation function, $\chi_{0\gamma}(r)/h_{0\gamma}(r) \to 0$, $r \to \infty$.

Since the moments exist, one may equate the coefficients of k^{-2} to obtain the solute electroneutrality condition

$$-q_0 = \underline{q}^T \underline{H}^{(0)} = \sum_\gamma q_\gamma \rho_\gamma \int h_{0\gamma}(r) \, d\mathbf{r} \qquad (3.4)$$

which confirms the result given above, Eq. (2.44). This shows that the charge on the solute is exactly canceled by the net charge in the double layer. This countercharge must have either dissociated from the originally neutral surface, or ions from the originally neutral bulk electrolyte must have adsorbed to the solute giving it its charge. The only way that the charge on the solute is not neutralized by the double layer is if the ion density profile about the solute were to be so long ranged as to be nonintegrable (and vice versa).

Turning now to the second moment, by equating the coefficients of k^0 in the Fourier transform of the Ornstein–Zernike equation one obtains

$$\underline{H}^{(0)} = \underline{\chi}^{(0)} + \underline{\underline{\chi}}^{(0)} \underline{H}^{(0)} - \underline{\underline{Q}} \underline{H}^{(2)} \qquad (3.5)$$

or, using the electroneutrality condition,

$$-q_0 = \underline{q}^T \underline{\chi}^{(0)} + \underline{q}^T \underline{\underline{\chi}}^{(0)} \underline{H}^{(0)} - \kappa_D^2 \underline{q}^T \underline{H}^{(2)} \qquad (3.6)$$

In this case the second moment of the ion–solute total correlation function depends on the zeroth moment of the short-range part of the ion–solute direct correlation function. The former is analogous to one of the individual second moments of the ion–ion total correlation functions of the bulk electrolyte, and it will be recalled that these were not individually constrained by the second moment condition. It was only their sum that was constrained, and because the solute is at infinite dilution, this last result can be added to the bulk second moment condition, Eq. (2.26), without effect. As in the case of the bulk electrolyte, one can rewrite Eq. (3.6) in terms of the zeroth moment of the mean electrostatic potential about the solute since the latter is related to the second moment of the ion–solute total correlation function by Poisson's equation. [Note that no new information is obtained by reversing the Ornstein–Zernike integrand, (i.e., replacing $c * h$ by $h * c$);

all that results are identities corresponding to the two moment conditions of the bulk electrolyte.]

The pairwise interaction between the solutes is given by the singlet approach when $\alpha = \gamma = 0$ in Eq. (2.3). The solute–solute Ornstein–Zernike equation for identical solute is

$$h_{00}(r) = c_{00}(r) + \int \underline{C}(s)^T \underline{H}(|\mathbf{r} - \mathbf{s}|) \, d\mathbf{s} \tag{3.7}$$

with Fourier transform

$$\hat{h}_{00}(k) = \hat{c}_{00}(k) + \underline{\hat{C}}(k)^T \underline{\hat{H}}(k)$$

$$= \hat{\chi}_{00}(k) - \frac{4\pi\beta q_0^2}{\varepsilon k^2} + \underline{\hat{\chi}}(k)^T \underline{\hat{H}}(k) - \frac{4\pi\beta q_0}{\varepsilon k^2} \underline{q}^T \underline{\hat{H}}(k) \tag{3.8}$$

Since the bulk closure remains the same, the exponential decay of the solute–solute interaction follows immediately. However, the coefficient of k^{-2} yields

$$-q_0 = \underline{q}^T \underline{H}^{(0)} \tag{3.9}$$

which contains nothing new, and the coefficient of k^0 yields

$$h_{00}^{(0)} = \chi_{00}^{(0)} + \underline{\chi}^{(0)T} \underline{H}^{(0)} - \frac{4\pi\beta q_0}{\varepsilon} \underline{q}^T \underline{H}^{(2)} \tag{3.10}$$

By using the result given above, the second moment could be eliminated and this could be reexpressed in terms of the zeroth moments of the various correlation functions. There is no electroneutrality condition for the solute–solute total correlation function; in this case the zeroth moment is nonuniversal and depends on the zeroth moments of various solute correlation functions.

2. Asymptotic Analysis

In the context of the singlet approach, the formal asymptotic analysis for the behavior of the double layer about a spherical solute is virtually unchanged from that for the bulk electrolyte (Section II.D). The solute is at infinite dilution (species 0, $\rho_0 \to 0$), and does not contribute to the solvent correlation functions because the Ornstein–Zernike convolution integral, Eq. (2.3), is multiplied by ρ_0 whenever the solute correlation functions appear in the integrand. That is, the properties of the bulk solvent are not affected by the addition of the infinitely dilute solute, and all of the preceding analysis for the ionic correlations remain. Hence, the decay length κ^{-1}, the effective charge on the ions $\tilde{q}_\gamma$, $\gamma > 0$, and the scale factor ν stay the same. The Fourier transform of the solute–ion Ornstein–Zernike equation (3.2) becomes

$$\hat{\underline{H}}(k) = (\underline{I} - \hat{\underline{C}}(k))^{-1}\hat{\underline{C}}(k) \tag{3.11}$$

The pole that determines the asymptotic behavior of the solute–ion total correlation function corresponds to the singularity in the inverted matrix. This consists solely of the bulk ion direct correlation functions, and hence may be explicitly seen to be unchanged by the solute; the determinant vanishes at the same $k = i\kappa$ as in the bulk electrolyte. Hence, the pole of the solute–solvent total correlation functions have the same residue as in the bulk electrolyte, with an effective solute charge $\tilde{q}_0$ appearing. One ends up with

$$\hat{\underline{H}}(k) \sim \frac{-4\pi\beta/\varepsilon}{k^2 + \Lambda(k)^2}\,\tilde{q}_0\underline{\tilde{q}} \qquad k \to i\kappa \tag{3.12}$$

where the effective charge on the solute is related to the short-range part of the solute–solvent direct correlation functions by

$$\tilde{q}_0 = \underline{\tilde{q}}^T\underline{\tilde{C}}(i\kappa)$$

$$= q_0 + \underline{\tilde{q}}^T\underline{\hat{\chi}}(i\kappa) \tag{3.13}$$

The asymptote is

$$h_{0\gamma}(r) \sim 2\,\mathrm{Re}'\left\{\frac{-\beta\tilde{q}_0\tilde{q}_\gamma}{\varepsilon\nu}\frac{e^{-\kappa r}}{r}\right\} \qquad r \to \infty \tag{3.14}$$

It is emphasized that κ, ν, and $\tilde{q}_\gamma$, $\gamma > 0$, are all properties of the bulk electrolyte and are unaffected by the solute. Only $\tilde{q}_0$ depends on the nature of the solute, via the short-ranged part of the solute–solvent direct correlation functions. The solute–solute total correlation function follows by setting $\gamma = 0$ in this result, and depends on the square of the effective solute charge.

A point worth emphasizing about this result is that the rate at which the ion densities in the double layer about an isolated solute decay to their bulk concentrations (the decay length κ) is the same as the decay length of the bulk electrolyte. One of the earliest analyses to show this effect was that of Stillinger and Kirkwood [65]. (This behavior generally is predicted by the singlet approach [70, 77, 78]; it holds also for the liquid–vapor interface [70, 71].) Among other things, this means that the ion profiles will become oscillatory at a point determined by the bulk electrolyte concentration, and similarly for the interaction between the solutes. This point has been emphasized in the context of the modified Poisson–Boltzmann theory [60, 79, 80].

The amplitude of decay of the ion densities in the double layer about a solute is determined by the nature of the solute via $\tilde{q}_0$, and the interaction between two identical solutes scales with the square of their effective charge. Consequently, the force between two identical macroions must be repulsive at large separations in the monotonic regime. This follows because all quantities are real, $\tilde{q}_0$ occurs as a square, and $\nu > 0$. [At low coupling, $\nu \to 1$, the Debye–Hückel result, and at the monotonic-oscillatory transition, $\nu \to 0$, Eq. (2.79).] In other words, asymptotically the force between identical charged solutes is either monotonic repulsive, or oscillatory. *It cannot be monotonically attractive.* In the oscillatory regime it is periodically attractive, and the attractive regions can be quite large since the period of oscillations becomes infinite as one approaches the bulk oscillatory–monotonic transition.

In the core-dominated regime, the analysis proceeds as in the bulk. Under the same assumption that the electrolyte ions have identical short-range interactions, one obtains for the spherical solute

$$\hat{\underline{H}}(k) \sim \frac{-2i\xi x(i\xi)}{\underline{\rho}^T \rho x'(i\xi)(k^2 + \xi^2)} \underline{\rho}\underline{\rho}^T \underline{\hat{\chi}}(i\xi) \qquad k \to i\xi \qquad (3.15)$$

with corresponding asymptote

$$h_{0\alpha}(r) \sim 2\,\mathrm{Re}'\left\{ \frac{-2i\xi}{4\pi\rho^2 x'(i\xi)} \frac{e^{-\xi r}}{r} \sum_{\gamma>0} \rho_\gamma \hat{\chi}_\gamma(i\xi) \right\} \qquad r \to \infty \qquad (3.16)$$

Since the right-hand side is independent of the index of the left-hand side, whatever the direct solute–ion on interaction for each species is, the solute–ion profiles become identical at large separations in this core-dominated regime (provided that the correlations of the ions themselves are asymptotically identical in the bulk).

The solute–solute total correlation function goes like

$$\hat{h}_{00}(k) = \hat{c}_{00}(k) + \hat{\underline{H}}^T(k)\hat{\underline{C}}(k)$$

$$\sim \frac{-2i\xi x(i\xi)}{\underline{\rho}^T \rho x'(i\xi)(k^2 + \xi^2)} \hat{\underline{\chi}}^T(i\xi)\underline{\rho}\underline{\rho}^T \underline{\hat{\chi}}(i\xi) \qquad k \to i\xi \qquad (3.17)$$

since the solute–solute direct correlation function is more short-ranged than the total correlation function, and since $\underline{\rho}^T \hat{\underline{C}} = \underline{\rho}^T \hat{\underline{\chi}}$ because $\underline{\rho}^T \underline{q} = 0$. One obtains the asymptote

$$h_{00}(r) \sim 2 \, \text{Re}' \left\{ \frac{-2i\xi}{4\pi\rho^2 x'(i\xi)} \frac{e^{-\xi r}}{r} \left(\sum_{\gamma > 0} \rho_\gamma \hat{\chi}_\gamma(i\xi) \right)^2 \right\} \qquad r \to \infty \quad (3.18)$$

The sum evidently represents the effective "charge" of the solute; it gives the magnitude of the ion density profile in the double layer about the isolated solute, and its square gives the magnitude of their pairwise interaction.

B. Isolated Planar Solutes

1. Wall–Ion Ornstein–Zernike Equation

Originally, the wall–solvent Ornstein–Zernike equation was derived from the large solute radius limit [81], and in the case of the electric double layer the derivation was modified to take care of the long-range Coulomb potential [18, 82–84]. Based on that experience, the singlet approach has come to be interpreted quite generally, and the existence of the solute–solvent Ornstein–Zernike equation is taken for granted, with the solute being characterized by the geometrical factors manifest in the symmetry of the arguments of the correlation functions. This is the approach taken here for a solute that represents a planar wall.

The wall–ion correlation functions depend only on the perpendicular distance from the surface of the wall, and the distance convention is such that contact occurs at $z = 0$ for all ions,

$$h_{0\alpha}(z) = -1 \qquad z < 0 \qquad\qquad (3.19)$$

Although a more realistic model could allow for ion-specific distances of closest approach to the surface, for example, due to variation in size or solvation, the consequent complication in the algebra obscures the subsequent analysis without any new features emerging. Therefore a single contact plane is assumed in all that follows. The wall here is a semiinfinite half-space; it is also possible to deal with a wall of finite thickness, but in this case one should be aware that correlations can occur between ions across the wall, and the properties of the coupled double layers are not identical to those at an isolated infinitely thick wall. Electroneutrality was shown above to hold for a spherical solute of arbitrary radius, and it also holds for the planar wall. In the present geometry the condition is

$$-\sigma = \int_0^\infty \underline{q}^T \underline{H}(z)\, dz \qquad (3.20)$$

This electroneutrality constraint is also known as the constant charge condition. Instead an alternative would be to specify the surface potential, since this mimics the effects of charge regulation at some or other level of approximation. The constant charge case is the easiest to deal with theoretically. One can in practice obtain any desired surface potential by empirical adjustments of the charge. Some care must be exercised when dealing with interacting double layers at constant potential because the free energy differs from that taken at constant charge due to the variation in charge with separation. Comparison between the two models at finite separations generally refer to double layers that are identical at infinite separation. The interaction pressure is the same at any separation provided the two models have the same surface potential and charge at that separation. In what follows, only the constant charge model will be treated.

In view of the symmetry, the Ornstein–Zernike convolution integral may be expressed in cylindrical coordinates,

$$\underline{H}(z) = \underline{C}(z) + 2\pi \int_0^\infty ds\, s \int_{-\infty}^\infty dz'\, \underline{\underline{C}}(\sqrt{(z'-z)^2 + s^2})\underline{H}(z') \qquad (3.21a)$$

$$= -\beta\psi(z)\underline{q} + \underline{\chi}(z) + 2\pi \int_0^\infty ds\, s \int_{-\infty}^\infty dz'\, \underline{\underline{\chi}}(\sqrt{(z'-z)^2 + s^2})\underline{H}(z') \qquad (3.21b)$$

Here the vector of wall–ion total correlation functions has components $\{\underline{H}(z)\}_\alpha = \rho_\alpha^{1/2} h_{0\alpha}(z)$, and similarly for the wall–ion direct correlation function $\underline{C}(z)$ and its short-ranged part $\underline{\chi}(z)$. As before, one has

$$c_{0\alpha}(z) = \chi_{0\alpha}(z) - \beta q_\alpha V_0^{\text{Coul}}(z) \qquad (3.22)$$

where the Coulomb part of the solute–ion potential may be written in general as

$$V_0^{\text{Coul}}(\mathbf{r}) = \int \frac{\sigma(\mathbf{s})}{\varepsilon|\mathbf{r} - \mathbf{s}|}\, d\mathbf{s} \qquad (3.23)$$

In the present particular, the wall charge density is $\sigma(\mathbf{r}) = \sigma\delta(z)$, where σ is the surface charge per unit area located at the plane $z = 0$. In view of

P. ATTARD

this, the mean electrostatic potential is defined by the passage from the
first to the second form of the Ornstein–Zernike equation. It is

$$\psi(z) = \frac{2\pi}{\varepsilon} \int_{-\infty}^{\infty} dz' \int_{0}^{\infty} ds\, s[\sigma\delta(z') + \underline{q}^T\underline{H}(z')]\frac{1}{\sqrt{s^2 + (z-z')^2}}$$

$$= \frac{-2\pi}{\varepsilon} \int_{-\infty}^{\infty} [\sigma\delta(z') + \underline{q}^T\underline{H}(z')]|z - z'|\, dz'$$

$$= \frac{2\pi}{\varepsilon} \int_{0}^{\infty} \underline{q}^T\underline{H}(z')z'\, dz' + \frac{4\pi}{\varepsilon} \int_{z}^{\infty} \underline{q}^T\underline{H}(z')(z - z')\, dz' \qquad z > 0$$

$$(3.24)$$

and $\psi(z) = \psi(0)$, $z < 0$. (The electroneutrality condition has been used to
cancel the upper limit of the first integration.) In this form the mean
electrostatic potential goes to a nonzero constant in the bulk electrolyte
far from the wall, $\psi(z) \to -\psi(0)$, $z \to \infty$. This is undesirable because in
order for the density to decay to its bulk value far from the wall, the
second form of the Ornstein–Zernike equation implies that $\chi(z) \to$
$-\beta q\psi(0)$, $z \to \infty$ (and the closure below implies the same limiting
behavior for the bridge function). By adding an appropriate constant one
can set the zero of the potential to be in the bulk,

$$\psi(z) = \begin{cases} \psi(0) & z \le 0 \\ \dfrac{4\pi}{\varepsilon}\sum_{\gamma} q_{\gamma}\rho_{\gamma} \displaystyle\int_{z}^{\infty} h_{0\gamma}(z')(z - z')\, dz' & z \ge 0 \end{cases} \qquad (3.25)$$

and now $\psi(z)$, $\chi(z)$, and $\underline{d}(z) \to 0$ all go to zero in the bulk far from the
wall. (It is possible to include a Stern layer, $-\delta < z < 0$, which is the
region between the plane of closest approach of the ions and the plane in
which the surface charge is located, in which the potential is linear. This
has no effect on the properties of the double layer, so it is not treated
here.) This choice of zero is equivalent to choosing the constant of the
wall–ion Coulomb potential such that

$$V_0^{\text{Coul}}(z) = -\frac{2\pi\sigma}{\varepsilon}|z| + \frac{2\pi\sigma}{\varepsilon}S + \frac{\psi(0)}{2} \qquad \text{all } z \qquad (3.26)$$

where $S \to \infty$ is the lateral extent of the wall. The potential drop across
the double layer is

$$\psi(0) = \frac{-4\pi}{\varepsilon}\sum_{\gamma} q_{\gamma}\rho_{\gamma} \int_{0}^{\infty} h_{0\gamma}(z')z'\, dz' \qquad (3.27)$$

and it is the variation of this with surface charge that is the differential capacitance. The first moment of the total correlation function represents the dipole moment of the double layer, and $\psi(0)$ is just the drop in electrostatic potential across a dipole layer.

The diagrammatic definitions of the wall–ion correlation functions are identical to the usual ion–ion ones, and the exact closure is

$$h_{0\alpha}(z) = -1 + \exp[-\beta q_\alpha V_0(z) + h_{0\alpha}(z) - c_{0\alpha}(z) + d_\alpha(z)] \quad (3.28a)$$

$$= -1 + \exp[h_{0\alpha}(z) - \chi_{0\alpha}(z) + d_{0\alpha}(z)] \quad (3.28b)$$

where $d_{0\alpha}(z)$ is the wall–ion bridge function. Here, and henceforth, it is assumed that the ions interact with the wall only via the Coulomb potential, for $z > 0$, and that there is an infinite repulsion (hard core) for $z < 0$. Note that the second form of the closure, Eq. (3.28b), with the second form of the Ornstein–Zernike equation, Eq. (3.21b), reduce to the nonlinear Boltzmann approximation when $\chi_{\alpha\gamma}(r)$ and $d_{0\alpha}(z)$ are set to zero. Also, the surface charge density does not explicitly appear in the expression for the mean electrostatic potential, Eq. (3.25), or in the alternate forms for the Ornstein–Zernike and closure equations (3.21b) and (3.28b); it must be determined by the electroneutrality condition, Eq. (3.20).

The wall–ion total correlation function that appears here may be identified with the solute–ion total correlation function of the preceding section in the limit that the radius of the macroionic solute goes to infinity,

$$\lim_{R \to \infty} h_{0\alpha}(R + z; R) = h_{0\alpha}(z) \quad (3.29)$$

where on the left-side macroion–ion total correlation function appears with its dependence on the macroion radius shown explicitly, and the right side defines the wall–ion total correlation function. An identical limit holds for $\chi_{0\alpha}(z)$, but in the case of $c_{0\alpha}(z)$, $V_0(z)$, and $\psi(z)$ the limiting equality only holds up to an arbitrary constant.

As in the case of the spherical solute, the short-range part of the wall–ion direct correlation function is exponentially short ranged, and assumed to have a shorter decay length than the total correlation function. This is true of the electrolyte side of the wall, but within the wall it goes to a constant. This can be seen from the Ornstein–Zernike equation (3.21b), where in the limit that $z \to -\infty$ the convolution integral is dominated by regions $z' \approx z$, because $\chi(r)$ is exponentially short ranged, and hence $\underline{H}(z')$ may be replaced by -1 and taken out of the integral. The result is

$$\lim_{z \to -\infty} \chi_{0\alpha}(z) = -1 + \beta q_\alpha \psi(0) + \sum_\gamma \rho_\gamma \hat{\chi}_{\alpha\gamma}(0) \qquad (3.30)$$

or using Eq. (2.21),

$$\underline{\rho}^T \underline{\chi}(z) \to -\beta \underline{\chi}_T^{-1} \qquad z \to -\infty \qquad (3.31)$$

The fact that $\chi_{0\alpha}(z)$ goes to a constant means that its one-dimensional Fourier transform is well defined in the upper half of the complex plane.

2. Asymptotics of the Isolated Planar Double Layer

The asymptotic behavior of the planar electric double can be obtained from the preceding analysis and the large radius limit of the solute asymptotes in Section III. The radial Fourier transforms for spherical symmetric functions become one-dimensional Fourier transforms in the planar case. In the large radius the relationship between them is

$$\hat{f}(k; R) = \frac{4\pi}{k} \int_0^\infty f(r; R) \sin kr \, r \, dr$$

$$= \frac{4\pi}{k} \int_{-R}^\infty f(R + z; R) \sin[k(R + z)](R + z) \, dz$$

$$= \frac{4\pi}{ik} \int_{-R}^\infty f(R + z; R) \sinh[ik(R + z)](R + z) \, dz$$

$$\sim \frac{-2\pi R e^{-ikR}}{ik} \int_{-\infty}^\infty f(z) e^{-ikz} \, dz \qquad R \to \infty \qquad \text{Im}\{k\} > 0$$

$$= \frac{-2\pi R e^{-ikR}}{ik} \bar{f}(k) \qquad (3.32)$$

which defines the one-dimensional Fourier transform, denoted by an overline.

The one-dimensional Fourier transform of the Coulomb part of the wall potential will also be required. This turns out to be a generalized function that may be treated by rewriting Eq. (3.26) in terms of Heaviside step functions

$$V_0^{\text{Coul}}(z) = \frac{-2\pi\sigma}{\varepsilon}[z\theta(z) - z\theta(-z)] + \frac{2\pi\sigma}{\varepsilon} S + \frac{\psi(0)}{2} \qquad (3.33)$$

Now the Fourier transform of $z\theta(z)$ is k^{-2}, provided $\text{Im}\{k\} < 0$, and the

Fourier transform of $-z\theta(-z)$ is also k^{-2}, but for $\mathrm{Im}\{k\} > 0$. Hence, analytic continuation is necessary, and the transform of the wall–ion Coulomb potential in the whole complex plane is

$$\bar{V}_0^{\mathrm{Coul}}(k) = \frac{4\pi\sigma}{\varepsilon k^2} + \left[\frac{2\pi\sigma}{\varepsilon} S + \frac{\psi(0)}{2} \right] 2\pi\delta(k) \qquad \text{all } k \qquad (3.34)$$

This result was obtained in [18] by using Poisson's equation, $V_0''(z) = -(4\pi\sigma/\varepsilon)\delta(z)$, and the fact that the Fourier transform of $f''(z)$ is $-k^2\bar{f}(k)$.

With these results the effective charge on the macroion, Eq. (3.13), becomes

$$\tilde{q}_0(R) = \tilde{\underline{q}}^T \hat{\underline{C}}(i\kappa; R)$$

$$\sim \frac{2\pi Re^{\kappa R}}{\kappa} [\tilde{\underline{q}}^T \underline{\tilde{\chi}}(i\kappa) - \beta \tilde{\underline{q}}^T \underline{q} \bar{V}_0^{\mathrm{Coul}}(i\kappa)] \qquad R \to \infty$$

$$= \frac{2\pi Re^{\kappa R}}{\kappa} [\sigma + \tilde{\underline{q}}^T \underline{\tilde{\chi}}(i\kappa)] \qquad (3.35)$$

where the fact that $\kappa^2 = (4\pi\beta/\varepsilon)\tilde{\underline{q}}^T \underline{q}$ has been used. In view of this, one defines the effective surface charge density [18]

$$\tilde{\sigma} = \tfrac{1}{2} [\sigma + \tilde{\underline{q}}^T \underline{\tilde{\chi}}(i\kappa)] \qquad (3.36)$$

where a factor of one-half is included in the definition to preserve the Poisson–Boltzmann form for the asymptote, and so that at low concentrations $\tilde{\sigma} = \sigma$. This last fact follows because in the Poisson–Boltzmann limit $\chi_{\alpha\gamma}(r) = 0$, and hence $\kappa = \kappa_D$ and $\tilde{\underline{q}} = q$. In this regime the linear Poisson–Boltzmann profile holds, $\underline{H}(z) = -\beta q\psi(z)$, $z > 0$, and Eq. (3.21b) shows that $\chi_{0\alpha}(z) = 0$, $z > 0$, and that $\chi_{0\alpha}(z) = -1 + \beta q_\alpha \psi(0)$, $z < 0$. In this limit $q^T \underline{\chi}(z) = \kappa_D^2 \varepsilon \psi(0)/4\pi = \kappa_D\sigma$, $z < 0$, and hence $q^T \underline{\chi}(i\kappa_D) = \sigma$. Note that this expression for the effective surface charge holds when there is no Stern layer; if one has a Stern layer of width δ, then one simply replaces σ by $\sigma e^{-\kappa\delta}$ in this result; the expressions given below remain formally unchanged.

Inserting these results into the solute–ion correlation function, Eq. (3.14),

$$h_{0\gamma}(R+z; R) \sim 2\,\mathrm{Re}' \left\{ \frac{-\beta\tilde{q}_0(R)\tilde{q}_\gamma}{\varepsilon\nu} \frac{e^{-\kappa(R+z)}}{R+z} \right\} \qquad z \to \infty \qquad (3.37)$$

one obtains in the planar limit

$$h_{0\gamma}(z) \sim 2\,\mathrm{Re}'\left\{\frac{-4\pi\beta\tilde{q}_\gamma\tilde{\sigma}}{\varepsilon\nu\kappa}\,e^{-\kappa z}\right\} \qquad z\to\infty \qquad (3.38)$$

The limiting procedure is $R\to\infty$, $z\to\infty$, $z/R\to 0$. [The factor of $4\pi/\kappa$ comes from the prefactor of $\tilde{q}_0(R)$ and the factor of $1/2$ in the definition of $\tilde{\sigma}$.] As before, $\mathrm{Re}'\{z\}$ means to take the real part of z if z is complex, and to take one-half of z if z is real, a consequence of the two mutually conjugate poles located in the upper half of the complex plane. The fact that the decay length for the planar double layer is the same as for the spherical double layer and for the bulk electrolyte is not unexpected, and obviously the transition from monotonic to oscillatory double-layer profiles will occur at the same point as in the bulk electrolyte. This was pointed out for the planar double layer by Stillinger and Kirkwood [65], who gave approximate expressions for the decay length, the transition point, and also the mean electrostatic potential drop across the double layer. (See also the modified Poisson–Boltzmann results [60, 79, 80].) Note that in planar geometry the Yukawa-form of the spherical case has become a pure exponential decay, a consequence of the fact that the three-dimensional Fourier transform of the short-ranged direct correlation function has been replaced by a one-dimensional one. Substituting this asymptote into the expression for the mean electrostatic potential, Eq. (3.25), one obtains

$$\psi(z) \sim 2\,\mathrm{Re}'\left\{\frac{4\pi\tilde{\sigma}}{\varepsilon\nu\kappa}\,e^{-\kappa z}\right\} \qquad z\to\infty \qquad (3.39)$$

Again, one sees that the effective charges give the response to the mean electrostatic potential, and that in the asymptotic regime the latter is proportional to the effective surface charge density.

Since electrostatics do not enter in the core-dominated regime, one can immediately take the planar limit of the core-dominated spherical solute result, Eq. (3.16), to obtain

$$h_{0\alpha}(z) \sim 2\,\mathrm{Re}'\left\{\frac{-2ie^{-\xi z}}{\rho^2 x'(i\xi)}\sum_{\gamma>0}\rho_\gamma\bar{x}_{0\gamma}(i\xi)\right\} \qquad \alpha>0 \qquad z\to\infty \qquad (3.40)$$

Despite the charge on the wall, this shows that all of the ion profiles are asymptotically identical in the core-dominated regime.

Figure 3 shows the asymptotic behavior of the isolated planar double layer in the monotonic regime. After several screening lengths the counterion total correlation function has gone over to its asymptotic

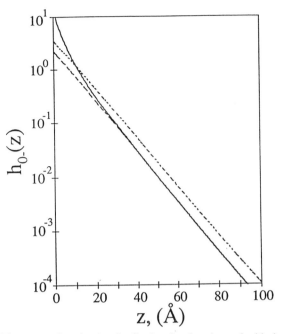

Figure 3. The counterion density distribution for the planar double layer for an area per surface charge of 250 Å^2, in contact with a 0.1 M monovalent restricted primitive model electrolyte ($\varepsilon_r = 78.358$, $T = 298.15$ K, $d = 4.25$ Å), with Debye length of 9.61 Å, and an actual decay length of 9.31 Å, and $\nu = 1.000$. The curve is the hypernetted chain result, which is well fitted by the straightline asymptote, Eq. (3.38), with effective change obtained from Eq. (3.36) equal to 405.2 Å^2 per unit surface charge. The remaining straight line is the linear Poisson–Boltzmann prediction using the actual surface charge density.

form, Eq. (3.28). In this case the Debye length, $\kappa_D^{-1} = 9.61$ Å is quite close to the actual screening length of the electrolyte $\kappa^{-1} = 9.32$ Å, and it would be hard to measure the discrepancy experimentally. (It would in fact be impossible to do so without a precise independent measure of the actual electrolyte concentration; the measured screening length is often used to determine the electrolyte concentration via the Debye–Hückel formula.) However, the magnitude of the density profile does show a significant discrepancy from the classical prediction. In this case the effective surface charge given by Eq. (3.36) in hypernetted chain approximation is 405.2 Å^2 (401.8 Å^2 when bridge diagrams are included), which is about two-thirds of the actual surface charge. In other words, if one were to analyze the measured profiles with the linear Poisson–Boltzmann theory, one would obtain quite a good fit to the data using a substantially decreased surface charge density, which one would presumably attribute to physicochemical binding of counterions to the surface, thereby deducing an erroneous binding constant. Similar conclusions hold for the nonlinear Poisson–Boltzmann theory, which could be used with an area per unit charge of 290 Å^2 to fit the asymptotic tail, about 15% too large. For a quantitative analysis of experimental data, one should use an accurate theory that includes ion size and correlations. Alternatively, one may fit the data with the mean-field Poisson–Boltzmann approximation, provided that one corrects the fitted surface charge density to obtain the actual surface charge density. Data for such a correction is given in Section V.

3. Algebraic Correlations along a Wall

The singlet approach that has been used so far gives the wall–ion distribution function, which corresponds to the density profile in the double layer. It does not directly give the correlation between ions in the presence of the wall (although these are implicitly contained in the diagrams comprising the wall–ion correlation functions). These ion–ion distribution functions near a wall are given by the inhomogeneous Ornstein–Zernike equation, and it will now be demonstrated that they decay as inverse cubics parallel to the wall, which contrasts with the exponential asymptotes of the bulk correlation functions and of the density profiles in the double layer.

For the planar geometry of an isolated charged wall, the density depends only on the distance from the wall, $\rho_\alpha(\mathbf{r}) = \rho_\alpha(z)$, and the pair correlation functions depend on the distances of the two ions from the wall and upon their separation in the direction parallel to the wall, $h_{\alpha\gamma}(\mathbf{r}_1, \mathbf{r}_2) = h_{\alpha\gamma}(s_{12}, z_1, z_2)$. The matrix of distribution functions becomes $\{\underline{\underline{H}}(s_{12}, z_1, z_2)\}_{\alpha\gamma} = \rho_\alpha^{1/2}(z_1) h_{\alpha\gamma}(s_{12}, z_1, z_2) \rho_\gamma^{1/2}(z_2)$. In this planar geometry the inhomogeneous Ornstein–Zernike equation is

$$\underline{\underline{H}}(s_{12}, z_1, z_2) = \underline{\underline{C}}(s_{12}, z_1, z_2) + \int d\mathbf{s}_3 \int_0^\infty dz_3 \, \underline{\underline{\hat{H}}}(s_{13}, z_1, z_3) \underline{\underline{\hat{C}}}(s_{32}, z_3, z_2)$$

$$(3.41a)$$

which partially factorizes upon Fourier transformation in the lateral direction,

$$\underline{\underline{\hat{H}}}(k, z_1, z_2) = \underline{\underline{\hat{C}}}(k, z_1, z_2) + \int_0^\infty \underline{\underline{\hat{H}}}(k, z_1, z_3) \underline{\underline{\hat{C}}}(k, z_3, z_2) \, dz_3 \quad (3.41b)$$

In this section the circumflex signifies the two-dimensional Fourier transform of a function with cylindrical symmetry, which is just the

Hankel transform of order zero,

$$\hat{f}(k) = 2\pi \int_0^\infty f(r) J_0(kr) r \, dr \qquad (3.42a)$$

with inverse

$$f(r) = \frac{1}{2\pi} \int_0^\infty \hat{f}(k) J_0(kr) k \, dk \qquad (3.42b)$$

In this cylindrical coordinate system the ion–ion potential is

$$\underline{\underline{u}}^{\text{Coul}}(s_{12}, z_1, z_2) = \frac{1/4\pi\beta}{\sqrt{s_{12}^2 + (z_1 - z_2)^2}} \underline{\underline{Q}}(z_1, z_2) \qquad (3.43a)$$

which has transform

$$\underline{\underline{\hat{u}}}^{\text{Coul}}(k, z_1, z_2) = \frac{e^{-k|z_1 - z_2|}}{2\beta k} \underline{\underline{Q}}(z_1, z_2) \qquad (3.43b)$$

(The charge matrix has a spatial dependence through the densities that it contains.) It is assumed without proof that the inhomogeneous ion–ion direct correlation function decays as the Coulomb potential, and hence that the function $\underline{\underline{\chi}}(s_{12}, z_1, z_2) = \underline{\underline{C}}(s_{12}, z_1, z_2) + \beta \underline{\underline{u}}^{\text{Coul}}(s_{12}, z_1, z_2)$ is at least integrable. The alternative form of the Ornstein–Zernike equation is

$$\underline{\underline{\hat{H}}}(k, z_1, z_2) = \underline{\underline{\hat{\chi}}}(k, z_1, z_2) + \int_0^\infty \underline{\underline{\hat{H}}}(k, z_1, z_3) \underline{\underline{\hat{\chi}}}(k, z_3, z_2) \, dz_3$$

$$- \underline{\underline{Q}}(z_1, z_2) \frac{e^{-k|z_1 - z_2|}}{2k} - \int_0^\infty \underline{\underline{\hat{H}}}(k, z_1, z_3) \underline{\underline{Q}}(z_3, z_2) \frac{e^{-k|z_3 - z_2|}}{2k} \, dz_3$$

$$(3.44)$$

The last two terms represent the transform of the fluctuation potential; $\phi_\alpha(s_{12}, z_1, z_2)$ is the change in the mean electrostatic potential at s_2, z_2 due to an ion of type α being at (s_1, z_1),

$$\phi_\alpha(s_{12}, z_1; z_2) = \frac{q_\alpha}{\varepsilon r_{12}} + \sum_\gamma \int h_{\alpha\gamma}(s_{13}, z_1, z_3) \frac{q_\gamma}{\varepsilon r_{32}} \rho_\gamma(z_3) \, d\mathbf{r}_3 \quad (3.45)$$

The fluctuation potential in the double layer plays the role of the mean electrostatic potential about an ion in the bulk electrolyte.

Now a moment expansion will be performed. As in the analysis of the bulk electrolyte (Section II.B), for an exponentially short-ranged function all moments exist and the odd moments are zero. Since in this planar geometry the transform of the Coulomb potential contains odd powers of k (whereas only $1/k^2$ appeared in the bulk) one anticipates that the correlation functions will exhibit power-law decay parallel to the wall. One defines the moments to be

$$\hat{\underline{\underline{H}}}(k, z_1, z_2) = \underline{\underline{H}}^{(0)}(z_1, z_3) + |k| \underline{\underline{H}}^{(1)}(z_1, z_3) + \cdots \qquad (3.46)$$

and similarly for $\hat{\underline{\underline{\chi}}}$. Note that the absolute value of k appears because by definition $\underline{\underline{H}}(r, z_1, z_2)$ is a real even function of r. One now takes the limit that $k \to 0$ in the Ornstein–Zernike equation and equates coefficients. (For finite $|z_1 - z_2|$ it is permissible to expand the various exponentials since z_3 is kept finite by the short range of the total correlation function in the integrand.) By equating the coefficients of k^{-1} one has

$$0 = 0 + 0 - \frac{1}{2} \underline{\underline{Q}}(z_1, z_2) - \frac{1}{2} \int_0^\infty \underline{\underline{H}}^{(0)}(z_1, z_3) \underline{\underline{Q}}(z_3, z_2) \, dz_3 \qquad (3.47a)$$

or

$$-q_\alpha = \sum_\gamma q_\gamma \int h_{\alpha\gamma}(s_{12}, z_1, z_2) \rho_\gamma(z_2) \, dz_2 \, ds_2 \qquad (3.47b)$$

This is just the local electroneutrality condition, Eq. (2.45b), in planar geometry.

Equating the coefficient of k^0 one obtains

$$\underline{\underline{H}}^{(0)}(z_1, z_2) = \underline{\underline{\chi}}^{(0)}(z_1, z_2) + \int_0^\infty \underline{\underline{H}}^{(0)}(z_1, z_3) \underline{\underline{\chi}}^{(0)}(z_3, z_2) \, dz_3$$

$$+ \underline{\underline{Q}}(z_1, z_2) \frac{|z_1 - z_2|}{2}$$

$$+ \int_0^\infty \underline{\underline{H}}^{(0)}(z_1, z_3) \underline{\underline{Q}}(z_3, z_2) \frac{|z_3 - z_2|}{2} \, dz_3$$

$$- \frac{1}{2} \int_0^\infty \underline{\underline{H}}^{(1)}(z_1, z_3) \underline{\underline{Q}}(z_3, z_2) \, dz_3 \qquad (3.48)$$

By premultiplying by $\underline{\underline{Q}}(z_0, z_1)$ and integrating over z_1, one can cancel most of these terms by the local electroneutrality condition, and one is left with

$$-\underline{\underline{Q}}(z_0, z_2) = \frac{-1}{2} \int_0^\infty dz_1 \int_0^\infty dz_3 \underline{\underline{Q}}(z_0, z_1) \underline{\underline{H}}^{(1)}(z_1, z_3) \underline{\underline{Q}}(z_3, z_2) \quad (3.49a)$$

or

$$1 = \frac{2\pi\beta}{\varepsilon} \sum_{\alpha\gamma} q_\alpha q_\gamma \int_0^\infty dz_1 \int_0^\infty dz_2 \rho_\alpha(z_1) \rho_\gamma(z_2) h_{\alpha\gamma}^{(1)}(z_1, z_2) \quad (3.49b)$$

This proves that the first moment of the total correlation function is nonzero. Now if $\hat{f}(k) \sim \text{constant} + A|k|$, $k \to 0$, then $f(r) \sim -A/2\pi r^3$, $r \to \infty$. That is,

$$h_{\alpha\gamma}(r, z_1, z_2) \sim \frac{-h_{\alpha\gamma}^{(1)}(z_1, z_2)}{2\pi r^3} \qquad r \to \infty \qquad (3.50)$$

where the numerator obeys the sum rule Eq. (3.49b). The asymptote and the sum rule for an electrolyte next to a planar wall were first given by Jancovici [85, 86] and have been discussed by others [87–89]. The fact that the ion–ion correlations decay as inverse cubics in the direction parallel to a planar wall is in marked contrast to the exponential decay of the bulk electrolyte. It is a consequence of the inability of the electrolyte to screen the charge on an ion equally on all sides due to the presence of the wall. The ion and its countercharge cloud has a net dipole moment, and the r^{-3} behavior results from the dipole–dipole interaction. Identical decay occurs for an electrolyte confined to a two-dimensional planar surface [90–94]; in this case the densities that appear in Eq. (3.49b) become δ functions and the two-dimensional charge–charge correlation function has a universal asymptote [93, 94]. Note that an r^{-3} tail is an integrable function in the two-dimensional subspace parallel to the wall.

This behavior does not depend on the charge on the wall but depends only on the fact that it excludes electrolyte from a semiinfinite half-space. For a wall of finite thickness t with electrolyte on both sides, it may be dominant behavior for $r \ll t$, but in the asymptotic limit the correlations parallel to the surface will decay exponentially with the bulk correlation length [88, 95]. For the seminfinite half-space treated in detail above, if the asymptotic limit is taken such that the perpendicular distance of the ions from the wall always remains much greater than their lateral separation, than one expects them to behave like the bulk correlation function and to decay exponentially. The existence of algebraic correlations depends on the inability of ions to be screened on all sides; the correlations between ions confined to an infinite cylinder decay as [96] $(z \ln|z|)^{-2}$, $|z| \to \infty$.

In addition to the local electroneutrality condition and the sum rule on the first moment of the inhomogeneous total correlation function, there is also the second moment condition, which can be written as a sum rule for the fluctuation potential [30, 31]

$$1 = -\beta \sum_\alpha q_\alpha \int ds\, dz_1\, \rho_\alpha(z_1)\phi_\alpha(s, z_1; z_2) \qquad (3.51)$$

which follows from the alternate form of the Ornstein–Zernike equation, Eq. (3.44), upon multiplication by the charge matrix, integration, k limitation, and invocation of the local electroneutrality condition, Eq. (3.47b). Finally, a dipole sum rule for the inhomogeneous total correlation function of the planar double layer was derived by Blum et al. [97] (see [6, 89] for an alternative derivation), and Carnie [98] showed that it was related to the second moment condition.

C. Analysis of Interacting Charged Walls

1. Wall–Wall Ornstein–Zernike Equation

The singlet approach can be applied to the problem of interacting planar walls [77]. In the general case of interacting solutes, the solute–solute Ornstein–Zernike equation exists [cf. Eq. (3.7)], as do the solute–solute pair correlation functions. However, for two walls of infinite cross-sectional area, the solute–solute pair correlation functions do not exist except as generalized functions (e.g., the pair distribution function is a set of delta functions). In this case the physical entity is the interaction free energy per unit area. In diagrammatic terms, this is the wall–wall potential, minus the connected pair diagrams that have the walls as nonadjacent root points not forming an articulation pair, per unit area. This simply corresponds to the solute–solute potential of mean force per unit area, which is just the exponent of the closure equation. One has

$$\beta w_{00}^{\text{int}}(z) = \beta V_{00}(z) - d_{00}(z) - \int_{-\infty}^{\infty} \underline{H}^T(z') \underline{C}(z - z')\, dz' \qquad (3.52)$$

where $\beta = 1/k_B T$, and where the wall–wall quantities per unit area symbolized by w, V, and d are the interaction free energy, potential, and the bridge function, respectively. The convolution integral, which is clearly just the series function per unit area, contains the wall–ion pair correlation functions discussed in Section III.B.1. It will be assumed that the walls interact only with the Coulomb potential, and that one is only interested in the region $z > 0$, where z represents the width of the region available to the centers of the ions [cf. Eq. (3.19)]. In order to express this interaction free energy in terms of short-range functions, one needs to remove the wall–ion Coulomb potential from the integral of the wall–ion direct correlation function [cf. Eq. (3.22)]. One has

$$\int_0^\infty \underline{H}^T(z') \underline{q} V_0^{Coul}(z - z') \, dz'$$

$$= \sigma\psi(z) + \frac{2\pi\sigma^2}{\varepsilon} z - \frac{2\pi\sigma^2}{\varepsilon} S - \sigma\psi(0) \quad z > 0 \qquad (3.53)$$

where the single-wall electroneutrality condition, Eq. (3.20), the mean electrostatic potential due to a charged wall, Eq. (3.25), the wall–ion Coulomb potential, Eq. (3.26), and the potential drop across an isolated double layer, Eq. (3.27), have all been used. In view of this, one chooses the arbitrary constants in the Coulomb part of the wall–wall potential to cancel these, leaving only the mean electrostatic potential outside of the integral. Accordingly, one chooses

$$V_{00}(z) = \frac{2\pi\sigma^2}{\varepsilon} S + \sigma\psi(0) - \frac{2\pi\sigma^2}{\varepsilon} z \quad z > 0 \qquad (3.54)$$

and the interaction free energy per unit area becomes

$$\beta w_{00}^{int}(z) = \beta\sigma\psi(z) - d_{00}(z) - \int_{-\infty}^\infty \underline{H}^T(z') \underline{\chi}(z - z') \, dz' \quad z > 0$$

$$(3.55)$$

In the case of a Stern layer of width $\delta > 0$, one replaces $\psi(z)$ by $\psi(z + \delta)$ in this expression. Note that $w_{00}^{int}(z) \to 0$, $z \to \infty$ (see below), and hence no surface or bulk contributions are included in this free energy expression. The net pressure (force per unit area) between the walls is

$$p^{net}(z) = \frac{-\partial w_{00}^{int}(z)}{\partial z} \qquad (3.56)$$

This corresponds to the total pressure of the confined electrolyte less the osmotic pressure of the bulk electrolyte, which is imagined to be pressing on the far sides of the walls (at infinity).

The above expressions are for two identical walls, but the generalization to the nonsymmetric situation is immediate. If one uses the subscripts "1" and "2" to denote the two walls, then one has

$$\beta w_{12}^{int}(z) = \beta \sigma_1 \psi_2(z) - d_{12}(z) - \int_{-\infty}^{\infty} \underline{H}_1^T(z') \underline{\chi}_2(z - z') \, dz'$$

$$= \beta \sigma_2 \psi_1(z) - d_{12}(z) - \int_{-\infty}^{\infty} \underline{H}_2^T(z') \underline{\chi}_1(z - z') \, dz' \quad (3.57)$$

2. The Derjaguin Approximation

The Derjaguin approximation [99] relates the total force between particles with curved surfaces to the interaction free energy per unit area between planar walls, and has found extensive application in colloid science [1], including nonspherical particles [100]. Its utility lies in the fact that it subsumes one degree of freedom, the curvature, into a single geometric parameter times the planar free energy. In general, the free energy is difficult to obtain, and considerable efficiency can result if one needs to calculate it only for the simplest case, namely, planar geometry, which has the greatest symmetry. The Derjaguin approximation allows the planar result to be applied to interacting particles of arbitrary curvature. What is perhaps obscure is the precise nature and regime of validity of the approximation. Attard and co-workers [77, 101] proceeded by comparing the formally exact Ornstein–Zernike results for planes and for spheres in the large radius limit of the latter. That approach is modified here for the electric double layer.

For macrospheres, the solute–solute potential of mean force may be written

$$\beta w_{00}^{int}(x + 2R; R) = \beta \int \sigma(s; R) \psi(|s - r|; R) \, ds - d_{00}(x + 2R; R)$$

$$- b_{00}^{sr}(x + 2R; R) \quad (3.58)$$

where the macroion has surface charge $\sigma(s; R) = \sigma \delta(s - R)$, and where $\mathbf{r} = (x + 2R)\hat{\mathbf{x}}$. The modified series function is

$$b_{00}^{sr}(r; R) = \int \underline{H}^T(s; R) \underline{\chi}(|\mathbf{r} - \mathbf{s}|; R) \, ds$$

$$= \frac{2\pi}{r} \int_0^{\infty} ds \, s \underline{H}^T(s; R) \int_{|r-s|}^{r+s} dt \, t \underline{\chi}(t; R) \quad (3.59)$$

In the last line bispherical coordinates are used, and s and t measure the radial distances from the centers of the macroions. Let the surface separation of the macroions be $x \equiv r - 2R$, and define $y \equiv s - R$ and

$z \equiv t - R$. The planar limit is obtained as $R \to \infty$, with the separations of interest being $0 < x \ll R$. Because the correlation functions are exponentially short ranged on the electrolyte side of the interface, the integral is dominated by regions $y \ll R$ and $z \ll R$. Changing the variables of integration, the lower limit of the t integral becomes $|2R + x - y - R| - R = x - y$, and the upper limit may be extended to infinity. This gives

$$b_{00}^{sr}(x + 2R; R) = \frac{2\pi}{x + 2R} \int_{-R}^{\infty} dy(y + R)\underline{H}^T(y + R; R)$$

$$\times \int_{x-y}^{\infty} dz(z + R)\underline{\chi}(z + R; R) \qquad (3.60)$$

Now, as $y \to -R$, $z \to \infty$, and $\chi_{0\alpha}(z + R; R) \to 0$. Therefore the integrals are dominated by $y \gg -R$ and $|z| \ll R$. Accordingly, the large radius limit yields

$$\lim_{R \to \infty} b_{00}^{sr}(x + 2R; R) = \pi R \int_{-\infty}^{\infty} dy\, \underline{H}^T(y) \int_{x-y}^{\infty} dz\, \underline{\chi}(z) \qquad (3.61)$$

where wall–ion correlation functions appear in the integrand. By differentiating one obtains

$$\lim_{R \to \infty} \frac{-1}{\pi R} \frac{\partial b_{00}^{sr}(x + 2R; R)}{\partial x} = \int_{-\infty}^{\infty} \underline{H}^T(y)\underline{\chi}(x - y)\, dy$$

$$\equiv b_{00}^{sr}(x) \qquad (3.62)$$

where the last quantity is the modified wall–wall series function per unit area. This result for interacting macrospheres was derived by Attard et al. [77] and by Henderson [102].

The treatment of the bridge function is very similar. Since it consists of $h_{0\gamma}(r; R)$ and $h_{\alpha\gamma}(r)$ bonds, which are exponentially short ranged, the convolution integrals are dominated by regions with the field points in the electrolyte between the two surfaces. One field point can be singled out, and the bridge function can be written as an integral over s_α and sum over α of $c_{00\alpha}(r, s, t; R)$, which is that part of the triplet direct correlation function with nonadjacent solutes not forming an articulation pair [101]. Transforming to bispherical coordinates, one concludes as above that

$|y| \ll R$ and that $|z| \ll R$, and hence that

$$\lim_{R \to \infty} \frac{-1}{\pi R} \frac{\partial d_{00}(x + 2R; R)}{\partial x} = \sum_\alpha \rho_\alpha \int_{-\infty}^{\infty} \underline{c_{00\alpha}}(y, x - y)\, dy$$

$$\equiv d_{00}(x) \qquad\qquad (3.63)$$

where the last quantity is the wall–wall bridge function per unit area.

The final contribution to the macrospherical solute–solute potential of mean force, Eq. (3.58), is the charge times the mean electrostatic potential, which is written as an integral over the surface charge density. Now the distance between the center of one macroion and the surface of the other at a perpendicular distance τ from the central axis is $t(\tau) = \sqrt{\tau^2 + (R + x + \tau^2/2R)^2} \to R + x + \tau^2/R + x^2/2R$, $R \to \infty$, and hence $\psi(t(\tau); R) \to \psi(x + \tau^2/R + x^2/2R)$, $R \to \infty$. One has

$$2\pi\sigma \int_0^\infty \psi(t(\tau); R)\tau\, d\tau \sim 2\pi\sigma \int_0^\infty \psi(x + \tau^2/R + x^2/2R)\tau\, d\tau$$

$$\sim \pi R\sigma \int_x^\infty \psi(z)\, dz \qquad R \to \infty \qquad (3.64)$$

When differentiated with respect to x this yields a contribution of the same form as for the bridge function and for the short-range part of the series function.

Now the negative derivative of the potential of mean force is the force between the solutes $F_{00}(x + 2R; R)$, and one concludes that

$$\lim_{R \to \infty} \frac{\beta}{\pi R} F_{00}(x + 2R; R) = \beta\sigma\psi(x) - d_{00}(x) - b_{00}^{\mathrm{sr}}(x)$$

$$= \beta w_{00}^{\mathrm{int}}(x) \qquad\qquad (3.65)$$

In words, the force between spherical solutes, divided by π times their radius, equals the interaction free energy per unit area between planar walls. This is the Derjaguin approximation, and the present analysis shows that it is exact in the asymptotic limit $R \to \infty$. Certain assumptions made in this derivation remain valid for solutes of large but finite radius if the surface separation is small compared to the radius, $x \ll R$, and, implicitly, if the range of the macroion–ion correlation functions in the electrolyte is less than the radius, $\kappa^{-1} \ll R$. (This last condition was used to find the dominant regions of the integrals, and to extend the ranges of integration.) Hence, one concludes that it is legitimate to apply the

Derjaguin approximation to colloid particles of finite radius whenever these two conditions are fulfilled.

This derivation of the Derjaguin approximation is based upon the formally exact Ornstein–Zernike expression for the solute–solute interaction free energy, and it is worthwhile to compare it with the more heuristic approach that is generally used [99, 100]. In this approach, one postulates that the potential of mean force between two solutes of large radius and close separation is the sum of a free energy per unit area acting between elements of area on their surface. Since the surfaces are nearly flat, this free energy density is taken to be that for infinite planar walls at the local surface separation. At a distance r from the central axis the latter is $h(r) = x + r^2/R$, where x is the surface separation on the central axis and R is the radius of the two spherical solutes. Consequently, one has

$$w_{00}^{\text{int}}(2R + x; R) = 2\pi \int_0^\infty w_{00}^{\text{int}}(x + r^2/R)r \, dr$$

$$= \pi R \int_x^\infty w_{00}^{\text{int}}(h) \, dh \qquad (3.66)$$

and differentiation yields Eq. (3.65). The quantitative accuracy of the Derjaguin approximation for interacting macroions has been explored within the context of the linear [51, 52] and the nonlinear [55, 53] Poisson–Boltzmann approximations. It has also been tested for hard-sphere fluids using the hypernetted chain approximation, with and without bridge diagrams [77, 103].

3. Asymptotics of Overlapping Planar Double Layers

The asymptotic behavior of two interacting walls is most easily extracted from the interaction free energy written in the form of Eq. (3.55). As previously for the bulk electrolyte and for the isolated double layer, one assumes the bridge function to be more short ranged than the rest. Consequently, the wall–ion mean electrostatic potential and the short-range series function dominate asymptotically, and both have range κ. From the transforms of Eq. (3.38) and of Eq. (3.39), one respectively obtains

$$\bar{H}(k) \sim \frac{-4\pi\beta\tilde{\sigma}}{\varepsilon\nu\kappa} \, \tilde{\underline{q}} \, \frac{1}{\kappa + ik} \qquad k \to i\kappa \qquad (3.67)$$

and

$$\bar{\psi}(k) \sim \frac{4\pi\tilde{\sigma}}{\varepsilon\nu\kappa} \frac{1}{\kappa + ik} \qquad k \to i\kappa \qquad (3.68)$$

assuming for the present that κ is real. Hence, near the pole the wall–wall interaction free energy per unit area, Eq. (3.55), goes as

$$\beta\bar{w}_{00}^{\text{int}}(k) \sim \left[\frac{4\pi\beta\sigma\tilde{\sigma}}{\varepsilon\nu\kappa} + \frac{4\pi\beta\tilde{\sigma}}{\varepsilon\nu\kappa} \underline{\tilde{q}}^T \underline{\tilde{\chi}}(i\kappa) \right] \frac{1}{\kappa + ik} \qquad k \to i\kappa$$

$$= \frac{8\pi\beta\tilde{\sigma}^2}{\varepsilon\nu\kappa} \frac{1}{\kappa + ik} \qquad (3.69)$$

where the expression for the effective surface charge density, Eq. (3.36), has been used. Consequently, the asymptote is [18]

$$w_{00}^{\text{int}}(z) \sim \frac{8\pi\tilde{\sigma}^2}{\varepsilon\nu\kappa} e^{-\kappa z} \qquad z \to \infty \qquad \text{Im}\{\kappa\} = 0 \qquad (3.70a)$$

and

$$w_{00}^{\text{int}}(z) \sim 2\,\text{Re}\left\{ \frac{8\pi\tilde{\sigma}^2}{\varepsilon\nu\kappa} e^{-\kappa z} \right\} \qquad z \to \infty \qquad \text{Im}\{\kappa\} \neq 0 \quad (3.70b)$$

This depends only on properties of the bulk electrolyte, κ and ν, and the effective surface charge of the isolated double layer, $\tilde{\sigma}$. This result reduces to the linear Poisson–Boltzmann theory in the Debye–Hückel limit, $\chi_{\alpha\gamma} = 0$. As for the interaction between identical macroions, because the effective surface charge occurs as a square, and because κ and ν are both positive, in the monotonic regime the force between identically charged walls is repulsive. That is, the interaction between two similar double layers is either monotonically repulsive, or oscillatory, and any attractions measured or predicted cannot persist for all separations (although there can be a very large period of oscillation near the bulk transition).

Finally, in the core-dominated bulk regime it follows from the wall–wall Ornstein–Zernike equation that the interaction free energy per unit area is

$$\beta w^{\text{int}}(z) \sim 2\,\text{Re}'\left\{ \frac{-2ie^{-\xi z}}{\rho^2 x'(i\xi)} \left(\sum_{\gamma>0} \rho_\gamma \bar{\chi}_{0\gamma}(i\xi) \right)^2 \right\} \qquad z \to \infty \quad (3.71)$$

where again the effective surface "charge" that described the amplitude

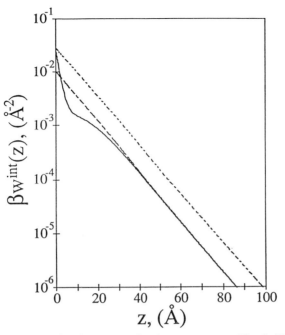

Figure 4. The interaction free energy for the same case as Fig. 3. The hypernetted chain curve goes rapidly to its asymptote, Eq. (3.70a), with $|e/\tilde{\sigma}| = 405.2 \text{ Å}^2$ from Eq. (3.36), but the linear Poisson–Boltzmann equation with the actual surface charge density $|e/\sigma| = 250 \text{ Å}^2$ significantly overestimates the repulsion.

of the decay of the density profiles away from the isolated wall appears as the square in their interaction.

Figure 4 shows the double-layer interaction between two identical planar walls that have the same parameters as the isolated wall described in Fig. 3. It is evident that the asymptotic regime extends to separations as small as several screening lengths, and that the effective charge for the isolated wall, Eq. (3.47b), describes the asymptote for two interacting walls accurately. The linear Poisson–Boltzmann theory using the actual area per unit surface charge of 250 Å^2 significantly overestimates the repulsion, but it would fit the data with the effective surface charge of 405.2 Å^2, from which one might erroneously conclude that one-third of the surface charge had been neutralized by bound counterions.

D.   Other Solute Geometries

One of the great advantages of the singlet method is the flexibility that it allows in the interpretation of a solute. So far results have been given for macrospherical solutes and for planar walls. In what remains of this section three generic types of solutes will be treated: cylinders, pores, and multimolecular species. These illustrate the range of problems that may be treated within the singlet approach.

1. Cylinders

As mentioned in the introduction to Section III.B for isolated walls, the modern single approach is predicated upon the assumption that the solute–ion Ornstein–Zernike equation exists, and the properties of the solute–ion correlation functions are deduced from the symmetry of the solute, and from the behavior of the Coulomb potential. The procedure holds for arbitrary shaped solutes, and here it is illustrated for cylinders. Previous work on the singlet method for the ion distributions about an isolated cylinder was in the context of the hypernetted chain approximation (i.e., neglect of the cylinder–ion bridge function), and Lozada–Cassou has reviewed these derivations and results [4]. Discussion of the approximate methods is deferred until Section IV, wheres in this section the object is to give formally exact singlet expressions for an isolated cylinder, and to extend them to two interacting cylinders. Asymptotic results are also given, including a formal microscopic expression for the amount of counterion condensation that occurs on a charged cylinder.

For an infinitely long cylinder of radius R, the solute–ion total correlation function depends only on r, the distance from the axis of the cylinder, and one has

$$h_{0\gamma}(r) = -1 \qquad r < R \tag{3.72}$$

where the surface of closest approach has been taken to be the same for all the ions. The cylinder–ion Ornstein–Zernike equation is

$$\underline{H}(r) = \underline{C}(r) + \int \underline{\underline{C}}(|\mathbf{r} - \mathbf{s}|)\underline{H}(s_r)\,d\mathbf{s}$$

$$= \underline{C}(r) + \int_{-\infty}^{\infty} dz \int_{0}^{2\pi} d\theta \int_{0}^{\infty} ds_r\, s_r \underline{\underline{C}}(\sqrt{z^2 + r^2 + s_r^2 - 2rs_r \cos\theta})\underline{H}(s_r) \tag{3.73}$$

where s_r is the radial component of s.

For a cylinder with surface charge density σ (i.e., with an axial length per unit charge of $|e/2\pi R\sigma|$, where e is the charge on an electron) the electroneutrality condition is

$$-\sigma = \frac{1}{R} \int_R^\infty \underline{q}^T \underline{H}(r) r \, dr \qquad (3.74)$$

and the cylinder–ion Coulomb potential is (up to an arbitrary constant)

$$V_0^{\text{Coul}}(r) = \begin{cases} \dfrac{-4\pi\sigma R}{\varepsilon} \ln r & r > R \\ \dfrac{-4\pi\sigma R}{\varepsilon} \ln R & r < R \end{cases} \qquad (3.75)$$

The mean electrostatic potential becomes

$$\psi(r) = V_0^{\text{Coul}}(r) - \frac{4\pi}{\varepsilon} \int_R^r \underline{q}^T \underline{H}(r') \ln(r) r' \, dr' - \frac{4\pi}{\varepsilon} \int_r^\infty \underline{q}^T \underline{H}(r') \ln(r') r' \, dr'$$

$$= \begin{cases} \psi(R) & r \leq R \\ \dfrac{4\pi}{\varepsilon} \int_r^\infty \underline{q}^T \underline{H}(r') \ln[r/r'] r' \, dr' & r \geq R \end{cases} \qquad (3.76)$$

Here the electroneutrality condition has been used, and it may be seen that as $r \to \infty$, $\psi(r)$ vanishes with the same decay length as $\underline{H}(r)$.

Analogously to the spherical and planar solutes treated above, the cylinder–ion Ornstein–Zernike equation may be rewritten in terms of short-range functions,

$$\underline{H}(r) = -\beta \underline{q} \psi(r) + \underline{\chi}(r) + \int \underline{\chi}(|\mathbf{r} - \mathbf{s}|) \underline{H}(s_r) \, d\mathbf{s} \qquad (3.77)$$

where $\underline{\chi}(r)$ is the short range part of the cylinder–ion direct correlation function. Similarly, the formally exact closure beyond contact is

$$h_{0\alpha}(r) = -1 + \exp[h_{0\alpha}(r) - \chi_{0\alpha}(r) + d_{0\alpha}(r)] \qquad (3.78)$$

where $d_{0\alpha}(r)$ is the cylinder–ion bridge function.

The asymptotic behavior of the ions far from the cylinder is obtained from the Fourier transform of the Ornstein–Zernike equation, Eq. (3.73),

$$\bar{\underline{H}}(k_r)\delta(k_z) = \bar{\underline{C}}(k_r)\delta(k_z) + \hat{\underline{C}}(\sqrt{k_r^2 + k_z^2})\bar{\underline{H}}(k_r)\delta(k_z) \qquad (3.79)$$

Here the overline denotes the two-dimensional Fourier transform of a circularly symmetric function (Hankel transform of order zero), k_r and k_z are the radial and axial components of the three-dimensional Fourier vector, and the Dirac deltas appear because of the constancy of the cylinder functions in the axial direction. Integrating both sides with respect to k_z, setting $k = k_r$, and rearranging, one obtains

$$\bar{H}(k) = (\underline{I} - \hat{\underline{C}}(k))^{-1}[\bar{\underline{\chi}}(k) - \beta \bar{V}_0^{\text{Coul}}(k)\underline{q}] \tag{3.80}$$

As always the transform of the cylinder–ion Coulomb potential is problematic. As in [18] [see also Eq. (3.34)] the analytic part can be extracted from Poisson's equation, $\nabla^2 V_0^{\text{Coul}}(r) = -(4\pi\sigma/\varepsilon)\delta(r - R)$, and the fact that the transform of $f''(r)$ is $-k^2\bar{f}(k)$. Hence,

$$-k^2\bar{V}_0^{\text{Coul}}(k) = 2\pi \int_0^\infty \frac{-4\pi\sigma}{\varepsilon} \delta(r - R)J_0(kr)r \, dr$$

$$= \frac{-8\pi^2\sigma R}{\varepsilon} J_0(kR) \tag{3.81}$$

One sees that the pole that occurs in the Ornstein–Zernike equation is just that of the bulk electrolyte, and hence the analysis of Section III.A.2 goes through unchanged. The result is

$$\bar{H}(k) \sim \frac{-4\pi\beta/\nu\varepsilon}{k^2 + \kappa^2} [\tilde{\underline{q}}^T \bar{\underline{\chi}}(i\kappa) + 2\pi\sigma RI_0(\kappa R)]\tilde{\underline{q}} \qquad k \to i\kappa \qquad \text{Im}\{\kappa\} = 0 \tag{3.82}$$

where $I_0(z) = J_0(iz)$ is the modified Bessel function of the first kind. Now

$$\frac{1}{2\pi} \int_0^\infty \frac{1}{k^2 + \kappa^2} J_0(kr)k \, dk = \frac{1}{2\pi} K_0(\kappa r) \tag{3.83}$$

where $K_0(z)$ is the modified Bessel function of the second kind, and hence

$$\underline{H}(r) \sim \frac{-2\beta}{\varepsilon\nu} [\tilde{\underline{q}}^T \bar{\underline{\chi}}(i\kappa) + 2\pi\sigma RI_0(\kappa R)]K_0(\kappa r)\tilde{\underline{q}} \qquad r \to \infty \qquad \text{Im}\{\kappa\} = 0 \tag{3.84}$$

Since

$$K_0(\kappa r) \sim \sqrt{\frac{\pi}{2\kappa r}} \left[1 - \frac{1}{8\kappa r} + \cdots\right]e^{-\kappa r} \qquad r \to \infty \tag{3.85}$$

one sees that the concentrations of ions in the double layer about a charged cylinder decay to their bulk values exponentially, over distances typified by the bulk screening length.

In order to identify the effective surface charge of the cylinder, the asymptote should be compared with the linear Poisson–Boltzmann result [104]

$$\underline{H}(r) = \frac{-4\pi\beta\sigma}{\varepsilon\kappa_D K_1(\kappa_D R)} K_0(\kappa_D r)\underline{q} \qquad (3.86)$$

where $K_1(z) = -K_0'(z)$ is the first-order modified Bessel function of the second kind. Hence, one should take

$$\tilde{\sigma} = \frac{\kappa K_1(\kappa R)}{2\pi} [\underline{\tilde{q}}^T \underline{\tilde{\chi}}(i\kappa) + 2\pi\sigma R I_0(\kappa R)] \qquad \mathrm{Im}\{\kappa\} = 0 \qquad (3.87)$$

as the effective surface charge. In the linear Poisson–Boltzmann limit, $[\underline{\chi}(r) = 0, \tilde{q} = q, \kappa = \kappa_D]$, one has $\underline{H}(r) = -\beta q\psi(r)$, and Eq. (3.77) shows that $\underline{q}^T \underline{\chi}(r) = 0$, $r > R$, and that $\underline{q}^T \underline{\chi}(r) = \kappa_D \sigma K_0(\kappa_D R)/K_1(\kappa_D R)$, $r < R$. It may be verified that the Hankel transform is $\underline{q}^T \underline{\tilde{\chi}}(i\kappa_D) = 2\pi\sigma R K_0(\kappa_D R) I_1(\kappa_D R)/K_1(\kappa_D R)$, and hence in this limit

$$\tilde{\sigma} = \sigma\kappa_D R[K_0(\kappa_D R)I_1(\kappa_D R) + I_0(\kappa_D R)K_1(\kappa_D R)]$$

$$= \sigma \qquad (3.88)$$

where the Wronskian for the modified Bessel functions has been used. Thus one recovers the Poisson–Boltzmann limit as one should, which persuades that the correct form for the effect surface charge density has been found. Manning [105] has put forward a theory for linear macroions based on counterion condensation in the cylindrical double layer; Eq. (3.87) represents a formal microscopic expression for the effective surface charge density that could presumably be used to characterize unambiguously the amount of condensation that occurs. With these definitions the asymptote is

$$\underline{H}(r) \sim \frac{-4\pi\beta\tilde{\sigma}}{\varepsilon\nu\kappa K_1(\kappa R)} K_0(\kappa r)\underline{\tilde{q}} \qquad r \to \infty \qquad \mathrm{Im}\{\kappa\} = 0 \qquad (3.89)$$

and presumably one takes twice the real part if it is complex.

The interaction between two parallel cylinders can also be treated at the singlet level. As in the case of walls, for finitely long cylinders the cylinder–cylinder pair correlation functions do not exist, and the quantity of physical import is the free energy per unit length. By analogy with the planar result this is

$$\beta w_{00}^{int}(r) = -d_{00}(r) + \beta \sigma R \int_0^{2\pi} \psi(\sqrt{r^2 + R^2 - 2rR \cos\theta}) \, d\theta$$

$$- \int_0^{2\pi} d\theta \int_0^\infty ds \, s \underline{H}^T(s) \underline{\chi}(\sqrt{r^2 + s^2 - 2rs \cos\theta}) \quad (3.90)$$

where $d_{00}(r)$ is the bridge function per unit length for parallel cylinders, and the other functions pertain to an isolated cylinder. For the case of nonparallel cylinders, the total interaction free energy exists, and hence so do the pair correlation functions. Although fundamentally no different from the case of aligned cylinders, geometric factors appear in the more general case that make the result appear more complicated. The result for identical cylinders, one on the z axis, and the other lying in the plane $x = r$ inclined at an angle $\theta \neq 0$ is

$$\beta w_{00}^{int}(r, \theta) = -d_{00}(r, \theta) + \beta \sigma R \int_{-\infty}^\infty dz' \int_0^{2\pi} d\phi \, \psi(t(x', y', z'))$$

$$- \int_{-\infty}^\infty dx \, dy \, dz \, \underline{H}^T(\sqrt{x^2 + y^2}) \underline{\chi}(t(x, y, z)) \quad (3.91)$$

where $x' = R \cos\phi$, $y' = R \sin\phi$, and

$$t(x, y, z) = ((x - r)^2 + y^2 \cos^2\theta - 2yz \sin\theta \cos\theta + z^2 \sin^2\theta)^{1/2} \quad (3.92)$$

is the perpendicular distance between a point (x, y, z) and the axis of the inclined cylinder. Differentiation of this result with respect to r gives the force between the cylinders, and differentiation with respect to θ give the torque. In both cases this free energy (potential of mean force) goes to zero at large separations because the cylinders are surrounded by electrolyte.

The final question to be addressed for cylinders is whether or not the Derjaguin approximation can be applied to two parallel cylinders. In Section III.C.2, it was shown that the force between two macrospheres, divided by π times their radius, equaled the interaction free energy between two walls in the limit that the radius of the spheres went to infinity. For finite radii this was the Derjaguin approximation, and for interacting nonparallel cylinders a corresponding result is known [99, 100]. The case of perfectly aligned cylinders is important because of its relevance to hexagonal phases of rodlike polyions and to the interaction between curved particles adsorbed on a surface or confined to a membrane. The simplest way to make the point is by the heuristic derivation,

Eq. (3.66). That is, one postulates that the potential of mean force (per unit length) for solutes of radius R can, at close separations, be written as the sum of free energy densities acting between elements of area on their surface. In the large radius limit the cylinders are virtually flat, and the free energy density is taken to be that between planar walls at a local separation of $h(y) = x + y^2/R$, where x is the closest separation of their surfaces, and y is the perpendicular distance from the plane containing the cylinders. One has

$$\beta w_{00}^{int}(x + 2R; R) = \int_{-\infty}^{\infty} w_{00}^{int}(x + y^2/R) \, dy$$

$$= \sqrt{R} \int_{x}^{\infty} \frac{w_{00}^{int}(h)}{\sqrt{h - x}} \, dh \tag{3.93}$$

This shows that the potential of mean force per unit length between parallel cylinders scales with the square root of their radius, and that it is a weighted integral of the interaction free energy per unit area of planar walls. (The weight function has an integrable singularity.) As such, it represents an efficient computational procedure, but unlike the inter-action between spheres, differentiation does not yield a simple expression for the force. This is rather different to the case of two spheres (or to a sphere and a plane, or to two crossed cylinders), where the Derjaguin approximation gives the total force as a geometric factor times the interaction free energy per unit area between planar walls. For parallel cylinders, or for particles adsorbed on a surface or confined to a membrane, the Derjaguin approximation only exists in the sense of Eq. (3.93).

The cylindrical double layer has been studied fairly extensively because of its relevance for rodlike polyelectrolytes such as DNA, and also because of its application to channels and pores in membranes and zeolites. The interaction between cylinders of finite length has been treated in linear Poisson–Boltzmann approximation by Halle [106]. For an isolated cylinder, it is possible to obtain an analytic solution to the nonlinear Poisson–Boltzmann equation for the case when there are only counterions in the double layer (no added salt) [107–109]. In the presence of added electrolyte, the linear Poisson–Boltzmann equation has solution in terms of modified Bessel functions [104, 110, 111]; solutions of the nonlinear version have also been discussed [112]. Anderson and Record [113] reviewed the cylindrical double layer as applied to DNA, and discussed the comparative advantages of the Poisson–Boltzmann theory and Manning's counterion condensation approach [105].

The singlet cylinder–ion hypernetted chain approximation was given by Lozada–Cassou [114]. Numerical results were obtained using the mean spherical approximation [115] and the hypernetted chain approximation [116] for the bulk ion–ion direct correlation function. Vlachy and co-workers [117–119] obtained simulation and integral equation results for cylindrical pores, and Yeomans et al. [120] recently compared the hypernetted chain/mean spherical approximation for that system with simulations and with the double layer at a planar wall. Das et al. [121] treated linear polyelectrolytes within the cell model by Monte Carlo, modified Poisson–Boltzmann (MBP5 [122]), and nonlinear Poisson–Boltzmann approximations. The cylindrical double layer has been reviewed [4, 123].

2. Membranes and Pores

The spherical, planar, and cylindrical macroions that have so far been analyzed illustrate the variety of solute geometries that can be utilized. A conceptually different system, namely, a confined electrolyte, can also be treated by the singlet method by considering the solute to be a pore. Spherical and planar geometries are the simplest. The planar slit pore will be treated in Section III.D.3, and this section will initially focus upon an isolated spherical pore of radius R, for which one has

$$\underline{H}(r) = -1 \qquad r > R \qquad (3.94)$$

Here $\{\underline{H}(r)\}_\alpha = \rho_\alpha^{1/2} h_{0\alpha}(r)$ is the pore–ion total correlation function. The ions are in equilibrium with a bulk electrolyte where they have number density ρ_α, and a common surface of closest approach, which defines the surface of the pore, has been used. If the pore has surface charge density σ, then the electroneutrality condition is

$$-4\pi R^2 \sigma = \int \underline{q}^T \underline{H}(r)\, d\mathbf{r} = 4\pi \sum_\alpha q_\alpha \rho_\alpha \int_0^R h_{0\alpha}(r) r^2\, dr \qquad (3.95)$$

In the event that one wants to model a fixed number of ions inside the pore, one can adjust the bulk concentrations until the integrals of the individual pore–ion distribution functions give the specified number of ions; average quantities (but not their fluctuations) will then be the same as if the canonical ensemble had been used.

As usual the solute–ion Ornstein–Zernike equation is

$$\underline{H}(r) = \underline{C}(r) + \int \underline{\underline{C}}(|\mathbf{r} - \mathbf{s}|)\underline{H}(s)\, d\mathbf{s} \qquad (3.96a)$$

$$= \underline{\chi}(r) - \beta \underline{q}\psi(r) + \int \underline{\underline{\chi}}(|\mathbf{r} - \mathbf{s}|)\underline{H}(s)\, d\mathbf{s} \qquad (3.96b)$$

The expression for the mean electrostatic potential follows from the integral of the Coulomb contribution to the bulk direct correlation function

$$-\beta \int \underline{q}\,\underline{q}^T \underline{H}(s) \frac{1}{\varepsilon |\mathbf{r} - \mathbf{s}|} \, ds = \frac{-4\pi\beta}{\varepsilon} \underline{q} \left[\frac{1}{r} \int_0^r \underline{q}^T \underline{H}(s) s^2 \, ds + \int_r^R \underline{q}^T \underline{H}(s) s \, ds \right]$$

$$(3.97)$$

Hence, the mean electrostatic potential may be written

$$\psi(r) = \begin{cases} \psi(0) + \dfrac{4\pi}{\varepsilon r} \displaystyle\int_0^r \underline{q}^T \underline{H}(s)[s - r]s \, ds & r \leq R \\[3mm] \psi(0) - \dfrac{4\pi R \sigma}{\varepsilon} - \dfrac{4\pi}{\varepsilon} \displaystyle\int_0^R \underline{q}^T \underline{H}(s) s \, ds & r \geq R \end{cases}$$

$$(3.98)$$

and consequently the pore–ion Coulomb potential is

$$V_0^{\text{Coul}}(r) = \begin{cases} \psi(0) - \dfrac{4\pi}{\varepsilon} \displaystyle\int_0^R \underline{q}^T \underline{H}(s) s \, ds + \dfrac{4\pi R \sigma}{\varepsilon} & r \leq R \\[3mm] \psi(0) - \dfrac{4\pi}{\varepsilon} \displaystyle\int_0^R \underline{q}^T \underline{H}(s) s \, ds + \dfrac{4\pi R^2 \sigma}{\varepsilon r} & r \geq R \end{cases}$$

$$(3.99)$$

There is an arbitrary constant here, signified by the mean electrostatic potential at the origin. In open geometries the zero of potential is chosen to lie in the bulk electrolyte far from the solute, and then all correlation functions decay to zero; no such principle exists for closed geometries. For analytic results the constant is immaterial, and it is convenient to choose the potential to be zero outside of the net neutral pore, but one does not have the same freedom in numerical approaches. Changing the value of the arbitrary constant of the potential implies corresponding changes in the constant contribution to the pore–ion short-ranged direct correlation function and the bridge function. However, approximate numerical approaches usually implicitly set the value of these functions (e.g., the hypernetted chain approximation discussed below sets the bridge function to zero), and hence one can no longer arbitrarily set the zero of the potential. In practice $\psi(0)$ can be found by satisfying the electroneutrality condition; if there are too many counterions it is decreased, and if too many co-ions it is increased (negative feedback). (This method is described in more detail in Section V.B.) These two equations for the mean electrostatic and Coulomb potentials combine to give the second form of the Ornstein–Zernike equation, Eq. (3.96b), with the short-range part of the pore–ion direct correlation function being $\underline{\chi}(r) = \underline{C}(r) + \beta \underline{q} V_0^{\text{Coul}}(r)$. One may verify that $\chi_\alpha(r) \to -1 + \beta q_\alpha \psi(R) +$

$\Sigma_\gamma\, \rho_\gamma\hat{\chi}_{\alpha\gamma}(0)$, $r\rightarrow\infty$. The exact closure is formally identical to those given above.

An electrolyte confined to spherical pores in this fashion is perhaps the simplest model of a reverse micelle. This consists of an aqueous core bounded by a spherical shell of amphiphilic molecules, all immersed in a bulk oil phase. (At this level image charges, which arise from the oil–water dielectric disparity, are ignored, which may be an unrealistic approximation in this system.) Interactions can occur between a pair of such reverse micelles, due to the correlated fluctuations of the ions they contain, even though the ions themselves are excluded from the oil phase and there is no net charge on the micelles [124, 125]. Now the pore–ion Ornstein–Zernike convolution integral consists of pore–ion correlation functions, which are constant as $r\rightarrow\infty$. One can subtract these constants to obtain the short-range contribution

$$\int \underline{H}^T(s)\underline{\chi}(|\mathbf{r}-\mathbf{s}|)\,ds = \int \underline{H}^{\mathrm{sr}\,T}(s)\underline{\chi}^{\mathrm{sr}}(|\mathbf{r}-\mathbf{s}|)\,ds + \underline{\hat{H}}^{\mathrm{sr}\,T}(0)\underline{\chi}(\infty)$$

$$+ \underline{H}^T(\infty)\underline{\hat{\chi}}^{\mathrm{sr}}(0) + \underline{H}^T(\infty)\underline{\chi}(\infty)V \qquad (3.100)$$

where V is the volume. Here $h_{0\alpha}(\infty)=-1$, and $h^{\mathrm{sr}}_{0\alpha}(r)=h_{0\alpha}(r)+1$ is just the pore–ion radial distribution function, which vanishes for $r>R$, and $\chi_{0\alpha}(\infty)=-1+\beta q_\alpha\psi(R)+\Sigma_\gamma\,\rho_\gamma\hat{\chi}_{\alpha\gamma}(0)$, with $\chi^{\mathrm{sr}}_{0\alpha}(r)=\chi_{0\alpha}(r)-\chi_{0\alpha}(\infty)$. Similarly, the pore–pore bridge function, which consists of pore–ion and ion–ion total correlation functions, goes to a constant at large separations. The constant can be eliminated by replacing two of the pore–ion h bonds by pore–ion g bonds (one connected to each pore, and not connected to the same field point), which leaves a short-ranged part that decays exponentially with the decay length of the bulk electrolyte. All of these constants can be discarded to give a formally exact expression for the pore–pore potential or mean force (for $r>2R$),

$$w^{\mathrm{int}}_{00}(r) = -d^{\mathrm{sr}}_{00}(r) - \int \underline{H}^{\mathrm{sr}\,T}(s)\underline{\chi}^{\mathrm{sr}}(|\mathbf{r}-\mathbf{s}|)\,ds$$

$$= -d^{\mathrm{sr}}_{00}(r) - 2\pi\sum_\alpha \rho_\alpha \int_0^R ds\, s^2 g_{0\alpha}(s) \int_0^\pi d\theta\, \chi^{\mathrm{sr}}_{0\alpha}(\sqrt{r^2+s^2-2rs\cos\theta})$$

$$(3.101)$$

There is no contribution to the interaction from the mean electrostatic potential because the micelle (pore plus confined ions) has no net charge. One expects that in the asymptotic limit the correlated fluctuations between the two confined electrolytes will give a van der Waals attraction [124, 125], $w_{00}^{int}(r) \sim \mathcal{O}(r^{-6})$, $r \to \infty$.

This method of treating the interaction between two pores can also be applied to the thickness-dependent free energy of a planar membrane. The wall–wall Ornstein–Zernike equation of Section III.C.1 gives the interaction free energy per unit area of two planar walls separated by an electrolyte (i.e., $z > 0$). But as in the case of reverse micelles there is no fundamental impediment to applying the result for $z < 0$, which could model a charged bilayer membrane with liquidlike interior, of thickness $|z|$ and with electrolyte on either side. Again, one has to account for the fact that the wall–ion correlation functions tend to constants deep inside the wall, and so one has to work with short-range functions. By using similar arguments to the above, the interaction free energy per unit area is

$$w_{00}^{int}(z) = d_{00}^{sr}(z) + \int_0^{\infty} \underline{H}^{srT}(z')\underline{\chi}^{sr}(z - z')\, dz' \qquad z < 0 \qquad (3.102)$$

where the first function in the integrand is the wall–ion distribution function, which vanishes for $z' < 0$, and the second term has been made short ranged by subtracting the constant given by Eq. (3.30). Again the bridge function has had two noncontiguous wall–ion h bonds replaced by g bonds, and it is exponentially short ranged. Note that there is no bulk free energy contribution to this, which would be just the osmotic pressure of the bulk electrolyte times the thickness of the membrane, nor is there any surface free energy part. This is purely the interaction free energy per unit area, which arises from correlations between ions in the electrolyte on the two sides of the membrane. It represents a van der Waals force, and hence for large thicknesses it should approach zero from below, $w_{00}^{int}(z) \sim \mathcal{O}(z^{-2})$, $z \to -\infty$.

In some ways this method that treats the membrane as two interacting walls at negative separations is not the most obvious way of analyzing the problem. What is perhaps more intuitive is to consider the membrane as a single planar wall of finite thickness t. Then all of the analysis for the isolated wall of Section III.B.1 holds, with minor modifications due to the replacement of the core exclusion condition, Eq. (3.19), by

$$h_{0\alpha}(z) = -1 \qquad |z| < t/2 \qquad (3.103)$$

This method and the one of the preceding paragraph treat the same problem. Each one has its own advantages. The membrane free energy is obtained directly in the first approach, but not in the second, whereas the ion profiles on both sides of the membrane are given directly by the second approach, but not the first. The two approaches illustrate another aspect of the singlet method, namely, that it is possible to regard two interacting solutes as a single species, and this is the next topic of discussion.

3. Dumbbells

The flexibility of the singlet Ornstein–Zernike approach is nowhere more evident then when the solute is a cluster of ions or particles. The membrane example at the end of the preceding section showed the fundamental equivalence of treating interacting walls either as two solutes or as a single molecular species. More generally, the solute can be the n ion of Section II.B.3, and then the solute–ion density profile simply corresponds to the $(n + 1)$ ion distribution function. When the solute is taken to be two particles, the method will be called the dumbbell singlet approach; it is based upon the solute–ion Ornstein–Zernike equation, and has to be distinguished from the method explored in detail above that utilizes the solute–solute Ornstein–Zenike equation for the same problem. The dumbbell approach is due to Lozada–Cassou (it is termed by him a three-point extension), and a number of applications have been reviewed [4]. Essentially, the dumbbell can be formed from any two particles at a fixed separation. A particularly interesting example is for a solute comprised of two ions, because this gives the triplet distribution function of the bulk electrolyte. In the dumbbell method one has to take care of additional geometric factors in formulating the solute–ion Ornstein–Zernike integral, which may then become complicated in detail though it remains conceptually straightforward. The simplest geometry is the planar case. This section will focus upon two infinitely thick walls, which corresponds to the wall–wall analysis of Section III.C.1, and which can also be considered as a slit pore. In Section V the merits of the two singlet methods for interacting solutes will be compared.

A solute comprised of two semiinfinite half-spaces separated by t has for the solute–ion total correlation function

$$h_{0\alpha}(z;t) = -1 \qquad |z| > t/2 \qquad\qquad (3.104)$$

which manifests the confinement of the electrolyte between the walls. That the solute actually consists of two walls will be explicitly signified by the separation appearing in the argument of the solute–ion correlation functions. Attention will be restricted to the symmetric system, $h_{0\alpha}(-z;t) = h_{0\alpha}(z;t)$, and if each wall has surface charge density σ, then the electroneutrality condition is

$$-2\sigma = \int_{-\infty}^{\infty} \underline{q}^T \underline{H}(z;t)\, dz = \sum_\alpha q_\alpha \rho_\alpha \int_{-t/2}^{t/2} h_{0\alpha}(z;t)\, dz \quad (3.105)$$

As usual the solute–ion Ornstein–Zernike equation is

$$\underline{H}(z;t) = \underline{C}(z;t) + \int \underline{\underline{C}}(|\mathbf{r}-\mathbf{r}'|)\underline{H}(r_z';t)\, d\mathbf{r}'$$

$$= \underline{C}(z;t) + 2\pi \int_0^\infty dr'\, r' \int_{-\infty}^{\infty} dz'\, \underline{\underline{C}}(\sqrt{r'^2 + (z-z')^2})\underline{H}(z';t)$$

$$(3.106)$$

From the Coulomb contribution to the bulk ion–ion direct correlation function one obtains the mean electrostatic potential,

$$\psi(z;t) = \psi(0;t) - \frac{4\pi}{\varepsilon} \int_0^z \underline{q}^T \underline{H}(z';t)(z-z')\, dz' \quad |z| \le t/2 \quad (3.107)$$

and the potential is constant and equal to its contact value within each wall. The arbitrary constant is signified by the appearance of the potential at the midplane, $\psi(0;t)$. In practice, its value may be determined from the electroneutrality condition; if neutralization is not satisfied, it should be increased or decreased in proportion to the extra co-ions or counterions, respectively (see Section V.B). In the limit $t \to \infty$, $\psi(0;t) \to 0$, since there is bulk electrolyte in the central region between the walls, and it may be verified by a shift in the coordinate system that this expression reduces to that for an isolated planar wall, Eq. (3.25). Now the Coulomb potential of an ion between two planes with identical charges is a constant, and this may be discarded along with the constant from the mean electrostatic potential to give the alternative formulation of the solute–ion Ornstein–Zernike equation,

$$\underline{H}(z;t) = -\beta\underline{q}\psi(z;t) + \underline{\chi}(z;t)$$

$$+ 2\pi \int_0^\infty dr'\, r' \int_{-\infty}^{\infty} dz'\, \underline{\chi}(\sqrt{r'^2 + (z-z')^2})\underline{H}(z';t) \quad (3.108)$$

As above, $\chi_{0\alpha}(z;t) \to -1 + \beta q_\alpha \psi(t/2;t) + \Sigma_\gamma \rho_\gamma \hat{\chi}_{\alpha\gamma}(0)$, $|z| \to \infty$. The formally exact closure equation is

$$h_{0\alpha}(z;t) = -1 + \exp[h_{0\alpha}(z;t) - \chi_{0\alpha}(z;t) + d_{0\alpha}(z;t)] \quad (3.109)$$

where $d_{0\alpha}(z;t)$ is the solute–ion bridge function. These equations correspond to a confined electrolyte in equilibrium with the bulk; by adjusting the bulk concentrations one can mimic a closed system with a specified number of ions, and the ion and potential profiles will then be identical to their counterparts in the canonical ensemble.

The equations above determine the ion and potential profiles between two charged walls separated by t. For the study of slit pores these are the two main quantities of interest. The wall–wall Ornstein–Zernike equation of Section III.C.1 did not determine the density profiles between the two walls, and so for the study of slit pores the dumbbell approach is preferable. What the wall–wall approach did give was the interaction free energy per unit area, a quantity that is not directly available from the formally exact dumbbell equations given above. The separation-dependent free energy is an important quantity in the study of solute–solute interactions (it determines, e.g., whether particles stick together or remain dispersed), and can be more or less directly measured. The pressure, which when integrated gives the free energy, can be obtained for the dumbbell geometry from the contact theorem, as is now shown.

As mentioned above, the density profile around a solute n ion corresponds to the $(n+1)$ particle distribution function. The Born–Green–Yvon hierarchy also relates these two levels of distribution functions. For the dumbbell solute the first member of the hierarchy is required. Usually, the Born–Green–Yvon equation expresses the gradient of the pair distribution function as the average direct force plus the indirect force mediated by a third particle. By dividing by the pair distribution function one can rewrite it as an expression for the gradient of the potential of mean force. For two spherical solutes it is

$$w_{00}'(r_{12}) = V_{00}'(r_{12}) + \sum_\alpha \rho_\alpha \int u_{0\alpha}'(r_{13}) \cos\theta \, \frac{g_{00\alpha}(r_{12}, r_{13}, \cos\theta)}{g_{00}(r_{12})} \, d\mathbf{r}_3$$

$$(3.110)$$

where V_{00} is the solute–solute potential, and $u_{0\alpha}$ is the isolated solute–ion potential. The ion density times the solute–solute–ion triplet distribution function divided by the solute–solute pair distribution function is the probability of finding an ion conditional upon the solutes being at $\mathbf{r}_1$ and $\mathbf{r}_2$. This is just the dumbbell-ion distribution function and one has

$$\frac{g_{00\alpha}(r_{12}, r_{13}, \cos\theta)}{g_{00}(r_{12})} = g_{0\alpha}(r_{13}) g_{0\alpha}(r_{23}) \exp - \beta\tau_{00\alpha}(r_{12}, r_{13}, \cos\theta)$$

$$= 1 + h_{0\alpha}(\mathbf{r}_3; r_{12}) \qquad (3.111)$$

Strictly speaking, the numerator and the denominator of the left side do not individually exist for planar walls of infinite area [cf. the discussion of the wall–wall Ornstein–Zernike equation, Eq. (2.52)], but the right sides are well defined for both finite and infinite solutes. (The function τ is the solute–solute–ion triplet potential of mean force, which is the set of connected, nonparallel diagrams with nonadjacent root points.) Inserting the dumbbell distribution function into the Born–Green–Yvon equation gives the mean solute–solute force (and upon integration their interaction free energy), as was originally given by Lozada-Cassou [126]. This enables the interaction of spherical and other finite solutes to be characterized.

For infinite planar solutes the analogue of the Born–Green–Yvon equation reduces to the contact theorem, as is now shown. Equation (3.110) says that the mean force between solutes is the sum of a direct part and an indirect part due to the mediation of the ions. In the planar case one has to deal with the force per unit area, but the same situation holds: The mean force per unit area is the direct pressure between the walls plus the force on one wall due to the ions, summed over species and integrated over space, weighted with their probability. The indirect contribution clearly scales with the area of the plates, and hence it is the axial integration that gives the indirect force per unit area. That is,

$$w'_{00}(t) = V'_{00}(t) + \sum_{\alpha} \rho_\alpha \int_{-t/2}^{t/2} u'_{0\alpha}(z)[1 + h_{0\alpha}(z;t)]\,dz \qquad (3.112)$$

Note that $\cos\theta$ has been set equal to 1 because in this planar geometry the component of force is always parallel to the line connecting the walls. Its worth mentioning that this result can be derived directly from Eq. (3.110) by taking the infinite solute radius limit, and by invoking the relationships of Section III.C.2. Because there is no electrolyte on the far side of these plates, the left side represents the total pressure of the double layer, $p^{\text{total}}(t) = -w'_{00}(t)$. To obtain the net pressure one substracts the osmotic pressure due to the bulk electrolyte, $p^{\text{net}}(t) = p^{\text{total}}(t) - p_\infty$, and hence the interaction free energy per unit area follows by integration,

$$w^{\text{int}}(t) = \int_t^\infty p^{\text{net}}(t')\,dt' \qquad (3.113)$$

with $w^{\text{int}}(t) \to 0$, $t \to 0$. For the electrolyte confined between walls, the wall–ion potential consists of the Coulomb part, Eq. (3.26), plus the hard-wall term, which gives a Dirac δ when differentiated,

$$u'_{0\alpha}(z) = -\frac{2\pi\sigma q_\alpha}{\varepsilon} - k_B T\delta\left(z - \frac{t}{2}\right) \qquad (3.114)$$

With the wall–wall Coulomb potential given by Eq. (3.54), the wall–wall Born–Green–Yvon equation becomes

$$p^{total}(t) = \frac{2\pi\sigma^2}{\varepsilon} - \sum_\alpha \rho_\alpha \int_{-\infty}^{\infty} \left[-\frac{2\pi\sigma q_\alpha}{\varepsilon} - k_B T\delta\left(z - \frac{t}{2}\right)\right][1 + h_{0\alpha}(z;t)]\,dz$$

$$= -\frac{2\pi\sigma^2}{\varepsilon} + k_B T \sum_\alpha \rho_\alpha \left[1 + h_{0\alpha}\left(\frac{t}{2};t\right)\right] \qquad (3.115)$$

where the electroneutrality condition, Eq. (3.105), has been used. This says that the total pressure (force per unit area) acting between the walls consists of the kinetic term due to the thermal motion of the ions in contact with the wall less a constant electrostatic contribution. This is known as the contact theorem [127–131]. Although formally exact, taking the difference of the two positive terms could be problematic in an approximate implementation of the dumbbell-ion scheme. The problems may be exacerbated by subtracting the bulk pressure to obtain the exponentially decaying net pressure; this will be tested by comparison with simulation results in Section V.B.

IV. APPROXIMATE METHODS

In Sections II and III the focus was on the formally exact expressions that could be used to analyze electrolytes and the electric double layer. Now attention shifts to numerical approaches, and in practice this means that one has to introduce approximations. The two computational schemes in current usage are simulation (molecular dynamics and Monte Carlo) and integral equation. The approximations are fundamentally different for the two approaches. Simulations set out to evaluate ensemble averages by generating all configurations of the ensemble. The limitations are the finite system size, the amount of equilibration, and the finite number of configurations used to generate the statistics, assuming numerical errors to be negligible. The accuracy of any simulation can be improved by running the simulation for a longer time with more particles. In contrast, integral equations are constructed to be at equilibrium and at the thermodynamic limit, and their limitation is due to having to close the set of formally exact relations between the correlation functions with an approximation. Whereas the numerical errors in solving an integral equation can be made negligible (by using finer grids and larger cut-offs), the approach itself is fundamentally approximate and can only be improved by changing the closure. Simulations are said to be exact (the above limitations notwithstanding), and they are used to generate benchmark results to test approximate theories. (The tests are carried out on well-defined model systems, which is the reason that one does not test the accuracy of an approximate theory by comparing it with experimental data.) Integral equations may be solved very efficiently, and results can be generated relatively quickly, which allows for direct application to experimental data in any regime where one is satisfied of the reliability of the approximation.

This section set out the various integral equation approximations—singlet, inhomogeneous, density functional—and describes the relationship between them. The discussion of density functional theory will be limited; although closely related to Ornstein–Zernike based theories, and although the results are often accurate, reliable and tractable, in some ways the approximations used in density functional theory are of a more ad hoc nature. The approximation in theories based upon the hypernetted chain closure is generally the neglect of the bridge function, and hence systematic improvements can be made by including bridge diagrams of increasing complexity, which seems an advantage. Inhomogeneous integral equations in general are the most accurate and the most reliable of the approximate methods. These equations have been used to generate benchmark results in cases where simulations are not feasible, such as fluids around macrospheres. However, they are rather complex to program and require heavy computational resources, in contrast to singlet integral equations. One is ultimately aiming for a practical theory that can be used for the quantitative interpretation of measurements, including data fitting. Accordingly, the major emphasis in this section will be on the solute–ion Ornstein–Zernike equation (singlet method) with hypernetted chain closure, and the systematic improvement to that approximation that includes the first bridge diagram.

A. Singlet Hypernetted Chain Approximation

The solute–ion Ornstein–Zernike equation, Eq. (3.2), relates the solute–ion total and direct correlation functions, and uses the bulk ion–ion direct correlation function. The latter may be obtained by any of the methods used to treat a bulk electrolyte, such as the hypernetted chain approximation, without [132] or with [133–137] the first bridge diagram. The mean spherical approximation is also used, since it has the advantage of providing an analytic polynomial expression for the ion–ion direct correlation function [62–64]. The second formally exact equation that relates the two solute–ion correlation functions is the closure equation. It is identical to that for the bulk electrolyte, Eq. (2.35), and requires a new unknown, the solute–ion bridge function. One needs to introduce an approximation, and most attention has focused upon the hypernetted chain closure, which appears the most reliable. (For example, the Percus–Yevick approximation gives negative solute–solute distribution functions for large macrospheres, wheres the hypernetted chain approximation is guaranteed to remain positive [138].) The bridge function consists of an infinite sum of highly connected cluster diagrams, which are difficult to calculate. The hypernetted chain approximation simply neglects the bridge function,

$$d_{0\alpha}^{\mathrm{HNC}}(r) = 0 \qquad (4.1)$$

This approximation is quite reliable and relatively accurate for bulk electrolytes, except possibly for divalent aqueous electrolytes at low concentrations. However, the electric double layer represents a more severe test, and it is known to be inaccurate for certain quantities and in certain regimes. For example, the contact theorem relates the ion density at a planar wall to the pressure of the bulk electrolyte [127–131],

$$k_B T \sum_\alpha \rho_\alpha [1 + h_{0\alpha}(0^+)] = p + \frac{2\pi\sigma^2}{\varepsilon} \qquad (4.2)$$

which may be compared to Eq. (3.115). However, the hypernetted chain approximation gives for the contact density [83]

$$k_B T \sum_\alpha \rho_\alpha [1 + h_{0\alpha}(0^+)] = k_B T \sum_\alpha \rho_\alpha - \frac{k_B T}{2} \sum_{\alpha\gamma} \rho_\alpha \rho_\gamma \hat{\chi}_{\alpha\gamma}(0) + \frac{2\pi\sigma^2}{\varepsilon}$$

$$(4.3)$$

Using the formal expression for the isothermal compressibility

$$\chi_T \equiv \rho \left(\frac{\partial p}{\partial \rho} \right)_T = k_B T \rho - k_B T \sum_{\alpha\gamma} \rho_\alpha \rho_\gamma \hat{\chi}_{\alpha\gamma}(0) \qquad (4.4)$$

where $\rho = \sum_\alpha \rho_\alpha$, the hypernetted chain result implies

$$p^{\mathrm{ex}} \equiv p - \rho k_B T = \frac{\rho}{2} \left(\frac{\partial p^{\mathrm{ex}}}{\partial \rho} \right)_T \qquad (4.5)$$

Hence, if the hypernetted chain contact densities were used to obtain the pressure, then this is exact for the second virial coefficient for a simple fluid; for an electrolyte this route does not give even the first correction to the ideal gas equation of state exactly. In contrast, the hypernetted chain approximation used in the virial theorem for a bulk simple fluid would yield the third virial coefficient exactly. The results persuade that the contact densities at a planar wall are not a good thermodynamic pathway to the bulk pressure.

A systematic correction to the hypernetted chain is to approximate the bridge function by the first bridge diagram. It is best to resume the f bonds, which are individually nonintegrable, in terms of the exponentially short-ranged h bonds, and the first resumed bridge diagram is [17]

$$d_{0\alpha}^{\text{HNCD}}(r_{12}) = \frac{1}{2} \sum_{\gamma\lambda} \rho_{\gamma}\rho_{\lambda} \int h_{0\gamma}(r_{13})h_{0\lambda}(r_{14})h_{\gamma\lambda}(r_{34})h_{\gamma\alpha}(r_{23})h_{\lambda\alpha}(r_{24}) \, d\mathbf{r}_3 \, d\mathbf{r}_4$$

$$(4.6)$$

This first bridge diagram alone significantly improves the hypernetted chain closure and extends its regime of applicability for the bulk electrolyte [133–137]. Usually one solves the hypernetted chain approximation for the total correlation functions, which are then used to evaluate this bridge diagram. One then iterates the Ornstein–Zernike and closure equations with fixed bridge function to obtain the new total correlation function; in most cases recalculating the bridge function to obtain self-consistency has negligible effect. The numerical evaluation of this bridge diagram is facilitated by expansion in Legendre polynomials [139, 140]

$$d_{0\alpha}^{\text{HNCD}}(r_{12}) = 2\pi^2 \sum_{\gamma\lambda} \rho_{\gamma}\rho_{\lambda} \sum_{n=0}^{\infty} \left(\frac{2}{2n+1}\right)^2 \int_0^{\infty} dr_3 \, r_3^2 \int_0^{\infty} dr_4 \, r_4^2$$

$$\times h_{\gamma\alpha}(r_3)h_{\lambda\alpha}(r_4)\hat{h}_{0\gamma}^{(n)}(r_1, r_3)\hat{h}_{0\lambda}^{(n)}(r_1, r_4)\hat{h}_{\gamma\lambda}^{(n)}(r_3, r_4) \quad (4.7)$$

where the ion α has been placed at the origin, $\mathbf{r}_2 = \mathbf{0}$. Here the solute has been taken to be spherically symmetric, and the Legendre coefficients are

$$\hat{f}^{(n)}(r, s) = \frac{2n+1}{2} \int_{-1}^{1} P_n(x)f(\sqrt{r^2 + s^2 - 2rsx}) \, dx \qquad (4.8)$$

where $P_n(x)$ is the Legendre polynomial of order n. For the case where the solute is a single planar wall one has

$$
d_{0\alpha}^{\mathrm{HNCD}}(z_2) = 2\pi^2 \sum_{\gamma\lambda} \rho_\gamma\rho_\lambda \sum_{n=0}^{\infty} \left(\frac{2}{2n+1}\right)^2 \int_0^\infty dr_3\, r_3^2 \int_0^\infty dr_4\, r_4^2
$$

$$
\times\, h_{\gamma\alpha}(r_3)h_{\lambda\alpha}(r_4)\hat{h}_{0\gamma}^{(n)}(z_2,r_3)\hat{h}_{0\lambda}^{(n)}(z_2,r_4)\hat{h}_{\gamma\lambda}^{(n)}(r_3,r_4) \tag{4.9}
$$

where the Legendre coefficients of the bulk ion–ion total correlation functions were given in Eq. (4.8), and those of the wall–ion total correlation function are [84]

$$
\hat{h}_{0\alpha}^{(n)}(z,r) = \frac{2n+1}{2} \int_{-1}^{1} P_n(x)h_{0\alpha}(z - rx)\, dx \tag{4.10}
$$

It is convenient to evaluate the Legendre coefficients by an orthogonal technique [141], which corresponds to a Gaussian quadrature; typically only 10–20 terms are needed in the expansion. Expressions for the Legendre expansion of the bridge diagrams with three field points have been given by Attard and Patey [138]. These authors found that the hypernetted chain approximation was increasingly improved as more bridge diagrams were added to the closure, especially for the solute–solute interaction [77, 138]. These results were for hard-sphere solutes in a hard-sphere fluid; to date bridge diagrams with three field points have not been used for electrolytes or for the electric double layer. An alternative expansion is the hydrostatic hypernetted chain approximation of Zhou and Stell [142]. The theory amounts to a formula for the bridge function that involves the bulk chemical potential and the bulk compressibility evaluated at either a local or a weighted density, and it has been shown to improve upon the singlet hypernetted chain approximation for hard spheres at and between hard walls [142].

The singlet hypernetted chain approximation has been used to obtain the properties of the electric double about an isolated solute, with much attention focused on a charged planar wall [2, 3]. At higher surface charge densities there occurs numerical instabilities in the solution of the approximation; the most efficient and robust algorithm appears to be the one given by Ballone et al. [143], which is based upon earlier work on the one-component plasma [144, 145]. The variational method of Feller and McQuarrie [146] also appears quite robust. Carnie and Torrie [2] give an extensive review and comparison of results for the isolated planar double layer prior to 1984, including the modified Poisson–Boltzmann approximation. Blum [3] compares various theories in detail for the case of $1\,M$ monovalent electrolyte ($d = 4.25\,\text{Å}$) against a wall with surface charge density $\sigma = 0.7e/d^2$. This is a demanding test case where the simulations [143, 147, 148] show a secondary peak in the counterion profile at about one diameter from contact, which is not predicted by the bare hypernetted chain approximation. However, inclusion of the first bridge diagram does give this peak [143]. Nielaba, Forstmann, and co-workers [149, 150] use the singlet hypernetted chain approach with a local density approximation that invokes the bulk ion–ion direct correlation function of a jellium and obtain the secondary peak. It is also given by certain density functional approximations; local [151] and weighted [152–154] densities have been used. Inhomogeneous integral equations (discussed below) are perhaps the most accurate of all approaches, and include the Born–Green–Yvon approach of Caccamo and co-workers [155, 156], and the inhomogeneous Ornstein–Zernike approach of Kjellander and Marčelja [157] and of Plischke and Henderson [158].

The singlet hypernetted chain approximation has been used in other geometries. Bratko [159] treated ions confined to a spherical pore, and results have also been obtained for cylinders and cylindrical pores [115–117, 120]. The dumbbell approach, which uses the hypernetted chain approximation for an isolated solute composed of two particles (Section III.D.3), has been used to obtain the interaction between planar walls [160–164].

The singlet equation for interacting solutes (e.g., the macrospheres of Section III.A, or the walls of Section III.C), can also be solved with the hypernetted chain approximation, with or without the first bridge diagram. Patey [165] obtained hypernetted chain results for charged spherical colloids at infinite dilution, and found an attraction at small separations between identically charged macroions with high surface charge densities. (Teubner's [166] argument that this attraction was spurious and an artifact of the hypernetted chain approximation is based on a theorem of Bell and Levine [167] that is only valid in Poisson–Boltzmann approximation.) Henderson [102] also used the hypernetted chain approximation for two charged spherical macroions, and others have studied highly asymmetric electrolytes [168–173]. For interacting walls, singlet hypernetted chain results have been obtained for hard walls in Lennard–Jones and in dipolar fluids [77], and for the electric double-layer interaction between charged walls without [174, 175], and with [84] the first bridge diagram.

1. Solvation Free Energy

A peculiar advantage of the hypernetted chain approximation is that one can perform the coupling constant integrations to obtain the chemical potential and the Helmholtz free energy as spatial integrals of the fully coupled pair correlation functions [176, 177]. This holds in a variety of applications besides bulk electrolytes, and this section derives the hypernetted chain approximation for the solvation free energy. This is a useful quantity that, for example, determines the electric double layer contribution to the solvation of charged particles and how it changes with curvature or surface charge. Also, for the case of the dumbbell singlet approach, the solvation free energy of the dumbbell gives the interaction free energy of the two solutes, and hence for the case of two planes the results of the present section provide an alternative to the contact theorem of Section III.D.3 for calculating the pressure. The following derivation of the hypernetted chain solvation free energy and its extension to include the first bridge diagram is based upon that of Kiselyov and Martynov [178] (see also [179]).

The solvation free energy is the change in the grand potential of the electrolyte when there is a solute fixed at the origin. Formally, this is

$$\mu_0 \equiv \Omega^{(1)}(\mu, V, T) - \Omega^{(0)}(\mu, V, T)$$

$$= \int_0^1 \frac{\partial \Omega^{(\lambda)}(\mu, V, T)}{\partial \lambda} \, d\lambda$$

$$= \mu_0^{\text{self}} + \int_0^1 \left\langle \sum_{\alpha,i} u_{0\alpha}(r_{\alpha i}) \right\rangle^{(\lambda)} d\lambda$$

$$= \mu_0^{\text{self}} + \int_0^1 d\lambda \sum_\alpha \rho_\alpha \int d\mathbf{r} \, u_{0\alpha}(r) g_{0\alpha}^{(\lambda)}(r) \tag{4.11}$$

Here the superscript indicates that the solute–ion distribution function is partially coupled; the solute–ion potential is $u_{0\alpha}^{(\lambda)}(r) = \lambda u_{0\alpha}(r)$. The quantity μ_0^{self} represents the self-energy of the solute, a zero-body term that must be added since the coupling integral represents only the contribution of the ions to the solvation free energy. For simplicity spherical solutes have been assumed, although the results will apply to any geometry in an obvious fashion, and a pairwise additive solute–ion potential has been invoked, which excludes image charge interactions. By using the exact closure [cf. Eq. (2.35)], one can perform two integrations by parts to obtain

$$-\beta \int_0^1 u_{0\alpha}(r) g_{0\alpha}^{(\lambda)}(r) \, d\lambda = h_{0\alpha}(r) - [h_{0\alpha}(r) - c_{0\alpha}(r) + d_{0\alpha}(r)] g_{0\alpha}(r)$$

$$+ \int_0^1 [h_{0\alpha}^{(\lambda)}(r) - c_{0\alpha}^{(\lambda)}(r) + d_{0\alpha}^{(\lambda)}(r)] \frac{\partial h_{0\alpha}^{(\lambda)}(r)}{\partial \lambda} \, d\lambda$$

$$\tag{4.12}$$

where the correlation functions without superscript are fully coupled.

Now for the hypernetted chain approximation, $d_{0\alpha}^{HNC}(r) = 0$, and the coupling constant integral involves the series function, which is just the Ornstein–Zernike convolution integral. One has

$$
\sum_\alpha \rho_\alpha \int d\mathbf{r}_\alpha \int_0^1 d\lambda [h_{0\alpha}^{(\lambda)}(r_\alpha) - c_{0\alpha}^{(\lambda)}(r_\alpha)] \frac{\partial h_{0\alpha}^{(\lambda)}(r_\alpha)}{\partial\lambda}
$$

$$
= \sum_{\alpha\gamma} \rho_\alpha \rho_\gamma \int d\mathbf{r}_\alpha \, d\mathbf{r}_\gamma \int_0^1 d\lambda \, h_{0\gamma}^{(\lambda)}(r_\gamma) c_{\gamma\alpha}(r_{\gamma\alpha}) \frac{\partial h_{0\alpha}^{(\lambda)}(r_\alpha)}{\partial\lambda}
$$

$$
= \sum_{\alpha\gamma} \rho_\alpha \rho_\gamma \int d\mathbf{r}_\alpha \, d\mathbf{r}_\gamma \int_0^1 d\lambda \, \frac{1}{2} \frac{\partial}{\partial\lambda} [h_{0\gamma}^{(\lambda)}(r_\gamma) c_{\gamma\alpha}(r_{\gamma\alpha}) h_{0\alpha}^{(\lambda)}(r_\alpha)]
$$

$$
= \frac{1}{2} \sum_{\alpha\gamma} \rho_\alpha \rho_\gamma \int d\mathbf{r}_\alpha \, d\mathbf{r}_\gamma \, h_{0\gamma}(r_\gamma) c_{\gamma\alpha}(r_{\gamma\alpha}) h_{0\alpha}(r_\alpha)
$$

$$
= \frac{1}{2} \sum_\alpha \rho_\alpha \int d\mathbf{r}_\alpha [h_{0\alpha}(r_\alpha) - c_{0\alpha}(r_\alpha)] h_{0\alpha}(r_\alpha) \tag{4.13}
$$

Here the product rule has been invoked to give the factor of one-half, because the right side is symmetric in the two solute–ion bonds. Note that the ion–ion correlation functions are always fully coupled and independent of λ. With this result the hypernetted chain solvation free energy may be written

$$
-\beta\mu_0^{HNC} = -\beta\mu_0^{self} + \sum_\alpha \rho_\alpha \int [c_{0\alpha}(r) - \tfrac{1}{2}\{h_{0\alpha}(r) - c_{0\alpha}(r)\}h_{0\alpha}(r)] \, d\mathbf{r}
$$

$$
\tag{4.14}
$$

In some cases (e.g., planar walls), it is desirable to replace the direct correlation function by its short-ranged counterpart, $\chi_{0\alpha}(\mathbf{r}) = c_{0\alpha}(r) + \beta q_\alpha V_0^{Coul}(\mathbf{r})$. For the most general solute with charge distribution $\sigma(\mathbf{r})$, one may write

$$
V_0^{Coul}(\mathbf{r}) = \text{const} + \int \frac{\sigma(\mathbf{s})}{\varepsilon|\mathbf{r} - \mathbf{s}|} \, d\mathbf{s} \tag{4.15}
$$

By using the definition of the mean electrostatic potential, $\psi(\mathbf{r}) = V_0^{Coul}(\mathbf{r}) + \Sigma_\alpha \rho_\alpha q_\alpha \int d\mathbf{s} \, h_{0\alpha}(\mathbf{s})/\varepsilon|\mathbf{r} - \mathbf{s}|$, it follows that the hypernetted chain solvation free energy may be rewritten

$$-\beta\mu_0^{\text{HNC}} = \frac{-\beta}{2} \int \sigma(\mathbf{s})\psi(\mathbf{s}) \, d\mathbf{s}$$

$$+ \sum_\alpha \rho_\alpha \int [\chi_{0\alpha}(\mathbf{r}) - \tfrac{1}{2}\{h_{0\alpha}(\mathbf{r}) - \chi_{0\alpha}(\mathbf{r})\}h_{0\alpha}(\mathbf{r})] \, d\mathbf{r} \qquad (4.16)$$

Note that the solute self-energy,

$$\mu_0^{\text{self}} = \frac{1}{2} \int \sigma(\mathbf{s}) V_0^{\text{Coul}}(\mathbf{s}) \, d\mathbf{s} \qquad (4.17)$$

is explicitly canceled in this final result.

It also possible to perform the coupling constant integral when one goes beyond the bare hypernetted chain approximation and includes bridge diagrams. The first solute–ion bridge diagram is given explicitly by Eq. (4.6), which can be represented as

$$d_{0\alpha}^{\text{HNCD}}(r) = \quad\text{}\quad \alpha \qquad (4.18)$$

Here the empty circles represent the solute and the ion α, the filled circles represent the ions that are integrated and summed over, and the lines represent total correlation function bonds. The solvation free energy in this approximation is

$$-\beta\mu_0^{\text{HNCD}} = -\beta\mu_0^{\text{HNC}} - \sum_\alpha \rho_\alpha \int d_{0\alpha}(r)g_{0\alpha}(r) \, d\mathbf{r}$$

$$+ \int_0^1 d\lambda \sum_\alpha \rho_\alpha \int d\mathbf{r} \, d_{0\alpha}^{(\lambda)}(r) \frac{\partial h_{0\alpha}^{(\lambda)}(r)}{\partial \lambda} \qquad (4.19)$$

As for the series function, which was treated in hypernetted chain approximation above, equivalence of the solute–ion bonds allows the coupling integrand to be expressed as the differential of a product. This is clear in the diagrammatic representation,

$$\text{}\,' \alpha = \frac{1}{3}\frac{\partial}{\partial \lambda}\,\text{}\, \alpha \qquad (4.20)$$

where the prime in the first diagram denotes the differentiated solute–ion h bond. The factor of $\tfrac{1}{3}$ arises because only the solute–ion bonds depend on λ, and the three of them are identical. There is no change in symmetry number in passing to the final diagram because of the label α; this integration will be now be done explicitly. The final result is

$$-\beta\mu_0^{HNCD} = -\beta\mu_0^{HNC} - \sum_\alpha \rho_\alpha \int [d_{0\alpha}^{HNCD}(r) + \tfrac{2}{3} d_{0\alpha}^{HNCD}(r) h_{0\alpha}(r)] \, d\mathbf{r}$$

(4.21)

This procedure may be extended to the higher order bridge diagrams [179, 180].

It is worthwhile to give explicit expressions for the solvation free energy for planar walls. For an isolated wall, $\sigma(\mathbf{r}) = \sigma\delta(z)$, and the solvation free energy per unit area becomes

$$-\beta\mu_0^{HNC} = \frac{-\beta\sigma\psi(0)}{2} + \sum_\alpha \rho_\alpha \int_{-\infty}^{\infty} [\chi_{0\alpha}(z) - \tfrac{1}{2}\{h_{0\alpha}(z) - \chi_{0\alpha}(z)\}h_{0\alpha}(z)] \, dz$$

(4.22)

This is in fact infinite because either correlation function vanishes inside the wall; $h_{0\alpha}(z) \to -1$ and $\chi_{0\alpha}(z) \to -1 - \beta q_\alpha \psi(0)/2 + \rho_\alpha \sum_\gamma \rho_\gamma \hat{\chi}_{\alpha\gamma}(0)$ as $z \to -\infty$. The physical origin of the divergence is that to make a wall of thickness L one has to displace bulk electrolyte and thus do work against the bulk pressure, which gives a contribution $\mu_0 \sim pL$. The divergent contribution in the above is

$$-\beta\mu_0^{HNC} \sim -\left[\sum_\alpha \rho_\alpha - \frac{1}{2}\sum_{\alpha\gamma} \rho_\alpha\rho_\gamma\hat{\chi}_{\alpha\gamma}(0)\right]L \qquad L \to \infty \qquad (4.23)$$

The bracketed term is just the hypernetted chain contact density expression for the bulk pressure, Eq. (4.3).

One can avoid this divergence by dealing with the electrostatic free energy, and the most convenient reference state is that of an uncharged wall. If one denotes the solute–ion correlation functions for an uncharged wall by the superscript zero, and if one defines difference functions such as $\Delta h_{0\alpha}(z) \equiv h_{0\alpha}(z) - h_{0\alpha}^{(0)}(z)$ that are short-ranged inside the wall, then one can form a well-defined expression for the surface charge dependent part of the free energy

$$-\beta[\mu_0^{HNC}(\sigma) - \mu_0^{HNC}(0)] = \frac{-\beta\sigma\psi(0)}{2} + \frac{1}{2}\sum_\alpha \rho_\alpha \int_0^{\infty} [2\Delta\chi_{0\alpha}(z) + \Delta\chi_{0\alpha}(-z)$$

$$- \Delta\{h_{0\alpha}(z)^2\} + \Delta\{\chi_{0\alpha}(z)h_{0\alpha}(z)\}] \, dz \qquad (4.24)$$

Because $\rho^T\chi(z) \to -\beta\chi_T^{-1}$, $z \to -\infty$, which is independent of surface charge, it is clear that the above integral is convergent. One has a similar situation for the hypernetted chain approximation that includes the first bridge diagram, and the solution is analogous. In this case deep inside the wall the bridge function goes to a constant that is given by

$$\lim_{z\to-\infty} d_{0\alpha}^{\text{HNCD}}(z) = \frac{1}{2} \sum_{\gamma\lambda} \rho_\gamma \rho_\lambda \int d\mathbf{r}\, d\mathbf{s}\, h_{\alpha\gamma}(r) h_{\alpha\lambda}(s) h_{\gamma\lambda}(|\mathbf{r} - \mathbf{s}|)$$

$$= \frac{1}{4\pi^2} \sum_{\gamma\lambda} \rho_\gamma \rho_\lambda \int_0^\infty \hat{h}_{\alpha\gamma}(k) \hat{h}_{\alpha\lambda}(k) \hat{h}_{\gamma\lambda}(k) k^2\, dk \quad (4.25)$$

which is clearly independent of the surface charge.

For the dumbbell solute that consists of two identical walls separated by t, $\sigma(\mathbf{r}) = \sigma[\delta(t/2 + z) + \delta(t/2 - z)]$, one has for the free energy per unit area

$$-\beta\mu_0^{\text{HNC}}(t) = -\beta\sigma\psi(t/2; t)$$

$$+ \sum_\alpha \rho_\alpha \int_{-\infty}^\infty [\chi_{0\alpha}(z; t) - \tfrac{1}{2}\{h_{0\alpha}(z; t) - \chi_{0\alpha}(z; t)\} h_{0\alpha}(z; t)]\, dz$$

$$(4.26)$$

Again this is divergent, and again only a free energy difference is meaningful. This time the zero of free energy is chosen to be at infinite separation, which corresponds to two isolated walls. One has

$$\lim_{t\to\infty} h_{0\alpha}(z; t) = h_{0\alpha}\left(\frac{t}{2} + z\right) + h_{0\alpha}\left(\frac{t}{2} - z\right) \quad (4.27a)$$

and

$$\lim_{t\to\infty} \chi_{0\alpha}(z; t) = \chi_{0\alpha}\left(\frac{t}{2} + z\right) + \chi_{0\alpha}\left(\frac{t}{2} - z\right) \quad (4.27b)$$

On the left dumbbell–ion functions occur, and on the right wall–ion functions. As usual one defines difference functions that vanish for large plate separations,

$$\Delta h_{0\alpha}(z; t) \equiv h_{0\alpha}(z; t) - h_{0\alpha}\left(\frac{t}{2} + z\right) - h_{0\alpha}\left(\frac{t}{2} - z\right) \quad (4.28a)$$

and

$$\Delta\chi_{0\alpha}(z; t) \equiv \chi_{0\alpha}(z; t) - \chi_{0\alpha}\left(\frac{t}{2} + z\right) - \chi_{0\alpha}\left(\frac{t}{2} - z\right) \quad (4.28b)$$

These also go to zero deep within the walls,

$$\Delta h_{0\alpha}(z;t) \to 0 \qquad |z| \to \infty \qquad (4.29a)$$

and

$$\Delta\{\underline{\rho}^T\underline{\chi}(z;t)\} \to 0 \qquad |z| \to \infty \qquad (4.29b)$$

Similarly, the difference in the surface potential vanishes for large separations,

$$\Delta\psi(t/2;t) \equiv \psi(t/2;t) - \psi(0) \to 0 \qquad t \to \infty \qquad (4.30)$$

Finally, the product of ion–wall total correlation functions from each isolated wall vanishes for large distances of separations,

$$h_{0\alpha}\left(\frac{t}{2}+z\right)h_{0\alpha}\left(\frac{t}{2}-z\right) \to 0 \qquad |z| \to \infty \qquad \text{or } t \to \infty \qquad (4.31)$$

and similarly for the product of h and χ. In view of these definitions it is straightforward to show that the dumbbell hypernetted chain approximation for the interaction free energy per unit area (see the corresponding wall–wall result in Section III.C), is

$$-\beta w_{00}^{\text{int}}(t) \equiv -\beta[\mu_0(t) - 2\mu_0]$$

$$= -\beta\sigma\,\Delta\psi(t/2;t) + \sum_\alpha \rho_\alpha \int_{-\infty}^{\infty} \Delta\chi_{0\alpha}(z;t)\,dz$$

$$-\frac{1}{2}\sum_\alpha \rho_\alpha \int_{-\infty}^{\infty} [h_{0\alpha}(z;t) - \chi_{0\alpha}(z;t)]\,\Delta h_{0\alpha}(z;t)\,dz$$

$$-\sum_\alpha \rho_\alpha \int_{-\infty}^{\infty} [\Delta h_{0\alpha}(z;t) - \Delta\chi_{0\alpha}(z;t)]h_{0\alpha}\left(\frac{t}{2}+z\right)dz$$

$$-\sum_\alpha \rho_\alpha \int_{-\infty}^{\infty} \left[h_{0\alpha}\left(\frac{t}{2}+z\right) - \chi_{0\alpha}\left(\frac{t}{2}+z\right)\right]h_{0\alpha}\left(\frac{t}{2}-z\right)dz$$

$$(4.32)$$

This expression contains both dumbbell–ion and wall–ion correlation functions and their differences. (Note that the dumbbell potential and the consistent short-ranged direct correlation function are given in Section III.D.3. Also, a formally identical expression holds for a nonzero Stern layer. It may be shown that the value of the interaction free energy is independent of the width of the Stern layer.) Differentiating this free energy provides an alternative to the contact theorem for calculating the net pressure between two walls in the dumbbell approach, Section III.D.3. This result holds for the hypernetted chain approximation, but an analogous result can be derived when the first bridge diagram is included. Feller and McQuarrie [163] formulated the singlet dumbbell hypernetted chain approximation as a variational principle. They did not use the free energy directly, but rather obtained the pressure from the contact theorem, and they avoided any divergences by dealing with plates of finite thickness.

B. Inhomogeneous Integral Equations

The singlet integral equation and the inhomogeneous integral equation represent two alternative viewpoints of the electric double layer. In the first, the charged particles are treated on the same footing as the ions of the electrolyte, and the solute–ion distribution function corresponds to the density profile of the double layer. In the second, the charged particles are treated as fixed and as the source of an external field, which causes the nonuniform density of the ions that is the double layer. In the inhomogeneous approach, one applies the closure approximation to the inhomogeneous ion–ion correlation functions, whereas in the singlet approach, the closure is applied to the density profile itself. Other things being equal one would expect a given closure approximation to be more accurate and reliable when used in the inhomogeneous method than when used in the singlet method. For example, it is necessary to include the first bridge diagram in the singlet approach to obtain results comparable to the bare hypernetted chain closure in the inhomogeneous approach. This is certainly true for the simplest singlet methods, such as the wall–ion Ornstein–Zernike equation, but the boundaries between the two approaches become blurred for dumbbells and n ions. For example, the ion density profile around a dumbbell consisting of a wall and a fixed ion corresponds to the inhomogeneous ion–ion correlation function next to a wall.

The inhomogeneous Ornstein–Zernike equation is

$$\underline{H}(\mathbf{r}_1, \mathbf{r}_2) = \underline{C}(\mathbf{r}_1, \mathbf{r}_2) + \int \underline{H}(\mathbf{r}_1, \mathbf{r}_3)\underline{C}(\mathbf{r}_3, \mathbf{r}_2) \, d\mathbf{r}_3 \qquad (4.33)$$

where the elements of the inhomogeneous total correlation function matrix are

$$\{\underline{H}(\mathbf{r}_1, \mathbf{r}_2)\}_{\alpha\gamma} = \rho_\alpha(\mathbf{r}_1)^{1/2} h_{\alpha\gamma}(\mathbf{r}_1, \mathbf{r}_2)\rho_\gamma(\mathbf{r}_2)^{1/2} \qquad (4.34)$$

and similarly for the inhomogeneous direct correlation function. One has $\underline{\underline{H}}(\mathbf{r}_1, \mathbf{r}_2) = \underline{\underline{H}}^T(\mathbf{r}_2, \mathbf{r}_1)$.

The diagrammatic definitions of the total and direct correlation functions are formally the same as for the uniform electrolyte; the densities that weight the field points are no longer constant but vary in space. Consequently, the formally exact closure remains

$$h_{\alpha\gamma}(\mathbf{r}_1, \mathbf{r}_2) = -1 + \exp - \beta u_{\alpha\gamma}(\mathbf{r}_1, \mathbf{r}_2) + h_{\alpha\gamma}(\mathbf{r}_1, \mathbf{r}_2) - c_{\alpha\gamma}(\mathbf{r}_1, \mathbf{r}_2) + d_{\alpha\gamma}(\mathbf{r}_1, \mathbf{r}_2)$$

$$(4.35)$$

where d is the inhomogeneous bridge function, and the pair potential u need not depend solely on separation (as do the hard-sphere and direct Coulomb potential), but can depend on position. Electrostatic images due to dielectric disparities are an example of a position-dependent contribution that is readily incorporated into the inhomogeneous approach; a discussion of these is deferred until the conclusion.

The remaining unknown function in the above is the density, and there are several formally exact expressions for its gradient. The oldest is the first member of the Born–Green–Yvon hierarchy for an inhomogeneous fluid, which balances the gradient of the density with the force due to an external potential, $V_\alpha(\mathbf{r})$, and the internal forces in the double layer [181–184],

$$\nabla \rho_\alpha(\mathbf{r}_1) = -\beta \rho_\alpha(\mathbf{r}_1) \nabla V_\alpha(\mathbf{r}_1) - \beta \rho_\alpha(\mathbf{r}_1) \sum_\gamma \int g_{\alpha\gamma}(\mathbf{r}_1, \mathbf{r}_2)$$

$$\times \nabla_1 u_{\alpha\gamma}(\mathbf{r}_1, \mathbf{r}_2) \rho_\gamma(\mathbf{r}_2) \, d\mathbf{r}_2 \qquad (4.36)$$

where the inhomogeneous pair distribution function is $g = h + 1$. An alternative result may be derived from the Born–Green–Yvon hierarchy for an homogeneous fluid mixture. For the case of a solute, the first member relates the solute–ion pair distribution function to the solute–ion–ion triplet distribution function,

$$\nabla_0 g_{0\alpha}(\mathbf{r}_{01}) = -\beta g_{0\alpha}(\mathbf{r}_{01}) \nabla_0 u_{0\alpha}(\mathbf{r}_{01}) - \beta \sum_\gamma \rho_\gamma \int \nabla_0 u_{0\gamma}(\mathbf{r}_{02}) g_{0\alpha\gamma}(\mathbf{r}_{01}, \mathbf{r}_{02}) \, d\mathbf{r}_2$$

$$(4.37)$$

If the solute is taken to be fixed and is considered as the source of an external field, then these may be rewritten in terms of the density profile and the inhomogeneous pair total correlation function (since the integral of the force over the unconnected diagrams vanishes) [185, 186],

$$\nabla \rho_\alpha(\mathbf{r}_1) = -\beta \rho_\alpha(\mathbf{r}_1) \nabla V_\alpha(\mathbf{r}_1) - \beta \rho_\alpha(\mathbf{r}_1) \sum_\gamma \int h_{\alpha\gamma}(\mathbf{r}_1, \mathbf{r}_2) \rho_\gamma(\mathbf{r}_2) \nabla V_\gamma(\mathbf{r}_2) \, d\mathbf{r}_2$$

(4.38)

This equation is eponymously referred to as the WLMB equation. Note that the external potential that appears here will usually contain in addition to the Coulomb potential a volume exclusion term, whose gradient contributes a δ function; this accounts for the boundary terms that some authors explicitly include in the WLMB equation [6, 187]. Actually, an equation equivalent to the WLMB equation was earlier given by Triezenberg and Zwanzig [188]

$$\nabla \rho_\alpha(\mathbf{r}_1) = -\beta \rho_\alpha(\mathbf{r}_1) \nabla V_\alpha(\mathbf{r}_1) + \rho_\alpha(\mathbf{r}_1) \sum_\gamma \int c_{\alpha\gamma}(\mathbf{r}_1, \mathbf{r}_2) \nabla \rho_\gamma(\mathbf{r}_2) \, d\mathbf{r}_2$$

(4.39)

(This equation, which with the external potential set to zero is Eq. (19) of [188], is also called by some the WLMB equation.) The two equations are related by the Ornstein–Zernike equation. Any one of these three equations for the density gradient, together with the Ornstein–Zernike and closure equations, are sufficient to determine the three unknown functions (the density profile, and the total and direct correlation functions), provided one has some prescription or approximation for the bridge function.

As in the bulk and singlet equations, the simplest approximation for the bridge function is to neglect it altogether, which is just the hypernetted chain closure applied to the inhomogeneous correlation functions,

$$d_{\alpha\gamma}^{\mathrm{HNC}}(\mathbf{r}_1, \mathbf{r}_2) = 0 \qquad\qquad (4.40)$$

As in the singlet case discussed in Section IV.A.1, within the inhomogeneous hypernetted chain approximation it is possible to perform the coupling integral, which gives an explicit expression for the local chemical potential. Enforcing the constancy of this yields an alternative to the three formally exact equations given above for determining the density profile [189]. The numerical evaluation of the first bridge diagram appears formidable for the inhomogeneous fluid, and to date systematic corrections to the hypernetted chain approximation have not been attempted. Various ansatz for the inhomogeneous bridge function have been ex-

plored for a Lennard–Jones fluid between two walls [190], and Percus–
Yevick hard-sphere bridge functions have been used for ions in the planar
electric double layer [191]. In the latter approximation, the results are
good but the computation appears rather demanding.

As in the bulk fluid, the direct correlation function decays as the pair
potential for large separations between the ions, as was assumed in
Section III.B.3. Hence, one defines the short-ranged part to be

$$\underline{\underline{\chi}}(\mathbf{r}_1, \mathbf{r}_2) = \underline{\underline{C}}(\mathbf{r}_1, \mathbf{r}_2) + \frac{1}{4\pi r_{12}} \underline{\underline{Q}}(\mathbf{r}_1, \mathbf{r}_2) \tag{4.41}$$

where the dyadic charge matrix now depends on the positions of the ions
due to densities that are incorporated into its definition, $\underline{\underline{Q}}(\mathbf{r}_1, \mathbf{r}_2) = (4\pi\beta/\varepsilon)\underline{q}(\mathbf{r}_1)\underline{q}^T(\mathbf{r}_2)$, with $\{\underline{q}(\mathbf{r})\}_\alpha = q_\alpha \rho_\alpha^{1/2}(\mathbf{r})$. (The pair potential due to
dielectric images, if present, should also be subtracted.) The mean
spherical approximation sets the short-ranged part of the direct correlation function to zero beyond contact,

$$\chi_{\alpha\gamma}^{\text{MSA}}(\mathbf{r}_1, \mathbf{r}_2) = 0 \qquad |\mathbf{r}_1 - \mathbf{r}_2| > d_{\alpha\gamma} \tag{4.42}$$

One of course enforces the exact condition $h_{\alpha\gamma}(\mathbf{r}_1, \mathbf{r}_2) = -1$, $|\mathbf{r}_1 - \mathbf{r}_2| < d_{\alpha\gamma}$. This corresponds to neglecting the bridge function and to linearizing
the hypernetted chain closure (apart from the mean electrostatic po-
tential). For the case of zero-size ions, $d_{\alpha\gamma} = 0$, this may be called the
Debye–Hückel approximation. Setting the inhomogeneous total correlation function to the exponential of the fluctuation potential is also known
as Loeb's closure [192]; it has been used to extract some analytic results
for the planar double layer [88, 193, 194]. The analysis of Attard et al.
[193, 194] was termed the extended Poisson–Boltzmann approximation,
which is the same name as that given to the earlier analysis at a similar
level of approximation by Podgornik and Žekš [195]. Solutions to Loeb's
closure (inhomogeneous mean spherical approximation for ions of zero
size) were obtained by Blum et al. [196] for the case when the density
profile can be expressed as a sum of exponentials.

The modified Poisson–Boltzmann theory, originally due to Bell and
Levine [197, 198], represents one of the earliest inhomogeneous integral
equation approaches. Loeb's closure was used, but instead of the
inhomogeneous Ornstein–Zernike equation the Kirkwood hierarchy was
invoked. The modified Poisson–Boltzmann theory has been used to study
the electric double layer at an isolated wall, and it includes volume
exclusion effects due to ion size, correlation effects due to the fluctuation
potential, and the effects of images due to the dielectric constant of the

wall. The various approximations have been refined several times (see [199] and references cited therein); the most recent version remains one of the more accurate theories for the isolated double layer at all but the highest surface charge densities [2].

The first member of the Born–Green–Yvon hierarchy for an inhomogeneous fluid, Eq. (4.36), was applied to the electric double layer by Croxton and McQuarrie [200], who approximated the inhomogeneous pair distribution function by the bulk one, with certain correction factors to ensure electroneutrality.

What makes the Ornstein–Zernike approach so popular for bulk fluids is that the convolution integral factorizes upon Fourier transformation; the consequent algebraic equation is readily "solved" for either the total or the direct correlation function. The availability of the fast Fourier transform is a significant advantage because one alternately iterates the closure in real space and the Ornstein–Zernike equation in Fourier space. For inhomogeneous fluids there are just two geometries in which the Ornstein–Zernike equation partially factorizes: planar, and spherical. As mentioned in Section III.B.3, in planar geometry a zero-order Hankel transform yields

$$\underline{\underline{\hat{H}}}(k, z_1, z_2) = \underline{\underline{\hat{C}}}(k, z_1, z_2) + \int_0^\infty \underline{\underline{\hat{H}}}(k, z_1, z_3)\underline{\underline{\hat{C}}}(k, z_3, z_2)\, dz_3 \quad (4.43)$$

The one-dimensional integral that remains is readily evaluated, and hence the transformed equation is in a form suitable for iteration. Because one has to transform back and forth from Hankel space many times, from the numerical point of view it is desirable to have a discrete orthogonal transform. Lado [201] has given an "almost" orthogonal transform, which appears to be quite robust in practice. Unfortunately, it does not appear possible to formulate a fast version of this. The density profile equations are also simplified by a Hankel transform. The planar inhomogeneous Ornstein–Zernike equation was solved by Sokolowski [202, 203] for a hard-sphere fluid with Percus–Yevick closure, and by Nieminen, Ashcroft, and co-workers [204–206] for a Lennard–Jones fluid with modified hypernetted chain closure. For the electric double layer Kjellander, Marčelja and co-workers [157, 189, 191, 207–214] used the hypernetted chain approximation for two walls (slit pore). Plischke and Henderson [158, 215, 216] used that approximation and also the mean spherical approximation for the double layer at an isolated wall.

For the case of a spherically inhomogeneous fluid, the density depends on the distance from the origin and the pair correlations depend on the two distances of the ions from the origin and the angle between them.

The Ornstein–Zernike equation is

$$\underline{\underline{H}}(r_1, r_2, \theta_{12}) = \underline{\underline{C}}(r_1, r_2, \theta_{12}) + \int \underline{\underline{H}}(r_1, r_3, \theta_{13})\underline{\underline{C}}(r_3, r_2, \theta_{32})\, d\mathbf{r}_3$$

$$(4.44)$$

which partially factorizes upon Legendre transformation [141, 217],

$$\underline{\underline{\hat{H}}}^{(n)}(r_1, r_2) = \underline{\underline{\hat{C}}}^{(n)}(r_1, r_2) + \frac{4\pi}{2n+1}\int_0^\infty \underline{\underline{\hat{H}}}^{(n)}(r_1, r_3)\underline{\underline{\hat{C}}}^{(n)}(r_3, r_2)r_3^2\, dr_3$$

$$(4.45)$$

(The equations for the density profile in spherical geometry also simplify upon Legendre transform [141].) The Legendre transforms are essentially the same as those used for the evaluation of the bridge function,

$$\hat{f}^{(n)}(r_1, r_2) = \frac{2n+1}{2}\int_0^\pi f(r_1, r_2, \theta)\sin\theta\, d\theta \qquad (4.46)$$

A discrete orthogonal version of the Legendre transform performs well in practice [141]. A quick (but not fast) version has been tested [218], and a fast (but not orthogonal) algorithm is known [219]. The spherically inhomogeneous Ornstein–Zernike equation has been solved for hard-sphere fluids with Percus–Yevick closure [141], and for Lennard–Jones fluids with hypernetted chain closure [218, 220]. Fushiki has also solved it for a charged spherical pore containing counterions only [159, 217]. Outhwaite and Bhuiyan [221] used the modified Poisson–Boltzmann theory (Kirkwood hierarchy with Loeb's closure) to study the electric double layer around a spherical macroion.

The planar and spherical geometries are the only two geometries in which the inhomogeneous pair correlation functions depends on three variables, one of them in an homogeneous fashion. In contrast, for example, for systems with cylindrical symmetry they depend on the two distances from the axis, the mutual angle, and the mutual separation along the axis. Currently, it does not appear feasible to solve the inhomogeneous Ornstein–Zernike equation in any but planar and spherical geometry (or circular or linear geometry in two dimension), although Bhuiyan and Outhwaite [122] have formulated their modified Poisson–Boltzmann theory in cylindrical geometry.

In concluding this section on the inhomogeneous approach, it is now shown that if the inhomogeneous direct correlation function is replaced by the bulk one in the equations for the density profile, then one recovers the singlet hypernetted chain approximation, as was pointed out by Badiali et al. [144]. By replacing the inhomogeneous $c_{\alpha\gamma}(\mathbf{r}_1, \mathbf{r}_2)$ by the bulk $c_{\alpha\gamma}(r_{12})$, the Triezenberg–Zwanzig equation for the density profile, Eq. (4.39), may be rewritten

$$\frac{1}{\rho_\alpha(\mathbf{r}_1)} \frac{\partial \rho_\alpha(\mathbf{r}_1)}{\partial \mathbf{r}_1} = -\beta \frac{\partial V_\alpha(\mathbf{r}_1)}{\partial \mathbf{r}_1} + \sum_\gamma \int \frac{\partial \rho_\gamma(\mathbf{r}_2)}{\partial \mathbf{r}_2} c_{\gamma\alpha}(r_{12}) \, d\mathbf{r}_2$$

$$= -\beta \frac{\partial V_\alpha(\mathbf{r}_1)}{\partial \mathbf{r}_1} - \sum_\gamma \int \rho_\gamma(\mathbf{r}_2) \frac{\partial c_{\gamma\alpha}(r_{12})}{\partial \mathbf{r}_2} \, d\mathbf{r}_2$$

$$= -\beta \frac{\partial V_\alpha(\mathbf{r}_1)}{\partial \mathbf{r}_1} + \sum_\gamma \int \rho_\gamma(\mathbf{r}_2) \frac{\partial c_{\gamma\alpha}(r_{12})}{\partial \mathbf{r}_1} \, d\mathbf{r}_2$$

$$= -\beta \frac{\partial V_\alpha(\mathbf{r}_1)}{\partial \mathbf{r}_1} + \frac{\partial}{\partial \mathbf{r}_1} \sum_\gamma \rho_\gamma \int h_{0\gamma}(\mathbf{r}_2) c_{\alpha\gamma}(r_{12}) \, d\mathbf{r}_2 \qquad (4.47)$$

The second equality follows an integration by parts, the integrated portion vanishing, the third equality arises because $\partial f(r_{12})/\partial \mathbf{r}_2 = -\partial f(r_{12})/\partial \mathbf{r}_1$, and the final equality is due to the fact that $\rho_\gamma(\mathbf{r}_2) = \rho_\gamma[1 + h_{0\gamma}(\mathbf{r}_2)]$, where the constant part gives a contribution that is independent of $\mathbf{r}_1$. Integrating this one obtains

$$\ln 1 + h_{0\alpha}(\mathbf{r}_1) = -\beta V_\alpha(\mathbf{r}_1) + \sum_\gamma \rho_\gamma \int h_{0\gamma}(\mathbf{r}_2) c_{\gamma\alpha}(r_{12}) \, d\mathbf{r}_2$$

$$= -\beta q_\alpha \psi(\mathbf{r}_1) + \sum_\gamma \rho_\gamma \int h_{0\gamma}(\mathbf{r}_2) \chi_{\gamma\alpha}(r_{12}) \, d\mathbf{r}_2 \qquad (4.48)$$

both sides vanishing as $\mathbf{r}_1 \to \infty$. This is just the singlet hypernetted chain approximation; in planar geometry it is equivalent to Eqs. (3.21b) and (3.28b), with $d_{0\alpha}(z) = 0$. Colmenares and Olivares [222] used this method to treat the electric double layer using hypernetted chain and mean spherical approximation bulk direct correlation functions.

For the Debye–Hückel closure to the bulk correlation functions, which is also the mean spherical approximation for ions with zero size, $\chi_{\alpha\gamma}(r) = 0$, this evidently reduces to the nonlinear Poisson–Boltzmann approximation for the density profile. This closure and profile approximation have been used in the inhomogeneous approach to analyze the electrostatic correlation contribution to the double layer interaction between planar surfaces. For the planar electric double with counterions only [193], and with a restricted primitive model electrolyte [194], Attard et al. solved the zero-sized mean spherical approximation for the inhomogeneous pair correlation functions using the nonlinear Poisson–Boltzmann density profile, and obtained analytic expressions for the double-layer interaction free energy. The same approximation was solved by Blum et al. [196] for exponential density profiles.

C. Density Functional Theory

The hypernetted chain solvation free energy, Eq. (4.14), may be rewritten

$$-\beta[\mu_0^{\text{HNC}} - \mu_0^{\text{self}}]$$

$$= -\sum_\alpha \rho_\alpha \int \{[h_{0\alpha}(r_1) - c_{0\alpha}(r_1)][1 + h_{0\alpha}(r_1)] - h_{0\alpha}(r_1)\} \, d\mathbf{r}_1$$

$$+ \frac{1}{2} \sum_\alpha \rho_\alpha \int [h_{0\alpha}(r_1) - c_{0\alpha}(r_1)] h_{0\alpha}(r_1) \, d\mathbf{r}_1$$

$$= -\sum_\alpha \rho_\alpha \int [\{\beta u_{0\alpha}(r_1) + \ln[1 + h_{0\alpha}(r_1)]\}[1 + h_{0\alpha}(r_1)] - h_{0\alpha}(r_1)] \, d\mathbf{r}_1$$

$$+ \frac{1}{2} \sum_{\alpha\gamma} \rho_\alpha \rho_\gamma \int d\mathbf{r}_1 \, d\mathbf{r}_2 \, h_{0\alpha}(r_1) c_{\alpha\gamma}(r_{12}) h_{0\gamma}(r_2) \qquad (4.49)$$

The solute–ion direct correlation function has been eliminated by using the hypernetted chain closure on the first term, and by using the Ornstein–Zernike equation on the second. One may confirm that this particular form of the free energy is optimized by the hypernetted chain approximation,

$$\frac{-\beta \delta \mu_0^{\text{HNC}}}{\rho_\alpha \delta h_{0\alpha}(r_1)} = -\beta u_{0\alpha}(r_1) - \ln[1 + h_{0\alpha}(r_1)] - 1 + 1$$

$$+ \sum_\gamma \rho_\gamma \int c_{\alpha\gamma}(r_{12}) h_{0\gamma}(r_2) \, d\mathbf{r}_2$$

$$= -[h_{0\gamma}(r_1) - c_{0\gamma}(r_1)] + h_{0\gamma}(r_1) - c_{0\gamma}(r_1)$$

$$= 0 \qquad (4.50)$$

In other words, minimizing the solvation free energy in the form of Eq. (4.49) with respect to the double layer about the solute is equivalent to solving the singlet hypernetted chain approximation. (Olivares and McQuarrie [223] give systematic methods for deriving variational principles for closure approximations.) The advantage of the variational approach is that one can use physically appropriate analytic functions to describe the density profile, and the free energy minimization by parameter variation is rather efficient. Moreover, the errors in the free energy are second order compared to the errors in the density profiles, and hence one only has to solve the minimization problem approximately.

By considering the solute as fixed and the source of an external field, the solute–ion distribution function may be replaced by the density profile. It is clear that Eq. (4.49) gives the grand potential as an explicit functional of the density. This is an example of a density functional approximation, in which the equilibrium density profile is given by the optimization of a grand potential functional. The various versions and applications of density functional theory have been reviewed by Evans [224, 225]. Normally, one writes the grand potential as a functional for the configurational part of the Hamiltonian (the pair potential, and the chemical potential and any other one-body external potentials). The functional derivative of the grand potential with respect to the one-body potential yields the density profile, and successive differentiation yields the density–density and higher order correlation functions. There is in fact a 1:1 relationship between the one-body potential and the density profile, and one may alternatively regard the free energy as a functional of the density. The Legendre transform of the grand potential yields the so-called intrinsic part of the Helmholtz free energy [224, 225],

$$\mathscr{F}[\underline{\rho}] = \Omega + \sum_\alpha \int [\mu_\alpha - V_\alpha(\mathbf{r})]\rho_\alpha(\mathbf{r})\, d\mathbf{r} \qquad (4.51)$$

For the case treated above of a spherical solute fixed at the origin, the external field is $V_\alpha(\mathbf{r}) = u_{0\alpha}(r)$ and the density profile is $\rho_\alpha(\mathbf{r}) = \rho_\alpha[1 + h_{0\alpha}(r)]$. Provided that one has a recipe for $\mathscr{F}$ as a functional of the density profile, one may regard this equation as giving the grand potential as a functional of the density, $\tilde{\Omega}[\underline{\rho}]$. This functional is minimized by the equilibrium density profile, at which point it equals the actual grand potential. (It is easiest to develop approximations directly for $\mathscr{F}$, which does not depend on the specific external field, and then to use these to obtain $\tilde{\Omega}$).

The intrinsic Helmholtz free energy may be split into ideal and excess parts. The former is

$$\mathscr{F}^{id}[\underline{\rho}] = \sum_\alpha \int f_\alpha^{id}(\rho_\alpha(\mathbf{r}))\, d\mathbf{r} \qquad (4.52)$$

where $f_\alpha^{id}(\rho) = k_B T \rho (\ln[\rho \Lambda_\alpha^3] - 1)$ is the ideal Helmholtz free energy density, and Λ_α is the de Broglie thermal wavelength. The functional derivatives of the excess part of the intrinsic Helmholtz free energy with respect to the density profile yield the hierarchy of direct correlation functions [224, 225]. Specifically, the one-body direct correlation function is

$$c_\alpha(\mathbf{r}_1; [\underline{\rho}]) = -\frac{\delta \mathscr{F}^{ex}[\underline{\rho}]}{\delta \rho_\alpha(\mathbf{r}_1)} \tag{4.53}$$

and the two-body direct correlation function is

$$c_{\alpha\gamma}(\mathbf{r}_1, \mathbf{r}_2; [\underline{\rho}]) = \frac{\delta c_\alpha(\mathbf{r}_1; [\underline{\rho}])}{\delta \rho_\gamma(\mathbf{r}_2)} = -\frac{\delta^2 \mathscr{F}^{ex}[\underline{\rho}]}{\delta \rho_\alpha(\mathbf{r}_1)\delta \rho_\gamma(\mathbf{r}_2)} \tag{4.54}$$

In view of these expressions, $\mathscr{F}$ as a functional of the density profile can be obtained by functional integration of the direct correlation function, and the resultant formally exact formulas provide a basis for a number of approximation schemes. If one introduces a coupling constant for the density profile, $\underline{\rho}^{(\lambda)}(\mathbf{r}) = \underline{\rho}^{(0)}(\mathbf{r}) + \lambda\Delta\underline{\rho}(\mathbf{r})$, then one obtains [224, 225]

$$\beta\mathscr{F}^{ex}[\underline{\rho}^{(1)}] = \beta\mathscr{F}^{ex}[\underline{\rho}^{(0)}] - \sum_\alpha \int_0^1 d\lambda \int d\mathbf{r}\, \Delta\rho_\alpha(\mathbf{r})c_\alpha(\mathbf{r}; [\underline{\rho}^{(\lambda)}]) \tag{4.55a}$$

$$= \beta\mathscr{F}^{ex}[\underline{\rho}^{(0)}] - \sum_\alpha \int d\mathbf{r}\, \Delta\rho_\alpha(\mathbf{r})c_\alpha(\mathbf{r}; [\underline{\rho}^{(0)}])$$

$$+ \sum_{\alpha\gamma} \int_0^1 d\lambda(\lambda - 1) \int d\mathbf{r}_1\, d\mathbf{r}_2\, \Delta\rho_\alpha(\mathbf{r}_1)\Delta\rho_\gamma(\mathbf{r}_2)c_{\alpha\gamma}(\mathbf{r}_1, \mathbf{r}_2; [\underline{\rho}^{(\lambda)}])$$

$$\tag{4.55b}$$

A convenient starting point for the functional integration is the uniform electrolyte, $\rho_\alpha^{(0)}(\mathbf{r}) = \rho_\alpha$, in which case the grand potential functional becomes

$$\beta\Omega[\underline{\rho}^{(1)}] - \beta\Omega[\underline{\rho}^{(0)}]$$

$$= \beta\sum_\alpha \int V_\alpha(\mathbf{r})\rho_\alpha(\mathbf{r})\, d\mathbf{r} + \sum_\alpha \int \left[\rho_\alpha(\mathbf{r})\ln\frac{\rho_\alpha(\mathbf{r})}{\rho_\alpha} - \rho_\alpha(\mathbf{r}) + \rho_\alpha\right] d\mathbf{r}$$

$$+ \sum_{\alpha\gamma} \int_0^1 d\lambda(\lambda - 1) \int d\mathbf{r}_1\, d\mathbf{r}_2\, \Delta\rho_\alpha(\mathbf{r}_1)\Delta\rho_\gamma(\mathbf{r}_2)c_{\alpha\gamma}(\mathbf{r}_1, \mathbf{r}_2; [\underline{\rho}^{(\lambda)}])$$

$$\tag{4.56}$$

When the external field represents a solute fixed at the origin, the left side is the solvation free energy, $\beta\mu_0$ (less the self-term).

This formally exact result gives the grand potential as a functional of the density profile. Obviously, one has to give some formula for the two-body direct correlation function that appears. The simplest approximation is to ignore the dependence on the coupling constant and to take it to be equal to the corresponding quantity of the bulk electrolyte,

$$c_{\alpha\gamma}(\mathbf{r}_1, \mathbf{r}_2; [\underline{\rho}^{(\lambda)}]) = c_{\alpha\gamma}(r_{12}; \underline{\rho}^{(0)}]) \tag{4.57}$$

In this case one obtains [224–226]

$$\beta\Omega[\underline{\rho}^{(1)}] - \beta\Omega[\underline{\rho}^{(0)}]$$

$$= \beta \sum_\alpha \int V_\alpha(\mathbf{r})\rho_\alpha(\mathbf{r})\,d\mathbf{r} + \sum_\alpha \int \left[\rho_\alpha(\mathbf{r}) \ln \frac{\rho_\alpha(\mathbf{r})}{\rho_\alpha} - \rho_\alpha(\mathbf{r}) + \rho_\alpha \right] d\mathbf{r}$$

$$- \frac{1}{2}\sum_{\alpha\gamma} \int d\mathbf{r}_1\,d\mathbf{r}_2\,[\rho_\alpha(\mathbf{r}_1) - \rho_\alpha][\rho_\gamma(\mathbf{r}_2) - \rho_\gamma]c_{\alpha\gamma}(r_{12}; [\underline{\rho}^{(0)}])$$

$$\tag{4.58}$$

For the case when the external field represents a solute fixed at the origin, this is evidently identical to Eq. (4.49). Hence, minimizing the approximate grand potential functional that has the inhomogeneous partially coupled direct correlation function replaced by the fully coupled bulk one is equivalent to solving the singlet hypernetted chain approximation.

An obvious improvement is to retain the next term in the functional Taylor expansion, which will give an expression that involves the triplet direct correlation function of the bulk electrolyte [227, 228]. That is equivalent to retaining bridge diagrams in the hypernetted chain approximation, since the bridge function can be written as an expansion over the three body and higher direct correlation functions [38, 39, 179].

Perturbation approximations are perhaps the most popular form of density functional theory [224, 225]. One typically uses the short-range repulsive part of the pair potential to define a reference fluid, which is predominantly responsible for the double-layer structure, and one usually treats the attractive part in mean field or other approximate fashion. The reference intrinsic free energy may be that of a bulk hard-sphere fluid using a simple local density approximation [229], or more sophisticated weighted versions [230–232]; see the review of Evans [225] for a discussion of these and other recipes for the reference free energy.

Density functional theory has been applied to the electric double layer at an isolated planar wall. Groot [151] used a density functional approximation that invoked the mean spherical bulk direct correlation function evaluated at a local density. A similar approach was earlier used by Nielaba, Forstmann, and co-workers [149, 150], who instead inserted a bulk direct correlation function, evaluated at a local nonneutral density, into the singlet hypernetted chain equation. Davis and co-workers [152, 233, 234] used a perturbation approach that invoked the hard-sphere free energy density evaluated at a density determined by a weighting due to Tarazona [232], together with a hard-sphere–Coulomb correction to the direct correlation function evaluated at the bulk density. They also studied interacting planar double layers in the primitive model [235] and with added hard sphere solvent [236]. Very similar approaches were used by Kierlik and Rosinberg [153], who give a transparent derivation of the electrostatic contributions, and by Patra and Ghosh [154], who evaluated the Coulomb correction at the effective rather than the bulk density. The latter authors tested as well the weighting due to Denton and Ashcroft [237], and obtained very good results for the restricted primitive model double layer [154, 238]. They also included a hard-sphere solvent [154], which is the solvent primitive model similarly treated by Tang et al. [234], and went on to treat an asymmetric binary electrolyte [239], the double layer at a metallic electrode with a jellium model [240], and the solvation force due to overlapping double layers [241]. Penfold et al. [242] obtained a correction to the Poisson–Boltzmann equation using a local density approximation and tested it for a spherical macroion (cell model). Feller and McQuarrie [146] formulated the singlet hypernetted chain approximation as a variational principle, and solved it for the isolated planar double layer.

For interacting walls, Stevens and Robins [243] obtained good results for the pressure for the counterions-only double layer using a local density approximation and the known free energy of the bulk, point-ion, one component plasma. Attard et al. [193, 194] obtained an analytic expression for the free energy of interacting planar double layers (counterions-only and restricted primitive model, including dielectric images), using the nonlinear Poisson–Boltzmann approximation for the density profile and the mean spherical approximation for ions of zero size for the inhomogeneous ion–ion direct correlation function. Podgornik and Žekš [195] used a similar level of approximation in the context of a functional integral description of the effects of ion correlations in the double layer. Feller and McQuarrie [163, 164] used variational techniques to obtain the interaction of two walls and the electric double layer between them (equivalent to the singlet dumbbell hypernetted chain approximation). As mentioned earlier, Davis and co-workers [235, 236] and Patra and Ghosh [241] used weighted density approximations for interacting planar double layers.

D. Simulations

The isolated planar double layer has been simulated for a restricted primitive model confined between two identically charged walls [244–246] and between one charged and one neutral wall [147, 247], which has the advantages of attaining bulk concentrations for smaller system sizes, and of avoiding time-consuming sums due to multiple image charges. (Ballone et al. [143] used two walls of equal and opposite charge, obtaining agreement with the results of Torrie and Valleau [147], which were for a smaller system.) Zhang et al. [236] simulated an electrolyte in a hard-sphere solvent confined between a neutral and a charged wall using (essentially) the canonical ensemble. Two-dimensional Ewald sums were used by these authors and by others [248–253]; the method of Lekner [251] appears most efficient. For conducting metal walls the usual three-dimensional Ewald sum can be used [254]. Torrie and Valleau [147] argue that Ewald sums exaggerate the correlations in the planar double layer, and that it is preferable to include only the mean-field contribution to the potential from ions external to the central simulation cell. Caillol and Levesque [148] avoided periodic boundary conditions in their simulations of the isolated planar double layer by placing the system on the surface of a 4-sphere.

The pressure due to the interacting of two planar double layers at small separations has been obtained for counterions only [214, 255–260], binary primitive model electrolytes [214, 259–262], and for molten salts [263]. The paper by Guldbrand et al. [256] is noteworthy for being the first simulation to show attractive double-layer forces between similarly charged surfaces. In some cases the canonical ensemble was used and Widom's method was applied to determine the chemical potential, and hence the concentration of the bulk electrolyte in equilibrium with the double layer. It is probably preferable to use grand canonical Monte Carlo, as this sets the desired equilibrium directly [147, 260]. For example, Bratko et al. [264] were able to study the adsorption excess of a restructed primitive model electrolyte confined between two neutral walls as a function of separation, and the consequent number of confined ions were used in canonical simulations for the net pressure between the neutral walls [265]. Grand canonical and isobaric Monte Carlo studies of relatively realistic models of a mica–clay slit pore with counterions in an aqueous solvent have been carried out [266–269]. Molecular dynamics have been performed in the canonical ensemble for water and mobile counterions between two ionic surfactant monolayers [270]. The solvation of sodium chloride at a platinum electrode has been reported [271], as has the structure of water between neutral and charged planar walls [272]. Results have also been reported for pure water between clay and surfactant lamellae [273–275]. Lee et al. [276] used molecular dynamics with Ewald summation to study a dipolar fluid confined between charged Lennard–Jones walls. The interaction of charged surfaces with grafted polyelectrolyte counterions has been simulated [277–281].

Simulations deal with finite sized samples, and except for fully closed pores one has to attend to the boundaries. Periodic boundary conditions with Ewald summation have ben used for bulk electrolytes. As mentioned above, Ewald summation has also been used for planar slits, although it has ben argued that mean-field corrections for the tail are more appropriate [147]. For the case of spherical macroions one typically encloses the macrosphere and surrounding electrolyte by a concentric, impenetrable, uncharged sphere; if the annular region is wide enough one will have bulk electrolyte in the system and so mimic an isolated spherical macroion. Obviously, for large radius macroions the situation is problematic, since the number of electrolyte ions required grows in proportion to the square of the radius; to date no periodic replication of an annular cell has been used. Some authors use this annular geometry to model interacting macroions, the so-called cell model; they invoke the electrolyte concentration at the outer boundary as the osmotic pressure of the dispersion [131]. The Monte Carlo tests of the cell model against isotropic solution results by Linse and Jönsson [282] show it to be inaccurate for the osmotic pressure except at low concentrations.

The double layer around a spherical macroion has been simulated within the cell model and in isotropic solution [131, 171, 242, 282, 283]. Degréve et al. [284] simulated the restricted primitive model double layer about small spherical macroions. Sloth and Sørenson [285] studied a primitive model electrolyte confined to a charged spherical pore using grand canonical Monte Carlo. Svennson and Jönsson [286] simulated the interaction of a spherical macroion with a similarly charged planar wall in the presence of counterions, and showed that the force could be attractive. Vlachy and co-workers [118, 119, 123] simulated the electric double layer inside a cylindrical pore. Cylindrical aggregates in the cell model have also been studied [121, 287–289].

V. NUMERICAL RESULTS

A. Effective Surface Charge

The various properties of an isolated planar double layer have been extensively reviewed [2, 3]. In particular, and among other things, Carnie and Torrie [2] cover the modified Poisson–Boltzmann approximation and singlet hypernetted chain approximations in some detail, and Blum [3] compares a number of theories for the ion profiles in $1\,M$ monovalent electrolyte at a high-surface charge density. However, since the asymptotic results for electrolytes and the electric double layer discussed in Sections II and III have begun to be emphasized only recently, it seems appropriate to begin with some numerical examples of these for an isolated wall. Perhaps the key result is that the ion density profiles behave as predicted by the Poisson–Boltzmann approximation, but with effective parameters. The screening length of the bulk electrolyte κ^{-1} is relatively close to the Debye length κ_D^{-1}, and the dielectric factor ν is almost unity, at least for electrolytes in the monotonic regime. Hence, the most important effective parameter is the surface charge density $\tilde{\sigma}$, which is given by Eq. (3.36). This effective surface charge density may also be called the fitted charge density because it is what one would obtain if one were to fit experimental measurements with the linear Poisson–Boltzmann approximation, at least in the asymptotic regime. As mentioned at the end of Section III.B.2, either one may use a sophisticated theory to describe the double layer, or one may use the simplest Poisson–Boltzmann approximation and correct this fitted surface charge to obtain the actual surface charge. The following data is presented with the latter procedure in mind. Note that the bulk electrolytes explored here are all in the monotonic regime, since this is the case where the effective surface charge can be given a physical interpretation.

Several methods will be used to obtain the effective surface charge. The most sophisticated use the singlet hypernetted chain approximation in conjunction with Eq. (3.36). This approximation is at least qualitatively reliable for the isolated planar double layer, and may be described as quantitatively accurate at not too high concentrations or surface charges [2, 3]. The accuracy of the single hypernetted chain approximation is improved by the inclusion of the first bridge diagram [143], and results for the effective surface charge with and without the bridge diagram are given. (In the data below the bridge function was recalculated to self-consistency, up to 10 times in the case of the divalent electrolyte at the highest surface charge density, in order to obtain a reliable value for the effective surface charge.) An analytic approximation will also be used, namely, the extended Poisson–Boltzmann theory of Attard et al. [194]. This is essentially the solution of the mean spherical closure to the inhomogeneous Ornstein–Zernike equation for ions of zero size; it thus includes the effects of ion correlations but not those of excluded volume. Attard et al. [194] gave a formula for converting surface charged fitted on the bases of the nonlinear Poisson–Boltzmann approximation to actual surface charge, but it is straightforward to reexpress this in terms of a fit to the linear Poisson–Boltzmann approximation. Assuming that to a good enough approximation $\kappa \approx \kappa_D$, $\nu \approx 1$, the result is

$$\tilde{\sigma} = \sigma \frac{A}{2s} \left[1 + \frac{\beta q^2 \kappa_D}{4\varepsilon} \{2I + \ln 2\} \right]^{1/2} \qquad (5.1a)$$

where $A \equiv 16(-1 + \sqrt{1 + s^2/4})/s$, the dimensionless surface charge is $s = 4\pi\beta q\sigma/\varepsilon\kappa_D$, and I is given by Eq. (4.23) of [194],

$$I = \frac{1}{2} \left(1 + \frac{2z^2 - 3}{(2z^2 - 1)^3} \right) \ln 2 + \frac{2 - 2z^3 + z}{2z(2z^2 - 1)^2}$$

$$- \frac{1}{2} \left(1 - \frac{2z^2 - 3}{(2z^2 - 1)^3} \right) \ln(z + z^2)$$

$$- \frac{\sqrt{z^2 - 1}}{z} \left(1 + \frac{2z^2 + 1}{(2z^2 - 1)^3} \right) \tan^{-1}\sqrt{\frac{z - 1}{z + 1}} \qquad (5.1b)$$

where $z \equiv \sqrt{1 + s^2/4}$.

Figure 5 shows the effective surface charge (i.e., that which would be fitted on the basis of the linear Poisson–Boltzmann approximation), for monovalent and divalent 1-mM electrolyte. The extended Poisson–Boltzmann approximation, Eq. (5.1), is compared to the hypernetted chain approximation with and without the first bridge diagram using Eq. (3.36). In general, the fitted surface charge is noticeably less than the actual surface charge, and the departure is larger for high-surface charge densities and in divalent electrolyte. If the two were equal (i.e., the linear Poisson–Boltzmann approximation were exact), one would get the straight line. At low surface charge this is indeed the case. As the surface charge is increased one finds a departure from linearity, which is the same in all three approximations. Evidently, the analytic extended Poisson–Boltzmann approximation provides the primary correction to the mean-field approximation. One therefore concludes that this departure is dominated by electrostatic correlations, and that excluded volume effects make a minor contribution in this regime (the extended Poisson–Boltzmann approximation includes the former but not the latter [194]). That the monovalent and divalent curves coincide here supports the use of the parameter $s \equiv 4\pi\beta q\sigma/\varepsilon\kappa_D$ as a dimensionless measure of surface charge. At higher surface charge densities the three approximations begin to disagree, particularly in the case of the divalent electrolyte, although qualitatively they behave similarly. The hypernetted chain approximation with the first bridge diagram is expected to be the most accurate of the

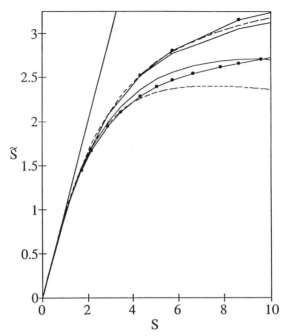

Figure 5. The effective or fitted surface charge density, $\tilde{s} \equiv 4\pi\beta q\tilde{\sigma}/\varepsilon\kappa_D$, as a function of the actual surface charge density, $s \equiv 4\pi q\sigma/\varepsilon\kappa_D$, for the restricted primitive model at a concentration of $10^{-3}\,M$ ($\varepsilon_r = 78.358$, $T = 298.15$ K, $d = 4.25$ Å). The upper triplet of curves is for monovalent ions, the lower triplet is for divalent ions, and the straight line is a guide to the eye. The solid curves with and without symbols are the hypernetted chain approximations, Eq. (3.36), with and without the first bridge diagram, and the dashed curve is the analytic result of the extended Poisson–Boltzmann approximation [194], Eq. (5.1).

three approximations, and the fitted surface charge calculated with it should be taken as definitive.

In the divalent data one sees that the curve actually has a maximum in the bare hypernetted chain and in the extended Poisson–Boltzmann approximations (these approximations also predict that a maximum occurs in the monovalent electrolyte at higher surface charges than those shown). This effect is clearer in the data for the higher concentration of 0.01 M shown in Fig. 6. Here again all the curves coincide at low-surface charge densities, and they would also coincide in this regime with the low-concentration data in Fig. 5, which lends further support to s as the appropriate dimensionless surface density. The divalent extended Poisson–Boltzmann data has a maximum at about $s = 2.85$ (it actually predicts an unphysical asymptotic attraction beyond $s = 5.9$), the bare hypernetted chain approximation has a maximum at about $s = 6.8$ (it did

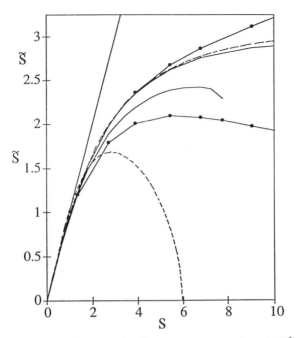

Figure 6. As in the preceding figure at a concentration of $10^{-2} M$.

not converge beyond $s = 8$), and including the first bridge diagram shifts the maximum to about $s = 5.5$ (and extends the regime of convergence). The nonmonotonic behavior of the fitted surface charge density is even more noticeable at the higher electrolyte concentration of $0.1 M$ (Fig. 7), and can also be seen in the monovalent data in the figure. The experimental consequence of this behavior is that for a given electrolyte the surface charge density will appear to saturate. Alternatively, there will be an ambiguity in its value that can only be resolved by additional information, such as data fitting in the nonasymptotic regime. This nonmonotonic behavior is both unusual and counterintuitive, but it is perhaps consistent with other hypernetted chain predictions for the double layer. In the case of the $0.1 M$ monovalent electrolyte, the maximum fitted charge occurs at an actual area per unit charge of $100–125$ Å^2. Shortly, it will be shown that the hypernetted chain approximation predicts that the electric double-layer interaction between two charged walls at small separations changes from repulsive to attractive at an area per unit surface charge of about 200 Å^2. Finally, the potential drop across the isolated double layer is predicted by the hypernetted chain approximation to begin to *decrease* with *increasing* surface charge;

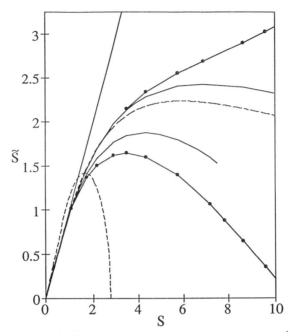

Figure 7. As in the preceding figure at a concentration of $10^{-1} M$.

the diffuse part at an area per unit surface charge of 50 $\overset{\circ}{A}^2$, and the total at about 10 $\overset{\circ}{A}^2$.

The rate of change of the electrostatic potential with surface charge density is called the differential capacitance, and the fact that the hypernetted chain approximation predicted it to be negative was at one time seen as a serious failing of that approximation. In fact there appears no fundamental proscription against a negative differential capacitance, and there is certainly no proof that it must be positive [290]. It has been shown that a negative differential capacitance could occur if the derivative of the charge density profile with respect to surface charge was anywhere positive, which appears to be what happens in the hypernetted chain calculations at high surface charge densities. Simulations of the restricted primitive model show a flattening of the potential-surface charge curve (without it actually turning over), but the diffuse part of the potential drop does turn over in simulations of the restricted primitive model, as does the total potential drop in simulations of an asymmetric electrolyte [291]. Hence, the fact that the hypernetted chain approximation predicts a negative differential capacitance for areas per unit charge less than 10 $\overset{\circ}{A}^2$, whereas simulations only show a small positive one in this case, is only a quantitative failing of that approximation, not a qualitative one. Similarly, the turnover in the effective surface charge shown in Figs. 5–7 is likely qualitatively correct, and the surface charge at which it occurs is probably accurately given by the hypernetted chain approximation that includes the first bridge diagram.

At high surface charge densities the effective surface charge continues to decrease, and for the 0.1 M divalent electrolyte it goes through zero and changes sign at $s \approx 11$. A layer of counterions forms at the wall with essentially close-packed density, excluding the co-ions, and at the same time overneutralizing the surface charge. Charge reversal occurs due to a combination of ion size and valence. It also depends on the bulk electrolyte concentration and surface charge density, which determine the concentration in the layer. The mechanism is clear in Fig. 8, where it can be seen that there is a dense layer of counterions next to the wall and that there are almost no co-ions in this region. Slightly beyond one diameter ($d = 4.25$ Å) from the wall an inversion occurs and the co-ion density

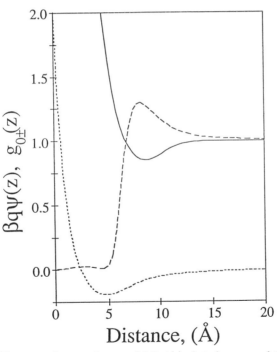

Figure 8. The mean electrostatic potential $\beta q \psi(z)$, dotted curve, and the counter- and co-ion distributions $g_{0\pm}(z)$, solid and dashed curves, respectively, for the 0.1 M divalent electrolyte with $e/\sigma = 75$ Å^2.

exceeds the counterion density. It is emphasized that this is a local charge oscillation induced by the high-surface charge on the wall. The bulk electrolyte is in the monotonic regime (with decay length $\kappa^{-1} = 5.4$ Å), and the density profiles and potential will decay monotonically beyond this surface region. In particular, although the surface is positively charged, the electrostatic potential turns negative at about 3 Å from the wall and remains negative. Measuring techniques that sample the double layer some distance from the surface will ascribe to the particle a charge with the wrong sign. For example, electrophoresis, in which the motion of a particle in an applied electric field depends on the electrostatic potential at the zeta plane away from the surface, may show a highly charged particle stop and reverse direction as the concentration is varied, even though its actual surface charge remains unchanged. The point of zero charge and the subsequent reversed effective charge are not due to physical binding of counterions to the surface, or to chemical association or dissociation, but rather to the electrostatic attraction and layering of the counterions and exclusion of the co-ions due to packing in the diffuse part of the double layer adjacent to the surface.

The modified Poisson–Boltzmann theory was perhaps the first to predict charge reversal, where it was seen in a primitive model electrolyte (monovalent co-ions and divalent counterions), at 0.15 M [292]. (These authors were careful to distinguish this surface-induced inversion of the ion profiles in the bulk monotonic regime from asymptotic oscillations present at higher concentrations.) The charge reversal, as evidenced by the inversion and the change in sign of the potential, were confirmed by simulations [246]. Similar behavior was also seen in a 0.5 M 2:2 electrolyte, which is very near the bulk monotonic-oscillatory transition, in Monte Carlo simulations by Torrie and Valleau [246], in the singlet mean spherical approximation by Feller and McQuarrie [146], and in density functional approaches by Mier-y-Teran et al. [233] and by Patra and Gosh [154]. Ennis [293] carried out an extensive study of the effective surface charge using the inhomogeneous hypernetted chain approximation (in contrast to the singlet approach implemented here), and systematically explored the reversal of the surface charge. That approximation has also shown charge reversal in a 2:1 primitive model electrolyte [294]. It appears that charge reversal occurs more readily in mixed valence (divalent counterions) that in symmetric electrolytes. Finally, charge reversal has also been seen in single hypernetted chain calculations for a *molecular* aqueous double layer [5, 295]. The solvent-induced fast screening and overcompensation of the surface charge in the 1 M monovalent electrolyte appears sensitive to the relative sizes of the solvent, counterions, and co-ions. The disparate ion diameters and consequent differential solvation doubtless quantitatively affect the point of zero charge. More generally, the values of the effective surface charge given above for the primitive model will be altered by a molecular solvent because they are a surface-sensitive property, although the asymptotic functional form will remain the same.

The agreement between the various approximations in Figs. 5–7 suggests that the effective surface charge data are fundamentally correct and even quantitatively accurate for the primitive model. That the extended Poisson–Boltzmann approximation agrees with the hypernetted chain based theories shows it to be soundly based. Furthermore, since the former neglects ion size, one concludes that the dominant contribution to the departure from the Poisson–Boltzmann approximation comes from the electrostatic correlations of the ions. In the context of the interactions between charged surfaces due to overlapping electric double layers, ion correlations add an attractive screened van der Waals component to the mean-field Poisson–Boltzmann repulsion [193, 194]. Hence, in order to describe double-layer force measurements with Poisson–Boltzmann theory, one needs to fit a smaller than actual surface charge, which gives a lower repulsion and effectively accounts for the otherwise neglected electrostatic fluctuations. This is precisely what is shown in the figures, where the effective surface charge density lies below the actual surface charge density.

Paradoxically, the hypernetted chain data in Fig. 9 appear to confirm the prediction made some time ago by Attard et al. [194] that at very low surface charge densities the fitted surface charge actually becomes greater than the actual surface charge. Those authors rationalized this behavior by arguing that ions are effectively repelled from an inert wall due to attractive correlations with ions in the bulk electrolyte (the Onsager–Samaris effect). At low surface charge densities this depletion of the electrolyte near the surface becomes important, and results in less screening of the surface charge by the double layer than is predicted by Poisson–Boltzmann theory, which gives rise to an effective surface charge that is greater than the actual surface charge. This effect was seen in every one of the six electrolytes analyzed in Figs. 5–7. In practice, this effect may be difficult to measure since it is relatively small and it occurs at extremely low surface charge densities.

An alternative analytic approximation for the effective surface charge may be developed on the basis of the self-consistent Debye–Hückel approach of Section II.C.1, as applied to the double layer. By using the formally exact asymptote for the ion profiles, Eq. (2.66), for all distances

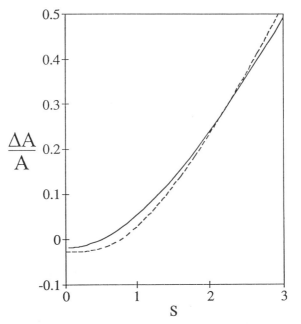

Figure 9. The relative change in the area per unit surface charge, $\sigma/\tilde{\sigma} - 1$, at low-surface charge densities in the 0.1 M monovalent electrolyte.

from the wall, and by using the electroneutrality condition, one obtains for all ions with the same diameter d

$$\tilde{\sigma} = \sigma \frac{\kappa^2}{\kappa_D^2} (1 + \kappa d) e^{-\kappa d} \qquad (5.2a)$$

from which the linear nature of the approximation is evident. (This approximation is similar in spirit to that used by Stillinger and Kirkwood [65].) At low concentrations the effective surface charge is less than the actual surface charge. A nonlinear version can be derived by applying the asymptote to the potential of mean force for all distances. For the restricted primitive model the effective surface charge satisfies

$$\frac{4\pi\beta q\sigma}{\varepsilon\kappa_D} = \kappa_D \int_0^\infty \sinh\left[\frac{4\pi\beta q\tilde{\sigma}e^{\kappa d}}{\varepsilon\kappa(1 + \kappa d)} e^{-\kappa z} \right] dz \qquad (5.2b)$$

This equation predicts that the effective surface charge will appear to saturate at high surface charge densities (more precisely it will increase logarithmically), but it does not turn over as is predicted by the hypernetted chain data. The linear result gives the tangent as $\sigma \to 0$. This analytic result is compared to the hypernetted chain data in Fig. 10. For the monovalent data it appears to give the primary departure correctly, but as the surface charge increases it underestimates the difference between the actual surface charge density and the effective surface charge density. This result is a consequence of the hypernetted chain data beginning to attain its maximum. In the divalent case, the analytic approximation performs poorly, and does not appear to have even the correct limiting tangent.

B. Interacting Double Layers

This section presents results for the double-layer interaction of two planar walls. The primary quantity of interest is the net pressure, since this, in

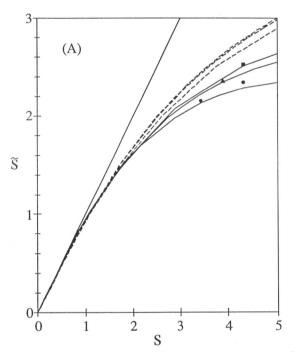

Figure 10. The effective surface charge density for, from bottom to top, 0.1, 0.01, and 0.001 M electrolyte (other parameters as in Fig. 5). The data represents the singlet hypernetted chain approximation with (symbols) and without (solid lines) the first bridge diagram, the dashed lines are the nonlinear self-consistent Debye–Hückel approximation, Eq. (5.2b), and the straight line is a guide to the eye. A. Monovalent electrolyte. B. Divalent electrolyte.

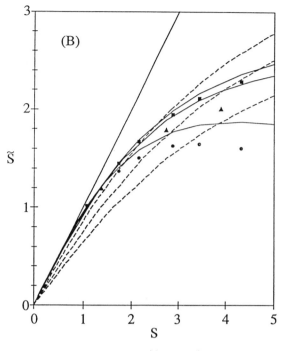

Figure 10. (*Continued*)

balance with the van der Waals attraction, determines whether charged particles in the electrolyte flocculate or remain dispersed and, in addition, it may be directly measured with molecular resolution. The net pressure is a rather sensitive property of overlapping double layers, and it provides a good test of the various approximations. Density functional theories for the pressure between planar walls have already been favorably compared to the simulation results for divalent electrolyte [235, 241]. Here the two singlet hypernetted chain methods, the dumbbell and the wall–wall, will be compared with inhomogeneous hypernetted chain and simulation data. Computational methods for the last three approximations have been given in the literature [84, 189, 260], and an algorithm for the planar dumbbell results presented here is now described.

The singlet Ornstein–Zernike equation for a planar dumbbell, Section III.D.3, was solved with the hypernetted chain closure without bridge diagrams. As usual the solute–ion total correlation functions were iterated using a simple Picard method with mixing. The Ornstein–Zernike convolution integral was evaluated by fast Fourier transform, using the transform of the bulk hypernetted chain short-ranged direct correlation function. The wall contributions to this integral were evaluated at the start of the program and stored. (The walls were of infinite thickness, and the limits discussed in Section III.D.3 were utilized.) The hypernetted chain closure requires the mean electrostatic potential, and the current iterate of this was evaluated by the robust algorithm of Badiali et al. [143, 144], modified for the present geometry, and with a few other changes. First, and probably of minor importance, the basic quantity $\psi''(z;t) - \kappa_D^2\psi(z;t)$ was replaced by $\psi''(z;t) - \kappa^2\psi(z;t)$, with the actual decay length obtained from the bulk electrolyte, and Eq. (2.66) being used. Second, the integration constant needed by the algorithm was the midplane potential $\psi_0(t)$ (since the integrals began at the midplane). The midplane potential cannot arbitrarily be set to zero, and its actual value was determined from negative feedback on the electroneutrality condition,

$$\psi_0^{(n+1)}(t) = \psi_0^{(n)}(t) + \frac{\sigma - \sigma^{(n)}}{q\Gamma^{(n)}} \tag{5.3}$$

where $\sigma^{(n)}$ is the charge density and $\Gamma^{(n)}$ is the total number density in the double layer on the nth iteration. This particular choice for the gain ensures that electroneutrality will be satisfied by the next iterate of the density profiles to linear order in the closure. Mixing (≈ 0.1) of old and new iterates was performed on the density profiles, but not on the potential except at large separations. Experience showed this procedure in combination with the algorithm of Badiali et al. [143, 144] to be quite robust, and it was confirmed that $\psi(0;t) \to 0$, $t \to \infty$. In the present calculations some 2^{13} grid points were used at a spacing of between 0.01 and 0.1 Å, depending on the concentration. Once electroneutrality was satisfied to 1 part in 10^7, the dumbbell–ion short-ranged correlation functions were evaluated, Eq. (3.108), and used in conjunction with the results for the isolated wall to evaluate the interaction free energy, Eq. (4.32). The net pressure was obtained by the first-order finite difference of this, and also from the contact theorem, Eq. (3.115). Whereas the former went smoothly to zero at large separations, the cancellation of various terms required in the latter gave rise to numerical errors at large separations. This appears to be related to the fact that the contact values in the singlet hypernetted chain approximation are related to the compressibility of the bulk electrolyte, and hence one should really subtract this quantity rather than the osmotic pressure. In the results given here, the net pressure from the contact pathway equals the calculated pressure less the pressure calculated at a separation beyond those shown.

1. *Monovalent Electrolyte*

Figure 11 shows the pressure between two walls with one charge every 250 Å^2 in a 0.01 M monovalent restricted primitive model electrolyte. The inhomogeneous hypernetted chain results are from [212] and are the benchmarks against which the simpler singlet results are to be judged.

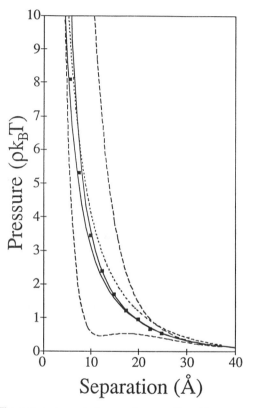

Figure 11. The net pressure between planar walls with an area per unit charge of 250 Å^2 in 0.1 M monovalent binary symmetric electrolyte ($T = 298$ K, $\varepsilon = 78.5$, $d = 4.25$ Å). The symbols represent inhomogeneous hypernetted chain results [213], and the dotted curve is the nonlinear Poisson–Boltzmann prediction. The pair of dashed curves are the singlet wall–wall hypernetted chain approximation [84] (the lower curve is the bare hypernetted chain approximation, and the upper curve includes the first bridge diagram). The pair of solid curves is the dumbbell hypernetted chain approximation (the lower curve utilises the contact theorem, and the upper curve comes from differentiating the interaction free energy). The separation here and in the remaining figures is the width of the region available to the centers of the ions; the walls may be taken to be an ion diameter further apart. The pressure is normalised by the kinetic pressure of the bulk electrolyte.

The inhomogeneous approach has been compared against simulations for the counterions only double layer [207], and for symmetric electrolytes [157, 214, 260]. In all cases the agreement has been very good, and in the absence of simulation data one may take the results of the inhomogeneous calculations to be exact. (The exception is the large separation regime, where both simulations and inhomogeneous integral equations become problematic because one requires the precise cancellation of several terms in order to obtain the net pressure.)

In this particular example the nonlinear Poisson–Boltzmann theory remains relatively close to the accurate results, but it does overestimate the repulsion. This is due to the neglect of correlations in that approximation, and is consistent with the data in Fig. 7, where in this regime the fitted surface charge is larger than the actual surface charge (see also Figs. 3 and 4 above). The singlet wall–wall hypernetted chain approximations bracket the inhomogeneous results. The bare hypernetted chain approximation overestimates the attractive component due to ion correlations; this deficiency occurs quite regularly, and will be seen in a number of the figures below. (The extended Poisson–Boltzmann approximation predicts an even greater attractive contribution, and again this is a general failing of that approximation [194].) The addition of the first bridge diagram overcorrects the hypernetted chain, and it predicts too high a repulsion at small separations. At large separations both wall–wall approximations appear accurate. The dumbbell hypernetted chain approximation is clearly the best of those compared in the figure; the contact route and the free energy route are in good agreement with each other and with the inhomogeneous hypernetted chain calculations.

The reason for the superior performance of the dumbbell compared to the wall–wall hypernetted chain approximation most likely lies in the fact that the two walls are treated as individual solutes in the latter, but as a single solute in the former. The diagrams with $n + 1$ field points in the wall–wall approach roughly correspond to diagrams with only n field points in the dumbbell approach because both root points are solutes in the former, whereas one is a solute and one is an ion in the latter. Similarly, the inhomogeneous approach outperforms the singlet approaches because both root points represent ions, and hence the equivalent diagrams only require $n - 1$ field points. Furthermore, since the solute–ion potential is more problematic than the ion–ion potential, it is an advantage to treat it via the formally exact equations for the density profiles, as in the inhomogeneous approach, rather than via the approximate closure, as in the singlet approaches.

Figure 12 shows the pressure at a higher concentration $(0.5 M)$ and surface charge density $(|e/\sigma| = 60 \text{ Å}^2)$. Again the bare wall–wall hy-

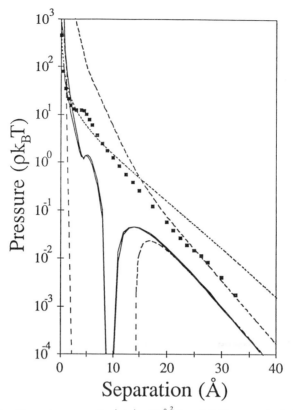

Figure 12. The net pressure for $|e/\sigma| = 60$ Å^2 in a $0.5\,M$ monovalent electrolyte with $d = 4.6$ Å. The remaining parameters are as in Fig. 11, and so are the various curves, except that the inhomogeneous hypernetted chain data is taken from [212].

pernetted chain approximation predicts an attractive regime that is not present in the inhomogeneous data [212], but which in this case also appears in the dumbbell approach. (The two dumbbell routes to the pressure are almost coincident.) When the first bridge diagram is added to the wall–wall hypernetted chain approximation the attraction disappears, but at small separations the bridge diagram once again overcorrects the bare hypernetted chain approximation and gives too large a repulsion. The bridge diagrams fail to give the hump shown by the inhomogeneous calculations at a separation of one diameter ($h \approx d = 4.6$ Å), but this is certainly present in the dumbbell calculations, even if the magnitude is underestimated. This peak is related to the (structural) free energy minimum that occurs when two integral layers of ions are confined between the surfaces. At large separations it is evident that the nonlinear Poisson–Boltzmann approximation does not have the correct decay length. The Debye length is here $\kappa_D^{-1} = 4.2$ Å, whereas the actual decay length is $\kappa^{-1} = 3.3$ Å. Evidently all the approximations based upon the hypernetted chain closure have the same decay rate, but it requires the inclusion of the first bridge diagram to obtain quantitative agreement with the inhomogeneous approach in the asymptotic regime. That is, the amplitude of the decay, which depends on the effective surface charge, is only given accurately when the first bridge diagram is included in the wall–wall approach. The bare hypernetted chain approximation for the effective surface charge would likely be inaccurate at this concentration; it is approximately 10% higher than is predicted by including the first bridge diagram at this actual surface charge density in the $0.1\,M$ monovalent electrolyte of Fig. 7.

As the separation goes to zero, the pressure becomes increasingly repulsive, as is evident in Figs. 11 and 12. In fact, the exact limiting result is [84]

$$p(h) \sim \frac{-2\sigma k_B T}{q_c h} + \mathcal{O}(h^0) \qquad h \to 0 \tag{5.4}$$

where q_c is the charge on the counterions (assuming a single species), and h is the width of the region accessible to the ion centers. This result arises from the electroneutrality condition and the contact theorem. At small separations only the counterions necessary to balance the surface charge remain in the double layer (or at least the contribution from those beyond these is bounded, see below). One may take the density profile of the counterions to be constant (equivalent to doing a Taylor expansion and keeping only the zeroth term because the separation, and hence the distance is going to zero), and hence the contact density is $\rho(0; h) \sim -2\sigma/q_c h + \mathcal{O}(h^0)$, $h \to 0$. Both remaining contributions to the net pressure, the wall–wall electrostatic interaction and the bulk osmotic pressure, are independent of separation, which gives the above result. All the numerical approaches except for the wall–wall hypernetted chain approximation appear to obey the limiting result; one can show analytically that the Poisson–Boltzmann approximation satisfies this limit for the counterions-only double layer.

This limiting result for the double layer is an extension of the corresponding result for uncharged walls [264, 297, 296]. For the restricted primitive model it is

$$\frac{p(h)}{\rho k_B T} \sim \gamma_\pm - \phi + \mathcal{O}(h) \qquad h \to 0 \tag{5.5}$$

where ϕ and $\gamma_{\pm}$ are the osmotic coefficient and the activity of the bulk electrolyte, respectively, and $\rho = \rho_{+} + \rho_{-}$ is its total number density. This result derives from the fact that the chemical potential, which is fixed, is comprised of an ideal part and an excess part. At small separations the excess part is just that of a two-dimensional fluid (plus terms of order h), and vanishes as the two-dimensional number density vanishes (because the three-dimensional number density remains finite). Hence, the contact density equals the fugacity (plus terms of order h) and the net pressure is simply this less that of the bulk electrolyte. Hence, the only difference between this result and that of the double layer is the two-dimensional number density, which in the latter case is fixed by the surface charge on the walls.

The divergence of the pressure at small separations, Eq. (5.4), is shown in the inset of Fig. 13 for the case of 1 M electrolyte at an area per unit surface charge of 85 Å^2. The nonlinear Poisson–Boltzmann approximation, the inhomogeneous hypernetted chain approximation, and the dumbbell hypernetted chain approximation for the contact pressure all quantitatively satisfy this result. [The neglected constant term in Eq. (5.4) becomes irrelevant on the logarithmic plot at small separations.] In contrast, the wall–wall hypernetted chain approximation does not diverge as $h \rightarrow 0$, but appears to have a finite contact value, and when the first bridge diagram is added this value actually decreases. The pressure via the free energy route in the dumbbell hypernetted chain approximation does diverge, but apparently somewhat faster than the limiting law.

More broadly, in Fig. 13 one sees once again that the wall–wall hypernetted chain approximation shows attractive double layer forces over much of the regime, and in this case the attractions persists over a limited range even when the first bridge diagram is included. (The attractive regime is signified on the logarithmic plot by a break in the curve, which is cut nearly perpendicularly by the abscissa.) The inhomogeneous data [212] does not show this small-separation attraction, but it does hint at a plateau at about a diameter separation, presumably due to the packing of two layers of ions. The dumbbell hypernetted chain approximation also shows this broad plateau, but underestimates its magnitude, as in Fig. 12. The two thermodynamic pathways to the pressure in the dumbbell hypernetted chain approximation are mutually consistent.

At larger separations once more it is apparent that the nonlinear Poisson–Boltzmann approximation has the wrong decay length, but what is interesting in this case is that the pressure is actually oscillatory in the asymptotic regime. This is manifest by the turn down in the wall–wall hypernetted chain data that include bridge diagrams, and as discussed in

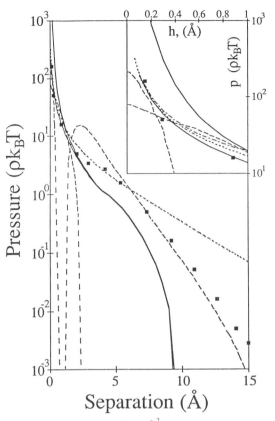

Figure 13. The net pressure for $|e/\sigma| = 85 \text{ Å}^2$ in a $1\,M$ monovalent electrolyte. The remaining parameters and curves are as in Fig. 11, and the symbols are inhomogeneous data from [212]. Inset. The dot–dashed curve shows the divergence of the pressure at small separations, Eq. (5.4), and almost coincident is the non-linear Poisson–Boltzmann approximation (dotted curve). The wall–wall hypernetted chain approximations (dashed curves) appear to have finite values at contact (the lower of the two at contact is the one with the first bridge diagram), whereas the dumbbell approximations (solid curves) appear to diverge (the contact route is almost coincident with the limiting equation).

Section III, the oscillations mimic those in the ion distribution functions of the bulk electrolyte. The bridge approximation is here quite accurate, and it gives 33 Å for the period of the oscillations, and 1.8 Å for the decay length. The dumbbell hypernetted chain approximation shows an attraction beyond about 10 Å, presumably due to asymptotic oscillations. On balance this approximation probably has the wrong phase, and the first asymptotic attraction likely begins at about 15 Å. There is perhaps a hint of this in the extreme inhomogeneous hypernetted chain data, but realistically it is not possible to extend that method into the asymptotic regime. At large separations one has problems in accurately calculating the net pressure because numerical errors mask the exact cancellation that is required. The situation is even worse for simulations, and at large separations the best approaches are the wall–wall hypernetted chain approximation with the first bridge diagram, or equivalently the asymptotic expressions using parameters calculated for the bulk electrolyte and for the isolated wall, again by the hypernetted chain approximation with the first bridge diagram.

In summary, the net double layer pressure for monovalent aqueous electrolytes shows a repulsion that increases as h^{-1} as the separation goes to zero, and a plateau at about one diameter separation due to the layering of ions. Attractions of the van der Waals type at small and intermediate separations are not present in the inhomogeneous data, but they are predicted by the singlet approaches. It is likely that such correlation attractions really do occur in monovalent aqueous electrolytes, but at higher surface charges than is indicated by the singlet approaches. At larger separations the double layer force is either monotonically repulsive or oscillatory, depending on the concentration of the bulk electrolyte. Of the approximate schemes tested, the nonlinear Poisson–Boltzmann approximation performs well at small separations, but has the wrong magnitude at intermediate concentrations, and the wrong decay length at high concentrations. The wall–wall hypernetted chain approximation is certainly the best in the asymptotic regime, particularly when the first bridge diagram is included, but it does not show the correct divergence at small separations, and it predicts attractions rather too readily. The dumbbell hypernetted chain approximation is remarkably self-consistent, it is the only approximate singlet theory to show the correct structure in the pressure at high concentrations, and it is quantitatively accurate at concentrations of $0.1\ M$.

2. Divalent Electrolyte

In general, divalent aqueous electrolytes show attractive double layer forces at lower surface charge densities than do monovalent aqueous electrolytes. This may be seen in Fig. 14, which is for a $0.1655\ M$ divalent electrolyte with one surface charge every $176\ \text{Å}^2$. The Monte Carlo data [260] at small separations have just begun to turn repulsive and one sees the limiting law, Eq. (5.4), begin to have an effect on the various approximations. The simulations show an attractive regime for separations between 5 and 15 Å. The dumbbell hypernetted chain approximation does not agree with the simulation data, and instead of an attraction, both pathways predict a positive plateau in the pressure. The

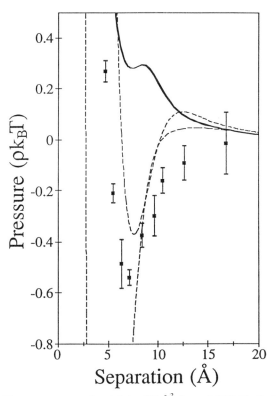

Figure 14. The net pressure for $|e/\sigma| = 176\,\text{Å}^2$ in a $0.1655\,M$ divalent electrolyte ($T = 298\,\text{K}$, $\varepsilon = 78.5$, $d = 4.2\,\text{Å}$). The symbols represent Monte Carlo data [260], the dashed lines are wall–wall hypernetted chain results [84] (the bare approximation shows more of an attraction than does the one that includes the first bridge diagram), and the nearly coincident solid curves result from the two pathways to the pressure in the dumbbell hypernetted chain approximation.

singlet wall–wall hypernetted chain data show an attractive regime for separations less than about $10\,\text{Å}$. As in the case of monovalents, including the first bridge diagram decreases the attraction predicted by the wall–wall hypernetted chain approximation, and in this case brings that theory into almost quantitative agreement with the simulation data. This bulk concentration corresponds to the monotonic regime ($\kappa^{-1} = 3.85\,\text{Å}$, compared to $\kappa_D^{-1} = 3.74\,\text{Å}$), and hence one knows that the attraction shown in Fig. 14 must be finite in extent, and that the pressure at large separations must be monotonic repulsive, as is predicted by all of the hypernetted chain-based theories. (The simulations evidently have difficulties in the large separation regime.) The attraction shown in the figure must therefore be of the van der Waals type, due to the fluctuations of the electrolyte confined between the walls.

The earliest prediction that the double-layer force between identically charged surfaces could be attractive was made by Oosawa [298, 299]. He made an approximate estimate of the influence of ion correlations and concluded that at high enough couplings, which in practice meant divalent aqueous electrolytes, the electrostatic fluctuations could give an attractive component that dominated the mean-field double layer repulsion. Oosawa clearly recognized that these attractions were properly considered as part of the van der Waals force. Patey [165] found attractions between highly charged macroions using the singlet hypernetted chain approximation, and they were also found for divalent counterions between charged planar walls by Guldbrand et al. [256] using Monte Carlo simulation, and by Kjellander and Marčelja [207] using the inhomogeneous hypernetted chain approximation. These numerical results were all rather sophisticated, but what became obscured was the physical interpretation that related the attraction to the van der Waals force. Although Guldbrand et al. [256] made connections to the work of Oosawa [298, 299], it was probably not until Attard et al. [193, 194] and Podgornik and Žekš [195] extended the Poisson–Boltzmann approximation to analyze the effects of correlations between ions in the double layer that the van der Waals origin of the attractions was once more explicit. Seen in this light attractive double layer forces between similarly charged surfaces can no longer be regarded as mysterious; it has always been known that the van der Waals attraction dominates the double layer repulsion at small separations. The strength of the attraction increases with ion coupling, principally characterized by valence and concentration, but also dependent in the general case on the dielectric constant and the temperature. Correlation effects are also important at higher surface charge densities because more ions are confined between the surfaces.

The case of zero surface charge, Fig. 15, is interesting because here there is no mean-field interaction, and the pressure is due solely to the correlated electrostatic fluctuations of the ions. In this case the pressure is monotonic attractive and its van der Waals origin is unmistakable. Because there is no net charge on the surfaces in Fig. 15, one expects that the net pressure will go like [194, 300–302]

$$p^{\text{net}} \sim -\alpha^2 e^{-2\kappa h} \qquad h \to \infty \qquad (5.6)$$

Such a fast decay is not evident in the simulation data, which is somewhat surprising because the separations are already greater than several decay lengths (the bulk electrolyte has a decay length of 2.28 Å, and a period of

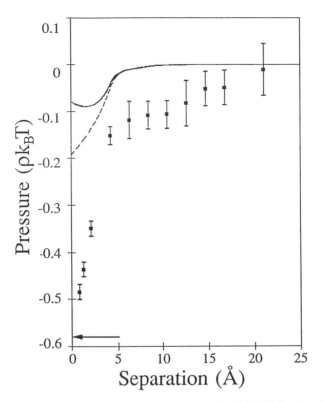

Figure 15.   The net pressure between uncharged walls in a 0.971 *M* divalent electrolyte. The remaining parameters and curves are as in Fig. 14. The bare wall–wall hypernetted chain approximation is coincident with the dumbbell results, and including the first bridge diagram makes the pressure more attractive at contact. The arrow denotes the limiting result, Eq. (5.5).

oscillation equal to 10.4 Å). It is possible that the range can be accounted for by a minimum from the asymptotic charge oscillations of the bulk electrolyte. The simulations by Bratko and Henderson [265] are in good agreement with the data of Valleau et al. [260] shown in the figure, and exhibit a similar range. In Fig. 15 the simulation data is tending to the limiting law, Eq. (5.5), but none of the hypernetted chain-based singlet approximations are particularly accurate here. The closest is the wall–wall approximation with bridge diagrams, and considerably less attractive at contact are the dumbbell and the bare wall–wall, which are coincident. Note that the wall–wall hypernetted chain results for this divalent 0.971 *M* electrolyte here and in the remaining figures are more accurate than those given in [84]. Also, the cycle of calculating wall–ion total

P. ATTARD

correlation functions and bridge functions was repeated up to five times in order to obtain self-consistency.

Figure 16 shows the pressure in the 0.971 M divalent electrolyte at the comparatively low-surface charge density of 1 charge every 1780 Å^2. The minimum in the simulation data at about 2.5 Å separation is present in all the singlet hypernetted chain approximations, albeit underestimated in magnitude. Including the first bridge diagram certainly improves the wall–wall approach in estimating the value of this maximal attraction. The simulation data has just begun to turn repulsive at small separations, in accord with the limiting divergence, Eq. (5.4). As in Fig. 13, the wall–wall hypernetted chain results appear to have a finite contact value, whereas the dumbbell approximation show the correct divergence. Once again the two routes to the pressure in the dumbbell approximation are in good agreement. As in Fig. 15 for neutral walls, the simulated attraction

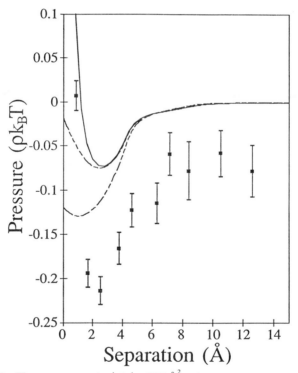

Figure 16. The net pressure for $|e/\sigma| = 1780$ Å^2, other parameters and curves as in Fig. 14. The wall–wall hypernetted chain approximation is made more attractive at small separations by the inclusion of the first bridge diagram.

appears strangely long ranged, particularly in comparison to the hypernetted chain approximations, which on this scale are indistinguishable from zero by about 10 Å separation.

At the higher surface charge density of 176 Å^2, Fig. 17, the small separation repulsion is more evident, and the position of the greatest attraction is shifted to a larger separation of about 5 Å. The wall–wall hypernetted chain approximations are in quite good agreement with each other and with the dumbbell approximation, where the two thermodynamic pathways are coincident. The approximations may be described as quantitatively reliable here, and they are certainly relatively more accurate in this case than in the same electrolyte at a lower surface charge density, Fig. 16. Compared to Fig. 14, which has the same surface charge density but lower concentration (giving a dimensionless surface charge 2.4 times larger), in the present case the approximations are much more accurate. Indeed, in Fig. 17 the dumbbell approximation almost passes through all of the simulation data, but in Fig. 14 it fails to predict any

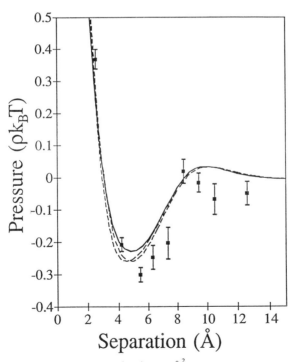

Figure 17. The net pressure for $|e/\sigma| = 176 \text{ Å}^2$, other parameters and curves as in Fig. 14.

attractions. What is also evident in all the approximations in Fig. 17 is the secondary maximum at about 8 Å separation, and which may just be made out in the simulation data. It is possible that this represents the first discernible peak of the oscillations in the pressure that are predicted to dominate asymptotically because the bulk electrolyte is here in the oscillatory regime (with a period of 10.4 Å).

In the case of the highest surface charge density for which simulation data is available (Fig. 18, $|e/\sigma| = 58.9$ Å^2), it was not possible to obtain solutions to the dumbbell hypernetted chain equations. The highest surface charge density at this concentration for which convergence could be obtained was 100 Å^2, which suggests that it may be worthwhile to include the first dumbbell–ion bridge diagram. It was, however, possible to solve the wall–wall hypernetted chain approximation with and without the bridge function, and the results are shown in the figure. (The bridge function and wall–ion total correlation functions were in this case iterated

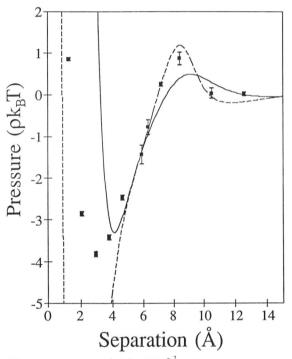

Figure 18.   The net pressure for $|e/\sigma| = 58.9$ Å^2, other parameters as in Fig. 14. The symbols represent Monte Carlo data [260], the dashed curve is wall–wall hypernetted chain data, and the solid curve includes the first bridge diagram. In this case the dumbbell hypernetted chain approximation did not converge.

to self-consistency five times.) The inclusion of the first bridge diagram certainly improves the bare hypernetted chain approximation, and it may be seen that the depth of the attraction is quite well estimated by that theory. Also evident is the small separation repulsion, and the secondary maximum at about 7-Å separation, which are present in both wall–wall hypernetted chain approximations.

In Fig. 19 the pressure given by density functional theories is compared to simulations [260] and to the wall–wall hypernetted chain approximation with the first bridge diagram [84]. The nonlocal density functional theory of Tang et al. [235] uses the Carnahan–Starling hard-sphere free energy, evaluated at an effective density using a weighting due to Tarazona [232], and the mean spherical approximation for a residual ion–ion direct correlation function of the bulk electrolyte. Patra and Ghosh [241] use the mean spherical approximation for the hard sphere and residual electrostatic ion–ion correlation function evaluated at an

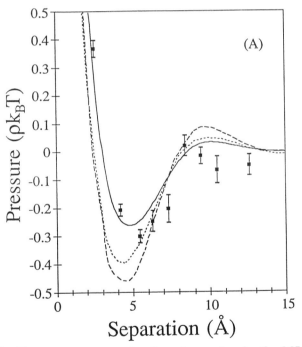

Figure 19. The net pressure as a function of separation for the $0.971\,M$ divalent electrolyte. The symbols represent simulations [260], the solid curve is the wall–wall hypernetted chain approximation with the first bridge diagram [84], and the dashed and the dotted curves are density functional results of Tang et al. [235] and of Patra and Ghosh [241], respectively. A. $|e/\sigma| = 176\,\text{Å}^2$. B. $|e/\sigma| = 58.9\,\text{Å}^2$.

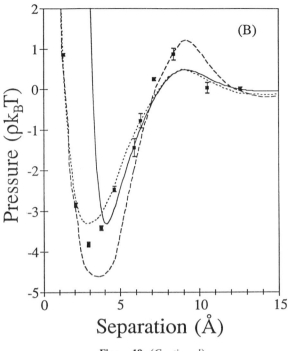

Figure 19. (*Continued*)

effective bulk density, obtained using a weighting due to Denton and Ashcroft [237]. The two approaches are similar in spirit, and it can be seen that they are of similar accuracy for the net double layer pressure. The approach of Tang et al. [235] possibly exaggerates the amplitude of the oscillations, but there is little to choose between the two recipes. The accuracy of the density functional theories is comparable to the wall–wall hypernetted chain approximation with first bridge diagram. For the highest surface charge density [Fig. 19B], the singlet approach predicts a repulsion that increases too rapidly at the smallest separations, whereas the density functional theories pass through the simulation datum.

In summary, the electric double layer force in divalent aqueous electrolyte shows small separation attractions due to ion fluctuations of the van der Waals type. At still smaller separations the pressure between charged surfaces becomes large and repulsive, whereas neutral surfaces show a finite adhesion in contact. The singlet hypernetted chain approximations appear relatively more accurate at higher surface charge densities and higher electrolyte concentrations, which is perhaps surprising. When structure is present in the double layer, the approximate theories appear capable of a quantitative description. At zero or low-surface charge densities, the simulation data [260, 265] appear to have a much longer range than one would have believed possible on the basis of the usual asymptotic notions.

VI. BEYOND THE MINIMAL MODEL

This chapter has been concerned with the behavior of electrolytes and the electric double layer, with particular emphasis on recent advances in theory that predict phenomena beyond the classical mean-field notions. The treatment was deliberately fundamental in an attempt to elucidate the general principles that apply to the various occurrences of the double layer, and it avoided as much as possible the specific details that make each system unique. It seems appropriate in this section to restore the balance by discussing some of the elaborations of the basic model, and the modifications that they necessitate. Specifically, concentrated dispersions, more realistic models for the charged surface, dielectric images, and molecular solvents will be mentioned.

A. Concentrated Dispersions

A large part of this chapter has been concerned with the interaction between particles due to the overlap of their double layers. This is an important topic because it is these forces that determine whether particles may approach, adhere, and fuse, and also the behavior of colloidal dispersions (flocculation and sedimentation or flotation, rheology). Are results obtained for the double-layer interaction between particles at infinite dilution applicable to dispersions of particles at finite volume fractions? One might guess that when the size of the particles is large compared to their separation and the decay length, the results given above for the pairwise double layer interaction will quantitatively hold. In the other extreme, the interactions between concentrated colloids which are small compared to the decay length, is not pairwise additive at larger separations, and one cannot apply the interactions calculated at infinite colloid dilution except in some qualitative sense. In these cases one needs to treat the multicomponent mixture explicitly.

There have been two approaches to this problem. In the cell model, one imagines space to be tiled with an infinite number of cells that have the same shape as the colloidal particle, and one calculates the properties of the double layer within a cell, subject to appropriate boundary conditions. The concentration of the dispersion is set by the size of the cell, and the interaction between particles is determined from the electrolyte concentration at the cell boundary [131]. The cell model has been used for spherical macroions [131, 171, 242, 282, 283] and for cylinders [121, 287, 288]. The cell model is probably of limited reliability; there are obvious questions about the physical realization of the tiling, and it appears to be inaccurate for the osmotic pressure of concentrated dispersions [282]. Probably preferable to the cell model are the integral equation methods for multicomponent mixtures. Here the colloid particle is just one of the components of the mixture, and it is treated on the same footing as the electrolyte ions. This is essentially the single approach at finite concentrations, and the hypernetted chain approximation and variants have been used to study concentrated dispersions of charged spherical colloids [168–173].

One long-running question that can be addressed in the context of the asymptotic analysis of this chapter concerns the decay length of the concentrated dispersion. Should one include the macroions and their counterions in addition to the electrolyte in the Debye length? On the basis of the results detailed above, one can say in the first place that the highly asymmetric electrolyte will have a unique decay length, but it will not be the Debye length. The macroions will appear in the formal expression for the decay length with an effective charge that depends on the direct correlation functions of the dispersion. This effective charge is likely much reduced from the bare macroionic charge, which reduction may be interpreted as counterion binding (which interpretation would be of limited value in the likely event of oscillatory correlations). For relatively dilute dispersions with low concentrations of added electrolyte, such that the correlations remain monotonic, the approximate formulas of Section V.A for the effective surface charge may be useful in giving the amount of bound and free counterions. The latter, the macroions with reduced charge, and the electrolyte may then be used to calculate the decay length of the dispersion, Eq. (2.66).

B.   Discrete and Regulated Surface Charges

The surface charge that gives rise to the electric double layer when a particle is immersed in an electrolyte arises either from dissociation of ionizable groups on the surface or from preferential adsorption of ions from the electrolyte. In both cases, the charges are discrete, although for utilitarian reasons the double layer is almost always modeled with a uniform surface charge. For the case of fixed charges, one expects that the charges will not be seen as individuals at distances large compared to their separation, so that discrete effects must decay at a rate at least in inverse proportion to their separation. In fact it is competition between the separation and the screening length that determines the rate of decay of discrete charge effects. For a periodic array of surface charges, the discrete contribution to the mean electrostatic potential goes as $\exp[-(G^2 + \kappa^2)^{1/2}z]$, where the smallest reciprocal lattice vector is related to the mean spacing between the surface charges by $G = 2\pi/a$. (The uniform or smeared out contribution corresponds to $G = 0$, and will only vanish if the surfaces with discrete charges are overall electroneutral.) The decay length for discrete charge effects is determined by the smaller of the bulk electrolyte screening length and the separation between the surface charge groups, and hence discrete charge effects may be expected to be important at low-electrolyte concentrations. The discrete contribution to the interaction between two surfaces can be attractive or repulsive depending on their juxtaposition, and any disorder in the discrete charge lattice or distribution of charge about the lattice sites diminishes the effects of discreteness.

Richmond [303] analyzed the interaction between net neutral surfaces with a periodic distribution of discrete charges on the basis of the linear Poisson–Boltzmann approximations, and found that the attraction due to perfectly misaligned lattices could be comparable to the van der Waals attraction for realistic parameters. Nelson and McQuarrie [304] also used the linear Poisson–Boltzmann approximation and explored the influence of discrete charges on the electrostatic potential near a membrane. Miklavic et al. [305] used both the linear and nonlinear Poisson–Boltzmann approximations to discuss the interaction between surfaces, and pointed out that thermal averaging in the lateral direction favors misalignment. This gives an attractive contribution similar to that found by Richmond [303], and to that found between dipolar lattices with [306] and without [307] electrolyte. (As mentioned above discrete charge effects are minimized if the lattices are not perfect or if there is any smearing of charge in the unit cell.) The simulations of a wall embedded with a periodic array of fixed charges reported by van Megan and Snook [244, 245] are vitiated by programming errors [2]. Zara et al. [258] performed simulations for counterions only confined between two planes bearing periodic distributions of discrete charges.

Kjellander and Marčelja [211] used the inhomogeneous hypernetted chain approximation to obtain the double layer force between walls bearing *mobile* surface charges. Although the charges are discrete, the fact that they form a two-dimensional ionic fluid on the surface means that *on average* they have a uniform charge distribution, which probably accounts for the rather small discrete charge effect found in this model [211]. The model, however, is of interest in its own right, and has been elaborated upon by Marčelja [213] to include specific adsorption of ions from the electrolyte by means of a short-ranged adsorption potential. The density of adsorbed ions changes as the separation is changed, and the model represents a realistic development of the two extreme boundary conditions commonly used in double layer theory—constant surface charge or constant surface potential. The nonlinear Poisson–Boltzmann equation can be solved in planar geometry in terms of elliptic functions for constant charge [308] and for charge regulation by ionizable surface groups [309]. This equation has been linearized and solved in cylindrical geometry in terms of Bessel functions for the case of charge regulation [104]. In this chapter attention has been focused on the constant charge condition because it is the easiest to treat both analytically and numerically. However, singlet hypernetted chain results have been presented for the constant surface potential case [4, 161, 162], and Spalla and Belloni [170] used that approximation to treat macroions with short-ranged adsorption potentials, which is similar to the model later used by Marčelja [213] to mimic charge regulation.

The surface charge may not only be nonuniform and nonstant, but the surface itself may have variable curvature. Surface roughness becomes particularly important for self-assembled systems such as micelles, membranes, and lamellae, where the energetics of undulation, protrusion, and curvature can be a crucial consideration. For small curvatures, where the departure from planarity may be treated as a perturbation, the Poisson–Boltzmann approximation has been used to analyze the double-layer contribution to the bending constants and curvature free energy of surfaces and membranes [310–315]. Goldstein et al. [316] found the electrostatic potential and free energy of isolated and interacting rough planar double layers in linearized Poisson–Boltzmann approximation using perturbative and iterative techniques. Blum [3] generalized the contact theorem for the pressure to a rough electrode. The double-layer interaction of a charged macrosphere and a *deformable* planar surface has been calculated self-consistently in linear Poisson–Boltzmann approximation [317].

Related to the possibility of surface roughness is the fact that even for a smooth surface not all ions may approach to within the same distance, due to their differing sizes or to their specific degree of hydration. In this chapter a single plane of closest approach was assumed since the additional complication does not appear to lead to any new principles. Nevertheless, a realistic model for the double layer in the region close to the surface may need to account not only for the ion exclusion region, but also the location of adsorbed surface charges and changes in the solvent dielectric constant (see below). In colloid science the simple division into the diffuse layer and the Stern layer (from which all ions are excluded) has been augmented with inner- and outer-Helmholtz planes, triple layers, and other embellishments for the compact inner layer (see [318] for a review of the traditional approaches to this region in the context of discrete surface charges). A distribution of surface charge over a region of finite thickness that is available to the electrolyte ions is important in modeling biological membranes [319, 320]. Surfaces with tethered poly-electrolytes are perhaps the extreme example; these may constitute the surface charge, or they may neutralize it, and their conformation is influenced by any interpenetrating electrolyte ions. The extent of the surface region about an isolated particle, and their interaction due to the combined effects of overlapping double layers and steric hindrance is of some practical importance [277–281, 321, 322].

In the context of the analysis of this chapter, any discreteness in the surface charge will not be evident at asymptotic separations, and the equations are formally unchanged. Obviously, however, the effective surface charge depends on the electrolyte in the vicinity of the surface and its quantitative value will change in the case of discrete charges (where it will likely *decrease*), charge thickness, and surface roughness. Asymptotically, the double-layer properties remain formally unchanged; in particular charge regulation becomes a second-order effect at large separations. The van der Waals attraction at short separations ought to be enhanced by mobile surface charges (because of the extra correlations), and also in the case of fixed discrete charges (because of the additional ions near the surface due to the more intense fields associated with each surface charge).

C. Dielectric Images

One of the major complications that was ignored in the minimal model of the electric double layer was that in general the charged particles have a different dielectric constant to the solvent, and at the continuum level this gives rise to image charges. These fictitious image charges represent the polarization responses of the media, and they ensure the continuity of the displacement electric field across the dielectric discontinuity at the boundary. When the dielectric constant of the particle is lower than that of the ions, which is the usual case for particles in aqueous electrolytes, the ions are repelled from the interface, and the coupling between them is increased. On the other hand, an image charge of opposite sign is induced in a metallic electrode, which attracts ions to the interface and reduces their correlations.

The image potential between two ions depends on their mutual separation and upon their distances from the particle. That is, the pair potential of the ions is noncentral and depends on the position of the particle. This poses no difficulties for simulations, where they have been included in simulations of a single wall [247], nor for inhomogeneous integral equations, because the image potential has the same symmetry as the pair correlation functions [189, 199]. An infinite sum of dielectric images arises from two walls, which represent a challenge for simulations that has to date been solved only approximately [257]. The sum can be expressed in closed form in Fourier space, which is suited to the inhomogeneous integral equation method [189]. Similarly, a Legendre transform renders the image potential tractable in spherical geometry.

Dielectric images are problematic for singlet integral equation approaches because they represent a three-body interaction. They have been approximately treated in the singlet method by including the self-image potential with a screening factor [247, 323]. Although seemingly ad hoc, the procedure is based upon the Onsager–Samaras analysis of an uncharged dielectric interface [324], and it avoids the spurious power law decay that would result from treating only part of the image interactions. (For example, it would be erroneous to attempt to modify Gouy–Chapman theory by including the one-body self-image interaction in the Boltzmann factor.) For simple liquids that interact with three-body potentials it is possible to formulate a type of hypernetted chain approximation that invokes an effective pair potential [325–330]. A linearized version of this effective pair potential has been used for a multipolar solvent without ions against a polarizable wall [331]. However, for the case of electrolyte, where in an exact treatment the density profile remains exponentially screened due to the cancellation of various diagrams, the hypernetted chain effective pair potential for the image interaction produces power law contributions to the profile because it only partially includes certain families of diagrams. A resummation to secure the correct screened form has not yet been carried out, and the double layer with dielectric images remains to be satisfactorily treated with singlet methods.

The effects of dielectric images are strongest at low surface charge densities and low electrolyte concentrations, and also in divalent electrolyte. Compared to no images, the ion densities near a wall are depleted when the wall has a lower dielectric constant than the solvent (repulsive self-image interaction) and is enhanced in the opposite extreme of a metallic electrode [247]. An adsorption decrement occurs in the electrolyte at an uncharged surface of low-dielectric constant that is believed to give a positive contribution to the surface tension of the air–electrolyte interface [199, 323, 324]. The modified Poisson–Boltzmann theory [199] has been shown to describe image charge effects at an isolated wall with high accuracy [247]. Vertenstein and Ronis [332] developed a cluster perturbation approximation for the planar double layer that included images, size, and correlations, and obtained good agreement with the simulation data [247] at low surface charge densities and low electrolyte concentrations; at higher couplings the theory overestimated the correlation effects. Spatial variation of the dielectric constant in nonplanar geometries frequently occur in modeling biological systems, and numerical algorithms have been given to solve the Poisson–Boltzmann equation [333–335].

For two interacting planar walls, dielectric images have been included using the inhomogeneous hypernetted chain approximation for counterions only [207], and for a symmetric electrolyte [209, 211], and approximately in simulations of the counterions-only double layer [257]. In calculating the net pressure between the walls in the presence of images, it is essential to include the zero frequency van der Waals force in Lifshitz approximation between the dielectric half-spaces, otherwise a spurious power law repulsion results [93, 94, 302]. For the usual case when the walls have a lower dielectric constant than the solvent, there is a depletion of the ion densities near the surfaces. At low-surface charge densities the images make the pressure even more repulsive than predicted by the Poisson–Boltzmann theory (see the discussion of Fig. 9), and at higher surface charges the images enhance the correlations between the ions leading to a larger attractive component. (For neutral surfaces, which have no mean-field repulsion, the van der Waals attraction is increased, cf. Fig. 15 and [194, 209, 260, 300–302].)

Qualitatively, dielectric images have little effect on the nonclassical behavior of the double layer described in this chapter. Any van der Waals fluctuation attraction at small separations is enhanced by the increased coupling between the ions due to the induced images. The asymptotic behavior of the profile and the interaction is still determined by the bulk electrolyte. Quantitatively, however, the effective surface charge will change, since images due to walls with a low-dielectric constant are expected to decrease the effective surface charge at higher surface charge densities, and to increase it as the surface charge goes to zero. The converse will occur for metallic walls. In general, however, the effects of dielectric discontinuities appear to be small, and ignoring image charges is probably less serious than the other physical simplifications or theoretical approximations that are made in describing the electric double layer.

D. Molecular Solvents

The dielectric constant represents the continuum contribution of the solvent, and in the primitive model for electrolytes and the electric double layer it appears in Coulomb's law for the interaction between charges *in media*. The solvent contributions have been integrated out of the problem (McMillan–Mayer representation), being subsumed into a linear polarization response. That Coulomb's law *in media* is really a solvent mediated interaction free energy can be seen from the fact that it is temperature dependent (via the dielectric constant). One expects the continuum approach to be exact in some asymptotic sense, where the potentials are weak and the response is averaged over many solvent molecules, but it can hardly be expected to hold in the close vicinity of a charged particle. The issue is to establish the regime of validity of the primitive model and the nature of its breakdown.

In a field with as many applications as the electric double layer there have been many efforts to incorporate noncontinuum contributions that account for various physical effects. The simplest such notion is to use a spatially varying dielectric constant, for example, one that is lower in an inner layer adjacent to the surface than it is in the bulk solvent. While no doubt the polarization response of the solvent is altered by the presence of a particle, the magnitude of the change and even the sign is difficult to estimate. Another development of the primitive model notes that the usual dielectric constant gives the polarization response to a uniform electric field, and attempts to account for the structure of the solvent by using a wavevector dependent dielectric tensor, $\underline{\underline{\varepsilon}}(\mathbf{k})$. Again the problem is that one has to postulate the functional form for such a dielectric function, and to guess the values of the adjustable parameters that occur. An example of the difficulties with this approach is the Yukawa form for the nonlocal dielectric function that was used by Kornyshev and co-workers for polar fluids [336]. This was later shown to violate certain formally exact requirements [337], and it is qualitatively inconsistent with accurate numerical results for polar fluids [337–340]. Both these examples—spatially varying and wave vector dependent dielectric functions—raise serious questions about the value of this type of elaboration of the continuum model. The postulated functions, although no doubt intuitively appealing, tend to be simplistic and to depend on unknown parameters. It is essential that they be tested against more sophisticated calculations if they are to be relied upon. Ultimately, progress beyond the continuum has to be at the level of a molecular model for the solvent, and it is these approaches that are now discussed.

Early results for the civilized model electric double layer were mostly for dipolar hard-sphere solvents at charged walls, and were obtained with the singlet mean spherical approximation [341, 342], with the linear [343], quadratic [344], and full [345, 346] singlet hypernetted chain approximation, and with density functional theory [347]. Sweeney et al. [348] used a gradient density functional approximation and the Clausius–Massoti formula to calculate the dielectric profile in a double layer at a hydrocarbon–water interface and to find the electrolyte dependence of the surface tension. Exact asymptotes for the dipole density and orientation profile are also known [349, 350]. Perhaps the most determined efforts to characterize solvent effects in the double layer has been the molecular hypernetted chain calculations of Torrie, Kusalik, and Patey [5, 295, 351, 352], and it is these results that will be summarized here. These authors concentrated on aqueous electrolytes and developed a relatively realistic model for the water molecule, namely, a hard sphere (diameter 2.8 Å) embedded with electrostatic multipoles (mostly in the symmetric tetrahedral quadrupole approximation, but also with the full C_{2v} water molecule that includes octapole moments and that preferentially solvates anions [353]), and with an enhanced permanent dipole moment determined self-consistently to account for the molecule's polarizability [354]. The solution of the molecular Ornstein–Zernike equation and hypernetted chain closure is accomplished by expanding the orientation-dependent pair distribution functions in rotational invariants, which are just generalized spherical harmonics. The reference bridge function approximates that of a pure hard-sphere fluid. The integral equation approximation has been shown to be accurate by testing against molecular dynamics simulations for the solvent–solvent and ion–solvent correlation functions for a bulk electrolyte [355]. The tetrahedral model appears to provide an acceptable description of water, giving a dielectric constant of $\varepsilon_r = 93.5$.

Calculations with this multipolar aqueous model of the bulk electrolyte yield information on ion solvation and other properties that have implications for the electric double layer. For the bulk electrolyte, Kusalik and Patey [46, 353] find that at the density of bulk water the ion–solvent distribution functions oscillate with period around the solvent diameter, and they have a large peak at contact. The primitive model seeks to account for this strong solvation shell of water about each ion by using a larger diameter than in the ionic crystal (e.g., the bare diameter of sodium is 2.4 Å, whereas a typical diameter used in the primitive model is 4.2 Å). In consequence of the behavior of the solvent, the ion–ion potentials of mean force at infinite dilution oscillate at separations of several solvent diameters, but are dominated by Coulomb's law *in media* asymptotically. The potential is either a maximum or a minimum at contact, depending on the relative signs of the two ions, and small unlike ions have a maximum at about one-half a solvent diameter from bare ion contact, whereas small like-charged ions have a primary minimum at this position. (This attractive well for similar ions has also been seen in reference interaction site calculations [356] and in simulations [357–359], which show it to be sensitive to the potentials and the boundary conditions.) Ions larger than the water molecule are less affected by solvation effects, and show smaller departures from Coulomb's law. Evidently, using a hydrated ion diameter that is about one-half a solvent diameter larger than the bare diameter roughly accounts for this solvation effect in the primitive model. At finite electrolyte concentrations the short-range structure in the ion–ion potentials is largely unchanged, but the screening length is observed to depart from the Debye length as the concentration is increased. In the context of the primitive model results of Section II.C.1, where for a monovalent electrolyte the screening length was mostly less than the Debye length in the monotonic regime, and where at a given concentration the departure increased as the ion diameter increased, Kusalik and Patey [46] find in the multipolar aqueous solvent that small ions such as sodium chloride (NaCl) indeed have a shorter screening length than the Debye length, but they also find that the departure *decreases* for larger sized ions such as potassium chloride (KCl), and for cesium iodide (CsI) ions, which are larger than the water molecule, the screening length is actually longer than the Debye length. Also found in the calculations is the decrease in dielectric constant with electrolyte concentration [46], which is observed experimentally, but which is not allowed for in the primitive model and which may contribute to the departure from the Debye length (calculated using the dielectric constant of pure water).

The multipolar model for water has been applied to the isolated electric double layer at the singlet hypernetted chain level of approximation by treating charged macrospheres at infinite dilution [5, 295, 351, 352]. (Results have also been obtained for dipolar fluids at planar walls [331, 360, 361].) Macroions with diameters up to 30 times that of the solvent, and with surface charge densities up to one charge per 91 Å^2 have been studied. At the surface of the macrosphere the water shows an icelike structure, similar to ice I with the c axis normal to the surface [351]. This gives one hydrogen bond per molecule dangling toward the surface, but maximizes the hydrogen bonds in the double plane of molecules adjacent to the surface. The structure is more developed for larger macroions, and the hypernetted chain results for the largest diameter are in good agreement with simulations for water at a planar wall [362–364]. The water near the surface remains mobile, but the average icelike arrangement is rather stable and hardly perturbed by the electric fields associated with a uniform surface charge [351, 364].

The induced structure and polarization of the water affect the distribution of ions in the double layer in a fashion that is sensitive to the relative size of the ions. Ions that are smaller than the water molecule experience a deep potential well out to about a solvent diameter from contact, whereas larger ions show a broad potential maximum out to about one and a half solvent diameters. This induced adsorption or desorption due to solvent structure is insensitive to the surface charge density on the macroion (but the solvent structure does depend on surface curvature), and holds for both counterions and co-ions; to a good approximation it may simply be added to Coulomb's law *in media* to obtain the total ion–macroion potential of mean force [5, 352]. Finite electrolyte concentrations have little effect on the structure of the solvent in the vicinity of the surface; at $4\,M$ KCl there is a relatively minor enhancement of the peak in the macroion–solvent distribution function [5, 295].

The solvent structure has a very great influence on the rate of neutralization of the macroion charge, and Torrie et al. [295] found that there is a fast screening regime induced very close to the surface. The rapid screening of the macroionic charge means that the solvent polarization oscillations induced at high surface charge densities in the pure solvent are not evident at finite electrolyte concentrations. In $0.1\,M$ KCl at the highest surface charge density, about 80% of the macroion charge is neutralized within about a solvent diameter of the surface. In part, no doubt, this occurs because ions see the truly bare surface charge that is screened by neither salt nor solvent. In addition, in this case the solvent-induced counterion adsorption may be enhanced by the fact that the potassium counterion ($3.0\,\text{Å}$) is slightly smaller than the chlorine co-ion ($3.2\,\text{Å}$); KCl is the only salt that has been studied at finite concentrations, and the effect of a larger counterion and/or a smaller co-ion is unclear. At $1\,M$ the preferential induced adsorption in KCl is so marked that about 20% overneutralization occurs so that beyond one solvent diameter from the macroion its charge appears to have the *opposite* sign. This is the charge reversal discussed in Section V.A. These results of Torrie et al. [5, 295] clearly indicate that surface-induced solvent structure has a major influence on the ion distribution in the double layer in the immediate vicinity of the surface, and that this can be quite dependent on the size of the specific ions. One expects that the asymptotic analysis of the primitive model described in detail in this chapter will hold in a formal sense, with the actual value of the effective surface charge being different from the numerical results given above.

The interaction between spherical macroions in the model multipolar aqueous solvent without added electrolyte has been characterized with the singlet hypernetted chain approximation [10]. (Dipolar fluids between planar walls and their interaction have been treated by density functional theory [347, 365, 366], and by the singlet wall–wall hypernetted chain approximation [77].) For small neutral macrospheres there is a short-ranged hydrophobic attraction, and for large macrospheres (30 times the diameter of the solvent) there is an oscillatory force that is due to the icelike structure induced at an isolated surface discussed above. What is remarkable about these oscillations is that they have period equal to almost twice the solvent diameter. This unusual phenomenon, which does not occur for a hard sphere or for a dipolar fluid, appears to be due to the fact that in the ice I structure the water molecules lie in pairs of planes perpendicular to the c axis, and it is energetically favorable for an integral number of these doubled planes to be between the surfaces. For charged macrospheres these oscillations at small separations persist. In addition, there is an r^{-4} cavity repulsion at intermediate separations and the continuum Coulomb asymptote at large [10]. Unpublished results by the authors of [10] show that the addition of 0.1 M KCl hardly changes the short-ranged structural interaction, and that the effective surface charge due to solvent-induced rapid neutralization at an isolated macroion is the quantity that characterizes their asymptotic interaction, which is of course exponentially screened.

This discussion of solvent effects in the double layer has been based on the singlet hypernetted chain studies of the multipolar aqueous solvent by Torrie, Kusalik, and Patey [5, 295, 351, 352]. As mentioned above, comparison with simulations and other approaches suggest that the model and approximation are reliable, particularly for the structure of the water at a wall [272, 362–364]. Oscillatory water density profiles have also been reported in simulations of slit pores with molecular structure with [266–271] and without [273–275] mobile counterions. In agreement with the hypernetted chain calculations, the former show that specific ion solvation has a strong influence on the location of the counterions. The solvent-induced rapid screening of the surface charge seen by Torrie et al. [352] does not disagree with the concept of effective surface charge discussed here on the basis of the asymptotics of the primitive model, although quantitative calculations remain to be made.

VII. CONCLUSION

In concluding this chapter of current theories for electrolytes and the electric double layer, it is appropriate to discuss the relevance and application to experiment, and the consequences of these recent advances in understanding for the measurement and interpretation of double-layer phenomena. There are two issues in applying theory to the real world: the model and the approximation used to treat it. Most of this chapter has been concerned with the minimal model of the double layer, which consists of the primitive model of electrolytes, and infinitely dilute particles with uniformly charged and curved surfaces. As discussed in

Section VI, in the real world one expects discrete surface charges, charge regulation, surface undulations and roughness, surface charge layers of finite and perhaps variable thickness, ion-specific surface contact distances, and of course dielectric image effects. Doubtless the most serious simplification is the neglect of the solvent molecules; although the continuum picture is valid in some asymptotic sense, it must break down for close approach of the ions to each other and to the surface.

So much has been left out of the minimal model that the relevance of results such as those reviewed here needs to be addressed. There are several arguments to support the basic philosophy of the present approach. The simplest model captures the essence of double layer phenomena, and as such it has a certain aesthetic appeal. It is preferred because it trims complications and other hirsute possibilities from consideration while accounting for most of the data. This allows a detailed analysis at the heart of the matter, and an understanding of the double layer at its most fundamental level. One avoids the mere empirical collection of data in favour of a unified description that rationalizes the data within a broad context; exceptional behavior can then be recognized and treated as such. The simplest model is also the most general. It ignores specifics and particulars, which may be quantitatively important in each system, but which limit interest to a specialized audience. By focusing on the common denominator, the minimal model remains relevant across a spectrum of fields and technologies.

The minimal model yields to sophisticated mathematical and computational analyses in ways that are not possible in more complicated models of the double layer. A number of limiting and benchmark results have been established that are suitable for well-defined and unambiguous tests of more tractable or more approximate theories, which themselves can be applied to experimental data or to more realistic models. Furthermore, the predictions of the sophisticated theories for the minimal model may be relied upon, and there is no doubt that the nonclassical phenomena described herein (effective surface charges, non-Debye lengths, attractions, and oscillations) really do occur in this model. And if they occur in the simplest model then they can also be expected in more elaborate models and in reality.

The dichotomy in theory between the model and the approximation is reflected in the two possible avenues for improving the Gouy–Chapman approach to the double layer. The alternative to the more advanced treatment of the minimal model that is advocated here is to apply the Poisson–Boltzmann theory to more developed models. While the motivation for such a program is firmly based on experimental pragmatism, the concern is that divergence of theory and experiment is as likely due to an artifact of the approximation as it is to a physical detail neglected in the model. A better fit obtained by embellishing the model may owe more to canceling the errors of the approximation than to accounting for some real physical effect; the model may be more realistic, but it is not described any more accurately than the original. The advantage of using sophisticated theories to treat the simplest model is that one can remove the uncertainty associated with the approximation itself. Any disagreement between theory and experiment can be reliably ascribed to the model, and it may be possible to anticipate which particular aspect requires elaboration. It is this ambiguity that prevents one testing an approximation scheme against experiment, and which has motivated the benchmarks being established for the simplest model. These benchmarks have enabled approximations such as the singlet integral equations that were discussed in some detail herein to be characterized, their reliability and regime of applicability has been established, and their feasibility for the analysis of experimental data remarked upon.

These then are the reasons for concentrating on the simplest possible model: Its simplicity captures the essence of the double layer, which can then be treated in a sophisticated and reliable fashion, and any discrepancy between theory and experiment can then be unambiguously ascribed to a breakdown of the model.

Although it is feasible to go beyond the Poisson–Boltzmann theory, most experiments continue to be described within that context, and a substantial amount of theoretical development follows the alternate path, namely, developing and using that approximation for more realistic models. The mean-field theory has been manifestly successful in describing the double layer for the better part of this century, and reports of its demise are perhaps premature. Why has it apparently succeeded, when the approximations inherent in the approach—the neglect of electrostatic and excluded volume correlations—can have such serious consequences, as detailed herein? Of course the Poisson–Boltzmann theory is expected to be valid at low-electrolyte concentrations and surface charge densities where the correlations are relatively small. However, it is often applied at surprisingly high concentrations without obvious contradictions. Part of the reason for this is that the Poisson–Boltzmann theory gives a formally exact description of an ideal gas in an external field, which happens to be the mean field. As such, it is an internally consistent theory, which partially explains its ability to account for a variety of data without discrepancies appearing. The asymptotic analysis detailed in this chapter also provides part of the answer. The electrostatic potential and ion profiles due to a charged particle have linear Poisson–Boltzmann form far from the surface, as does the interaction between surfaces. Measurements that sample this regime will appear to confirm the applicability of that approximation, and the fitted surface charge will be taken as the actual surface charge. Hence, a certain amount of ion binding or dissociation will be said to have been measured, and an independent measurement of surface charge may be difficult if not impossible to obtain. In other words, the Poisson–Boltzmann theory apparently describes the data, although in fact its error has been quantitatively incorporated into the fit.

Of course, it is not always possible to reconcile the Poisson–Boltzmann theory of the minimal model with measured data, even with effective parameters. In these cases it is tempting to conclude that the model has broken down, and to incorporate some extra physical effect or other. However, this chapter has stressed that the advanced treatments of recent years show that even the minimal model can display behavior beyond the classical Poisson–Boltzmann prediction. For the case of an isolated planar double layer, the effective surface charge increases less rapidly than the actual surface charge, until it reaches a maximum. Thus the surface charge that can be supported by a given electrolyte appears to saturate. As the surface charge is further increased the effective charge begins to decrease. It would be difficult to account for this effect with a Poisson–Boltzmann theory for an embellished counterion binding model. At even higher charge densities the effective surface charge passes through zero and changes sign, due to the packing of the finite-sized ions against the surface. This overscreening in the vicinity of the surface would appear as charge reversal in any measurement that sampled the double-layer beyond about one diameter from the surface. (Of course, the force between identical particles would still be asymptotically repulsive if both their effective charges are reversed and the bulk electrolyte is in the monotonic regime.)

Another nonclassical effect appears in the bulk electrolyte, namely, the departure of the screening length from the Debye length. However, this effect is relatively small, and it would take a rather precise measurement to quantify it. The departure does increase as the electrolyte increases, but the double layer shrinks even faster, which may again preclude measurement. One case where one may possibly see the effect is for ions with a large size at moderate decay lengths, since the departure depends on $\kappa_D d$.

For even higher electrolyte concentrations the bulk ion correlation functions become oscillatory, and consequently so do the double-layer ion profiles, potential, and interaction. This certainly contradicts the Poisson–Boltzmann picture, but the decay length is so short that it would be difficult to measure directly. However, there are likely measurable indirect consequences, particularly for the flocculation and coagulation of colloids, or for the interaction between charged surfaces. The traditional picture of colloid stability, due to Derjaguin, Landau, Verwey, and Overbeek [1], relies upon the barrier to the primary minimum that arises from the balance of the van der Waals attraction and the Poisson–Boltzmann repulsion. In contrast, the modern theories predict that at high concentrations of the bulk electrolyte multiple oscillations in the double layer occur, which considerably complicates the calculation of the most likely particle separation and the behavior of the colloid.

Double layer interactions do provide one of the most direct tests of theory, since the forces between charged surfaces can be directly measured. The net force is of course a surface phenomenon; it has the bulk electrolyte contribution automatically subtracted, and hence the entire signal yields information about the double layer. Even at moderate electrolyte concentrations nonclassical behavior has been predicted. Modern theories—simulation, singlet and inhomogeneous integral equation, density functional—all show that at intermediate to small separations the double layer interaction between *like*-charged surfaces can be *attractive*. This is the van der Waals regime, and it occurs at high ion couplings, which in practice means divalent aqueous electrolytes. (It can also be induced by high-surface charge densities.) As contact is approached there is a repulsive pressure that diverges as the reciprocal of the separation (at least for fixed surface charge), and at large separation the force between like-charged surfaces must be repulsive (in the bulk monotonic regime). But there is also this intermediate regime where the double-layer force can be attractive due to the electrostatic fluctuations of the ions (part of the zero-frequency van der Waals force), which can dominate the osmotic repulsion that is calculated in the mean-field Poisson–Boltzmann approximation.

Double layer attractions are now well established in theory, and there is some experimental support for the idea. As one qualitative example, the correlation between divalent ions and double-layer attractions may provide new insight into the role of calcium and magnesium in colloid and soil science, and in cell biology. Thus irreversible adhesion and coagulation in the presence of these particular cations need not be ion bridging in a literal sense, but may in large part be due to a deep van der Waals minimum arising from the mechanism discussed above. More quantitatively, inhomogeneous hypernetted chain calculations have been compared with the attractive double layer interactions measured between charged mica surfaces [294, 367, 368] and between clay platelets [369] in divalent electrolyte. The attractive regime predicted by the theory agrees relatively well with that inferred from the measurements, and the experiments clearly show nonclassical behavior.

The prediction of double-layer attractions is one of the high points of recent double-layer theory, and its apparent confirmation is a tribute to the sophistication of the experiments. At the same time the limitations of the continuum picture are becoming increasingly evident. The same experiments that support the above interpretation [294, 367–369] show oscillations with period several angstroms in the measured forces. The primitive model double-layer calculations do not exhibit oscillations (they mostly correspond to the monotonic regime of the bulk electrolyte), and the experimental data therefore provides relatively direct evidence of molecular solvent effects. This is a case where the sophisticated treatment of the minimal model has heralded reliable experimental evidence for its breakdown. It is the precision of these and other contemporary measurements that provides strong motivation for theory to include the solvent as a distinct molecular species and to go beyond the continuum model of electrolytes and the electric double layer.

ACKNOWLEDGMENTS

I thank Emma Chorley and Clem Colman for assistance with the dumbbell hypernetted chain and the effective surface charge calculations, respectively. I thank Stan Miklavic and Stjepan Marčelja for critical comments on the manuscript.

REFERENCES

1. E. J. W. Verwey and J. Th. G. Overbeek, *Theory of the Stability of Lyopholic Colloids*, Elsevier, Amsterdam, 1948.
2. S. L. Carnie and G. M. Torrie, *Adv. Chem. Phys.*, **56**, 141 (1984).
3. L. Blum, *Adv. Chem. Phys.*, **78**, 171 (1990).
4. M. Lozada-Cassou, in *Fundamentals of Inhomogeneous Fluids*, D. Henderson, Ed., Marcel Dekker, New York, 1992, Chapter 8.
5. G. N. Patey and G. M. Torrie, *Chem. Scr.*, **29A**, 39 (1989).
6. Ph. A. Martin, *Rev. Mod. Phys.*, **60**, 1075 (1988).
7. A. Alustuey, in *Strongly Coupled Plasma Physics*, F. J. Rogers and H. E. Dewitt, Eds., NATO Advanced Science Institute Series, No. B 154, Plenum, New York, 1987, p. 331.
8. M. Parrinello and M. P. Tosi, *Riv. Nuovo Cim.*, **2** (6) (1979).
9. J. A. Barker, R. A. Fisher, and R. O. Watts, *Mol. Phys.*, **21**, 657 (1971).
10. P. Attard, D. Wei, G. N. Patey, and G. M. Torrie, *J. Chem. Phys.*, **93**, 7360 (1990).
11. A. E. Sherwood, A. G. de Rocco, and E. A. Mason, *J. Chem. Phys.*, **44**, 2984 (1966).
12. D. C. Brydges, *Commun. math. Phys.*, **58**, 313 (1978).
13. D. C. Brydges and P. Federbush, *Commun. math. Phys.*, **73**, 197 (1980).

14. T. Kennedy, *Commun. math. Phys.*, **92**, 269 (1983).
15. J. Z. Imbrie, *Commun. math. Phys.*, **87**, 515 (1983).
16. P. Federbush and T. Kennedy, *Commun. math. Phys.*, **102**, 361 (1985).
17. J. P. Hansen and I. R. McDonald, *Theory of Simple Liquids*, Academic, New York, 1976.
18. P. Attard, *Phys. Rev. E*, **48**, 3604 (1993).
19. H. L. Friedman and P. S. Ramanathan, *J. Phys. Chem.*, **74**, 3756 (1970).
20. P. V. Giaquinta, M. Parrinello, and M. P. Tosi, *Phys. Chem. Liq.*, **5**, 305 (1976).
21. M. Rovere, M. Parrinello, M. P. Tosi, and P. V. Giaquinta, *Philos. Mag. B*, **39**, 167 (1979).
22. P. Vieillefosse, *J. Phys. Lett.*, **38**, L43 (1977).
23. C. N. Likos and N. W. Ashcroft, *J. Chem. Phys.*, **97**, 9303 (1992).
24. L. G. Suttorp and A. J. van Wonderen, *Physica*, **145A**, 533 (1987).
25. A. J. van Wonderen and L. G. Suttorp, *Physica*, **145A**, 557 (1987).
26. J. G. Kirkwood and F. P. Buff, *J. Chem. Phys.*, **19**, 774 (1951).
27. F. H. Stillinger and R. Lovett, *J. Chem. Phys.*, **48**, 3858 (1968).
28. F. H. Stillinger and R. Lovett, *J. Chem. Phys.*, **49**, 1991 (1968).
29. D. J. Mitchell, D. A. McQuarrie, A. Szebo, and J. Groenevold, *J. Stat. Phys.*, **17**, 15 (1977).
30. C. W. Outhwaite, *Chem. Phys. Lett.*, **24**, 73 (1974).
31. S. L. Carnie and D. Y. C. Chan, *Chem. Phys. Lett.*, **77**, 437 (1981).
32. D. Pines and Ph. Noziéres, *The Theory of Quantum Liquids*, Benjamin, New York, 1966.
33. P. C. Martin, *Phys. Rev.*, **161**, 143 (1967).
34. P. Vieillefosse and J. P. Hansen, *Phys. Rev. A*, **12**, 1106 (1975).
35. M. Baus, *J. Phys. A*, **11**, 2451 (1978).
36. G. Stell and J. L. Lebowitz, *J. Chem. Phys.*, **48**, 3706 (1968).
37. R. Kjellander and D. John Mitchell, *J. Chem. Phys.*, **101**, 603 (1994).
38. G. Stell, in *Statistical Mechanics Part A: Equilibrium Techniques*, B. J. Berne, Ed., Plenum, New York, 1977, p. 47.
39. P. Attard, *J. Chem. Phys.*, **93**, 7301 (1900); **94**, 6936 (1991).
40. G. Stell, *Phys. Rev. B*, **1**, 2265 (1970).
41. J. E. Enderby, T. Gaskell, and N. H. March, *Proc. Phys. Soc.*, **85**, 217 (1965).
42. G. Nienhuis and J. M. Deutch, *J. Chem. Phys.*, **56**, 1819 (1972).
43. J. S. Høye and G. Stell, *J. Chem. Phys.*, **61**, 562 (1974).
44. J. S. Høye and G. Stell, *J. Chem. Phys.*, **68**, 4145 (1978).
45. G. Stell, G. N. Patey, and J. S. Høye, *Adv. Chem. Phys.*, **38**, 183 (1981).
46. P. G. Kusalik and G. N. Patey, *J. Chem. Phys.*, **88**, 7715 (1988).
47. C. Gruber, J. L. Lebowitz, and P. A. Martin, *J. Chem. Phys.*, **75**, 944 (1981).
48. L. Blum, C. Gruber, J. L. Lebowitz, and P. A. Martin, *Phys. Rev. Lett.*, **48**, 1769 (1982).
49. P. Debye and E. Hückel, *Z. Phys.*, **24**, 185 and 305 (1923).

50. S. Marčelja, D. J. Mitchell, B. W. Ninham, and M. J. Sculley, *J. Chem. Soc. Faraday Trans. 2*, **73**, 630 (1977).
51. A. B. Glendinning and W. B. Russel, *J. Colloid Interface Sci.*, **93**, 95 (1983).
52. S. L. Carnie and D. Y. C. Chan, *J. Colloid Interface Sci.*, **155**, 297 (1993).
53. E. Barouch and E. Matijević, *J. Chem. Soc. Faraday Trans. 1*, **81**, 1797 (1985).
54. J. E. Sánchez-Sánchez and M. Lozada-Cassou, *Chem. Phys. Lett.*, **190**, 202 (1992).
55. S. L. Carnie, D. Y. C. Chan, and J. Stankovich, *J. Colloid Interface Sci.*, **165**, 116 (1994).
56. J. Ennis, R. Kjellander, and D. John Mitchell, *J. Chem. Phys.*, **102**, 975 (1995).
57. D. J. Mitchell and B. W. Ninham, *Phys. Rev.*, **174**, 280 (1968).
58. D. J. Mitchell and B. W. Ninham, *Chem. Phys. Lett.*, **53**, 397 (1978).
59. R. Kjellander and D. John Mitchell, *Chem. Phys. Lett.*, **200**, 76 (1992).
60. C. W. Outhwaite, *Statistical Mechanics: A Specialist Periodical Report*, **2**, 188 (1975).
61. R. Lovett and F. H. Stillinger, *J. Chem. Phys.*, **48**, 3869 (1968).
62. E. Waisman and J. L. Lebowitz, *J. Chem. Phys.*, **56**, 3086 and 3093 (1972).
63. L. Blum, *Mol. Phys.*, **30**, 1529 (1975).
64. L. Blum and J. S. Høye, *J. Phys. Chem.*, **81**, 1311 (1977).
65. F. H. Stillinger and J. G. Kirkwood, *J. Chem. Phys.*, **33**, 1282 (1960).
66. J. G. Kirkwood, *Chem. Rev.*, **19**, 275 (1936).
67. J. G. Kirkwood and J. C. Poirier, *J. Phys. Chem.*, **58**, 591 (1954).
68. M. E. Fisher and B. Widom, *J. Chem. Phys.*, **50**, 3756 (1969).
69. G. A. Martynov, *Fundamental Theory of Liquids: Method of Distribution Functions*, Hilger, Bristol, 1992.
70. J. R. Henderson and Z. A. Sabeur, *J. Chem. Phys.*, **97**, 6750 (1992).
71. R. Evans, J. R. Henderson, D. C. Hoyle, A. O. Parry, and Z. A. Sabeur, *Mol. Phys.*, **80**, 755 (1993).
72. R. Evans, R. J. F. Leote de Carvalho, J. R. Henderson, and D. C. Hoyle, *J. Chem. Phys.*, **100**, 591 (1994).
73. R. J. F. Leote de Carvalho and R. Evans, *Mol. Phys.*, **83**, 619 (1994).
74. R. H. Fowler, *Statistical Mechanics* Chapter XIII, Cambridge University Press, New York, 1929.
75. L. Onsager, *Chem. Rev.*, **13**, 73 (1933).
76. J. G. Kirkwood, *J. Chem. Phys.*, **2**, 767 (1934).
77. P. Attard, D. R. Bérard, C. P. Ursenbach, and G. N. Patey, *Phys. Rev. A*, **44**, 8224 (1991).
78. P. Attard, C. P. Ursenbach, and G. N. Patey, *Phys. Rev. A*, **45**, 7621 (1992).
79. C. W. Outhwaite, *J. Chem. Soc. Faraday Trans. 2*, **74**, 1214, 1670 (1978).
80. C. W. Outhwaite, L. B. Bhuiyan, and S. Levine, *J. Chem. Soc. Faraday Trans. 2*, **76**, 1388 (1980).
81. D. Henderson, F. F. Abraham, and J. A. Barker, *Mol. Phys.*, **31**, 1291 (1976).
82. D. Henderson and L. Blum, *J. Chem. Phys.*, **69**, 5441 (1978).
83. S. L. Carnie, D. Y. C. Chan, D. J. Mitchell, and B. W. Ninham, *J. Chem. Phys.*, **74**, 1472 (1981).
84. P. Attard and S. J. Miklavic, *J. Chem. Phys.*, **99**, 6078 (1993).
85. B. Jancovici, *J. Stat. Phys.*, **28**, 43 (1982).

86. B. Jancovici, *J. Stat. Phys.*, **29**, 263 (1982).
87. M. Baus, *Mol. Phys.*, **48**, 347 (1983).
88. S. L. Carnie and D. Y. C. Chan, *Mol. Phys.*, **51**, 1047 (1984).
89. B. Jancovici, J. L. Lebowitz, and Ph. A. Martin, *J. Stat. Phys.*, **41**, 941 (1985).
90. F. Stern and W. E. Howard, *Phys. Rev.*, **163**, 816 (1967).
91. A. L. Fetter, *Phys. Rev. B*, **10**, 3739 (1974).
92. V. M. Muller and B. V. Derjaguin, *J. Colloid Interface Sci.*, **61**, 361 (1977).
93. P. Attard, R. Kjellander, and D. J. Mitchell, *Chem. Phys. Lett.*, **139**, 219 (1987).
94. P. Attard, R. Kjellander, D. J. Mitchell, and Bo Jönsson, *J. Chem. Phys.*, **89**, 1664 (1988).
95. W. J. Ellis, *Mol. Phys.*, **82**, 973 (1994).
96. B. Jancovici and X. Artru, *Mol. Phys.*, **49**, 487 (1983).
97. L. Blum, D. Henderson, J. L. Lebowitz, Ch. Gruber, and Ph. A. Martin, *J. Chem. Phys.*, **75**, 5974 (1981).
98. S. L. Carnie, *J. Chem. Phys.*, **78**, 2742 (1983).
99. B. V. Derjaguin, *Kolloid Z.*, **69**, 155 (1934).
100. L. R. White, *J. Colloid Interface Sci.*, **95**, 286 (1983).
101. P. Attard and J. L. Parker, *J. Phys. Chem.*, **96**, 5086 (1992).
102. D. Henderson, *J. Chem. Phys.*, **97**, 1266 (1992).
103. P. Attard and G. N. Patey, *J. Chem. Phys.*, **92**, 4970 (1990).
104. S. L. Brenner and D. A. McQuarrie, *J. Theor. Biol.*, **39**, 343 (1973).
105. G. S. Manning, *J. Chem. Phys.*, **51**, 924, 3249 (1969); G. S. Manning, *Biophys. Chem.*, **7**, 95 (1977).
106. B. Halle, *J. Chem. Phys.*, **102**, 7238 (1995).
107. R. M. Fuoss, A. Katchalsky, and S. Lifson, *Proc. Natl. Acad. Sci. USA*, **37**, 579 (1951).
108. T. Alfrey, P. W. Borg, and H. Morawitz, *J. Polymer Sci.*, **7**, 543 (1951).
109. H. P. Gregor and J. M. Gregor, *J. Chem. Phys.*, **66**, 1934 (1977).
110. F. Booth, *J. Chem. Phys.*, **19**, 821 (1951).
111. T. L. Hill, *Arch. Biochem. Biophys.*, **57**, 229 (1955).
112. C. P. Woodbury, *J. Chem. Phys.*, **82**, 1482 (1985).
113. C. F. Anderson and M. T. Record, *Annu. Rev. Phys. Chem.*, **33**, 191 (1982).
114. M. Lozada-Cassou, *J. Phys. Chem.*, **87**, 3729 (1983).
115. E. González-Tovar, M. Lozada-Cassou, and D. Henderson, *J. Chem. Phys.*, **83**, 361 (1985).
116. R. Bacquet and P. J. Rossky, *J. Phys. Chem.*, **88**, 2660 (1984).
117. V. Vlachy and D. A. McQuarrie, *J. Phys. Chem.*, **90**, 3248 (1986).
118. V. Vlachy and A. D. J. Haymet, *J. Am. Chem. Soc.*, **111**, 477 (1989).
119. V. Vlachy and A. D. J. Haymet, *J. Electroanal. Chem.*, **283**, 77 (1990).
120. L. Yeomans, S. E. Feller, E. Sánchez, and M. Lozada-Cassou, *J. Chem. Phys.*, **98**, 1436 (1993).
121. T. Das, D. Bratko, L. B. Bhuiyan, and C. W. Outhwaite, *J. Phys. Chem.*, **99**, 410 (1995).
122. L. B. Bhuiyan and C. W. Outhwaite, *Philos. Mag. B*, **69**, 1051 (1994).
123. V. Vlachy and A. D. J. Haymet, *Aust. J. Chem.*, **43**, 1961 (1990).

124. C. E. Woodward, Bo Jönsson, and T. Åkesson, *J. Chem. Phys.*, **89**, 5145 (1988).

125. A. Luzar and D. Bratko, *J. Chem. Phys.*, **92**, 642 (1990).

126. M. Lozada-Cassou, *J. Chem. Phys.*, **80**, 3344 (1984).

127. D. Henderson and L. Blum, *J. Chem. Phys.*, **69**, 5441 (1978).

128. D. Henderson, L. Blum, and J. L. Lebowitz, *J. Electroanal. Chem.*, **102**, 315 (1979).

129. S. L. Carnie and D. Y. C. Chan, *J. Chem. Phys.*, **74**, 1293 (1981).

130. S. L. Carnie and D. Y. C. Chan, *J. Chem. Phys.*, **78**, 3348 (1983).

131. H. Wennerström, Bo Jönsson, and P. Linse, *J. Chem. Phys.*, **76**, 4665 (1982).

132. J. F. Springer, M. A. Pokrant, and F. A. Stevens, *J. Chem. Phys.*, **58**, 4863 (1973).

133. P. J. Rossky, J. B. Dudowicz, B. L. Tembe, and H. L. Friedman, *J. Chem. Phys.*, **73**, 3372 (1980).

134. R. Bacquet and P. J. Rossky, *J. Chem. Phys.*, **79**, 1419 (1982).

135. H. Iyetomi and S. Ichimaru, *Phys. Rev. A*, **27**, 1241 (1983).

136. P. Ballone, G. Pastore, and M. P. Tosi, *J. Chem. Phys.*, **81**, 3174 (1984).

137. J. Wiechen, *J. Chem. Phys.*, **85**, 7364 (1986).

138. P. Attard and G. N. Patey, *J. Chem. Phys.*, **92**, 4970 (1990).

139. J. A. Barker and J. J. Monaghan, *J. Chem. Phys.*, **36**, 2564 (1962).

140. A. D. J. Haymet, S. A. Rice, and W. G. Madden, *J. Chem. Phys.*, **74**, 3033 (1981).

141. P. Attard, *J. Chem. Phys.*, **91**, 3072 (1989); **92**, 3248 (1990).

142. Y. Zhou and G. Stell, *J. Chem. Phys.*, **92**, 5533 (1990).

143. P. Ballone, G. Pastore, and M. P. Tosi, *J. Chem. Phys.*, **85**, 2943 (1986).

144. J. P. Badiali, M. L. Rosinberg, D. Levesque, and J. J. Weis, *J. Phys. C*, **16**, 2183 (1983).

145. A. Alastuey and D. Levesque, *Mol. Phys.*, **47**, 1349 (1982).

146. S. E. Feller and D. A. McQuarrie, *J. Phys. Chem.*, **96**, 3454 (1992).

147. G. M. Torrie and J. P. Valleau, *J. Chem. Phys.*, **73**, 5807 (1980).

148. J. M. Caillol and D. Levesque, *J. Chem. Phys.*, **94**, 597 (1991).

149. P. Nielaba and F. Forstmann, *Chem. Phys. Lett.*, **117**, 46 (1985).

150. B. D'Aguanno, P. Nielaba, T. Alts, and F. Forstmann, *J. Chem. Phys.*, **85**, 3476 (1986).

151. R. D. Groot, *Phys. Rev. A*, **37**, 3456 (1988).

152. Z. Tang, L. Mier-y-Teran, H. T. Davis, L. E. Scriven, and H. S. White, *Mol. Phys.*, **71**, 369 (1990).

153. E. Kierlik and M. L. Rosinberg, *Phys. Rev. A*, **44**, 5025 (1991).

154. C. N. Patra and S. K. Gosh, *J. Chem. Phys.*, **100**, 5219 (1994).

155. C. Caccamo, G. Pizzimenti, and L. Blum, *J. Chem. Phys.*, **84**, 3327 (1986).

156. E. Bruno, C. Caccamo, and G. Pizzimenti, *J. Chem. Phys.*, **86**, 5101 (1987).

157. R. Kjellander and S. Marčelja, *Chem. Phys. Lett.*, **127**, 402 (1986).

158. M. Plischke and D. Henderson, *J. Chem. Phys.*, **88**, 2712 (1988).

159. D. Bratko, *Chem. Phys. Lett.*, **169**, 555 (1990).

160. M. Lozada-Cassou and D. Henderson, *Chem. Phys. Lett.*, **127**, 392 (1986).

161. M. Lozada-Cassou and E. Díaz-Herrera, *J. Chem. Phys.*, **92**, 1194 (1990).

162. M. Lozada-Cassou and E. Díaz-Herrera, *J. Chem. Phys.*, **93**, 1386 (1990).

163. S. E. Feller and D. A. McQuarrie, *J. Phys. Chem.*, **97**, 12083 (1993).

164. S. E. Feller and D. A. McQuarrie, *J. Colloid Interface Sci.*, **162**, 208 (1994).
165. G. N. Patey, *J. Chem. Phys.*, **72**, 5763 (1980).
166. M. Teubner, *J. Chem. Phys.*, **75**, 1907 (1981).
167. G. M. Bell and S. Levine, *Trans. Faraday Soc.*, **54**, 785 (1958).
168. L. Belloni, *Chem. Phys.*, **99**, 43 (1985).
169. L. Belloni, *J. Chem. Phys.*, **88**, 5143 (1988).
170. O. Spalla and L. Belloni, *J. Chem. Phys.*, **95**, 7689 (1991).
171. P. Linse, *J. Chem. Phys.*, **93**, 1376 (1990).
172. P. Linse, *J. Chem. Phys.*, **94**, 3817 (1991).
173. D. Forciniti and C. K. Hall, *J. Chem. Phys.*, **100**, 7553 (1994).
174. S. J. Miklavic and P. Attard, *J. Phys. Chem.*, **98**, 4320 (1994).
175. S. E. Feller and D. A. McQuarrie, *Mol. Phys.*, **80**, 721 (1993).
176. T. Morita, *Prog. Theor. Phys.*, **23**, 829 (1960).
177. L. Verlet and D. Levesque, *Physica*, **28**, 1124 (1962).
178. O. E. Kiselyov and G. A. Martynov, *J. Chem. Phys.*, **93**, 1942 (1990).
179. P. Attard, *J. Chem. Phys.*, **94**, 2370 (1991).
180. L. L. Lee, *J. Chem. Phys.*, **97**, 8606 (1992).
181. J. Yvon, "La Théorie Statistique des Fluides et l'Equation d'Etat", *Actualités Scientiques et Industrielles*, Vol. 203, Hermann, Paris, 1935.
182. J. G. Kirkwood, *J. Chem. Phys.*, **3**, 300 (1935).
183. N. N. Bogoliubov, *J. Phys. (Moscow)*, **10**, 256 (1946).
184. M. Born and H. S. Green, *Proc. R. Soc. London Ser. A*, **188**, 10 (1946).
185. M. S. Wertheim, *J. Chem. Phys.*, **65**, 2377 (1976).
186. R. A. Lovett, C. Y. Mou, and F. P. Buff, *J. Chem. Phys.*, **58**, 1880 (1976).
187. L. Blum, Ch. Gruber, D. Henderson, J. L. Lebowitz, and Ph. A. Martin, *J. Chem. Phys.*, **78**, 3195 (1983).
188. D. G. Triezenberg and R. Zwanzig, *Phys. Rev. Lett.*, **28**, 1183 (1972).
189. R. Kjellander and S. Marčelja, *J. Chem. Phys.*, **82**, 2122 (1985).
190. R. Kjellander and S. Sarman, *Mol. Phys.*, **70**, 215 (1990).
191. H. Greberg and R. Kjellander, *Mol. Phys.*, **83**, 789 (1994).
192. A. L. Loeb, *J. Colloid Sci.*, **6**, 75 (1951).
193. P. Attard, D. J. Mitchell, and B. W. Ninham, *J. Chem. Phys.*, **88**, 4987 (1988); **92**, 3248 (1990).
194. P. Attard, D. J. Mitchell, and B. W. Ninham, *J. Chem. Phys.*, **89**, 4358 (1988).
195. R. Podgornik and B. Žekš, *J. Chem. Soc. Faraday Trans. 2*, **84**, 611 (1988).
196. L. Blum, J. Hernando, and J. L. Lebowitz, *J. Phys. Chem.*, **87**, 2825 (1983).
197. S. Levine and G. Bell, *J. Phys. Chem.*, **64**, 1188 (1960).
198. S. Levine and G. Bell, *Disc. Faraday Soc.*, **42**, 69 (1966).
199. C. W. Outhwaite and L. B. Bhuiyan, *J. Chem. Soc. Faraday Trans. 2*, **79**, 707 (1983).
200. T. L. Croxton and D. A. McQuarrie, *Mol. Phys.*, **42**, 141 (1981).
201. F. Lado, *Phys. Rev. B*, **17**, 2827 (1978).
202. S. Sokolowski, *J. Chem. Phys.*, **73**, 3507 (1980).
203. S. Sokolowski, *Mol. Phys.*, **49**, 1481 (1983).
204. R. M. Nieminen and N. W. Ashcroft, *Phys. Rev. A*, **24**, 560 (1981).

205. S. M. Foiles and N. W. Ashcroft, *Phys. Rev. B*, **25**, 1366 (1982).
206. K. Hillebrand and R. M. Nieminen, *Surf. Sci.*, **147**, 599 (1984).
207. R. Kjellander and S. Marčelja, *Chem. Phys. Lett.*, **112**, 49 (1984).
208. R. Kjellander and S. Marčelja, *J. Phys. Chem.*, **90**, 1230 (1986).
209. R. Kjellander and S. Marčelja, *Chem. Phys. Lett.*, **142**, 485 (1987).
210. R. Kjellander, *J. Chem. Phys.*, **88**, 7192 (1988).
211. R. Kjellander and S. Marčelja, *J. Chem. Phys.*, **88**, 7138 (1988).
212. R. Kjellander and S. Marčelja, *J. Phys. (Paris)*, **49**, 1009 (1988).
213. S. Marčelja, *Biophys. J.*, **61**, 1117 (1992).
214. R. Kjellander, T. Åkesson, Bo Jönsson, and S. Marčelja, *J. Chem. Phys.*, **97**, 1424 (1992).
215. D. Henderson and M. Plischke, *J. Phys. Chem.*, **92**, 7177 (1988).
216. M. Plischke and D. Henderson, *J. Chem. Phys.*, **90**, 5738 (1989).
217. M. Fushiki, *Chem. Phys. Lett.*, **154**, 77 (1989).
218. P. Attard, *J. Chem. Phys.*, **95**, 4471 (1991).
219. B. K. Alpert and V. Rokhlin, *SIAM J. Sci. Stat. Comput.*, **12**, 158 (1991).
220. M. Fushiki, *Mol. Phys.*, **74**, 307 (1991).
221. C. W. Outhwaite and L. B. Bhuiyan, *Mol. Phys.*, **74**, 367 (1991).
222. P. J. Colmenares and W. Olivares, *J. Chem. Phys.*, **88**, 3221 (1988).
223. W. Olivares and D. A. McQuarrie, *J. Chem. Phys.*, **65**, 3604 (1976).
224. R. Evans, *Adv. Phys.*, **28**, 143 (1979).
225. R. Evans, in *Fundamentals of Inhomogeneous Fluids*, D. Henderson, Ed., Marcel Dekker, New York, 1992, Chapter 8.
226. M. J. Grimson and G. Richayzen, *Mol. Phys.*, **42**, 47 (1981).
227. J. R. Henderson, *Mol. Phys.*, **52**, 1467 (1984).
228. G. Rickayzen and A. Augousti, *Mol. Phys.*, **52**, 1355 (1984).
229. S. Toxvaerd, *J. Chem. Phys.*, **55**, 3116 (1971).
230. S. Nordholm, M. Johnson, and B. C. Freasier, *Aust. J. Chem.*, **33**, 2139 (1980).
231. P. Tarazona, *Mol. Phys.*, **52**, 81 (1984).
232. P. Tarazona, *Phys. Rev. A*, **31**, 2672 (1985); **32**, 3148 (1985).
233. L. Mier-y-Teran, S. H. Suh, H. S. White, and H. T. Davis, *J. Chem. Phys.*, **92**, 5087 (1990).
234. Z. Tang, L. E. Scriven, and H. T. Davis, *J. Chem. Phys.*, **97**, 494 (1992).
235. Z. Tang, L. E. Scriven, and H. T. Davis, *J. Chem. Phys.*, **97**, 9258 (1992).
236. L. Zhang, H. T. Davis, and H. S. White, *J. Chem. Phys.*, **98**, 5793 (1993).
237. A. R. Denton and N. W. Ashcroft, *Phys. Rev. A*, **44**, 8242 (1991).
238. C. N. Patra and S. K. Gosh, *Phys. Rev. A*, **47**, 4088 (1992).
239. C. N. Patra and S. K. Gosh, *J. Chem. Phys.*, **101**, 4143 (1994).
240. C. N. Patra and S. K. Gosh, *J. Chem. Phys.*, **102**, 2556 (1995).
241. C. N. Patra and S. K. Gosh, *Phys. Rev. E*, **49**, 2826 (1994).
242. R. Penfold, S. Nordholm, Bo Jönsson, and C. E. Woodward, *J. Chem. Phys.*, **92**, 1915 (1990).
243. M. J. Stevens and M. O. Robins, *Europhys. Lett.*, **12**, 81 (1990).
244. W. van Megan and I. Snook, *J. Chem. Phys.*, **73**, 4656 (1980).

245. I. Snook and W. van Megan, *J. Chem. Phys.*, **75**, 4104 (1981).
246. G. M. Torrie and J. P. Valleau, *J. Phys. Chem.*, **86**, 3251 (1982).
247. G. M. Torrie, J. P. Valleau, and G. N. Patey, *J. Chem. Phys.*, **76**, 4615 (1982).
248. D. M. Heyes, M. Barber, and J. H. R. Clarke, *J. Chem. Soc. Faraday Trans. 2*, **73**, 1485 (1977).
249. S. W. de Leeuw and J. W. Perram, *Mol. Phys.*, **37**, 1313 (1979).
250. H. Totsuji, *J. Phys. C*, **19**, L573 (1986).
251. J. Lekner, *Physica*, **176**, 485 (1991).
252. L. Zhang, H. S. White, and H. T. Davis, *Mol. Simul.*, **9**, 247 (1992).
253. J. Hautman and M. L. Klein, *Mol. Phys.*, **75**, 379 (1992).
254. J. Hautman, J. W. Halley, and Y.-J. Rhee, *J. Chem. Phys.*, **91**, 467 (1989).
255. Bo Jönsson, H. Wennerström, and B. Halle, *J. Phys. Chem.*, **84**, 2179 (1980).
256. L. Guldbrand, Bo Jönsson, H. Wennerström, and P. Linse, *J. Chem. Phys.*, **80**, 2221 (1984).
257. D. Bratko, Bo Jönsson, and H. Wennerström, *Chem. Phys. Lett.*, **128**, 449 (1986).
258. S. J. Zara, D. Nicholson, N. G. Parsonage, and J. Barker, *J. Colloid Interface Sci.*, **129**, 297 (1989).
259. B. Svensson, Bo Jönsson, and C. E. Woodward, *J. Phys. Chem.*, **94**, 2105 (1990).
260. J. P. Valleau, R. Ivkov, and G. M. Torrie, *J. Chem. Phys.*, **95**, 520 (1991).
261. T. Åkesson and Bo Jönsson, *Electrochim. Acta*, **36**, 1723 (1991).
262. P. G. Bolhuis, T. Åkesson, and Bo Jönsson, *J. Chem. Phys.*, **98**, 8096 (1993).
263. D. Bratko, L. Blum, and L. B. Bhuiyan, *J. Chem. Phys.*, **94**, 597 (1991).
264. D. Bratko, D. J. Henderson, and L. Blum, *Phys. Rev. A*, **44**, 8235 (1991).
265. D. Bratko and D. Henderson, *Phys. Rev. E*, **49**, 4140 (1994).
266. N. T. Skipper, A. K. Soper, J. D. C. McConnell, and K. Refson, *Chem. Phys. Lett.*, **166**, 141 (1990).
267. N. T. Skipper, K. Refson, and J. D. C. McConnell, *J. Chem. Phys.*, **94**, 7434 (1991).
268. A. Delville, *Langmuir*, **8**, 1796 (1992).
269. A. Delville, *J. Phys. Chem.*, **97**, 9703 (1993).
270. Z. Gamba, J. Hautman, J. C. Shelley, and M. L. Klein, *Langmuir*, **8**, 3155 (1992).
271. D. A. Rose and I. Benjamin, *J. Chem. Phys.*, **95**, 6856 (1991).
272. S.-B. Zhu and G. W. Robinson, *J. Chem. Phys.*, **94**, 1403 (1991).
273. R. Kjellander and S. Marčelja, *Chem. Scr.*, **25**, 73 (1985).
274. R. Kjellander and S. Marčelja, *Chem. Phys. Lett.*, **120**, 393 (1985).
275. M. L. Berkowitz and K. Raghavan, *Langmuir*, **7**, 1042 (1991).
276. S. H. Lee, J. C. Rasaiah, and J. B. Hubbard, *J. Chem. Phys.*, **86**, 2383 (1987).
277. S. J. Miklavic, C. E. Woodward, Bo Jönsson, and T. Åkesson, *Macromolecules*, **23**, 4149 (1990).
278. M. K. Granfeldt, S. J. Miklavic, S. Marčelja, and C. E. Woodward, *Macromolecules*, **23**, 4760 (1990).
279. M. K. Granfeldt, Bo. Jönsson, and C. E. Woodward, *J. Phys. Chem.*, **95**, 4819 (1991).
280. L. Sjöström, T. Åkesson, and Bo Jönsson, *J. Chem. Phys.*, **99**, 4739 (1993).

281. C. E. Woodward, T. Åkesson, and Bo Jönsson, *J. Chem. Phys.*, **101**, 2569 (1994).

282. P. Linse and Bo Jönsson, *J. Chem. Phys.*, **78**, 3167 (1983).

283. P. Linse, G. Gunnarsson, and Bo Jönsson, *J. Phys. Chem.*, **86**, 413 (1982).

284. L. Degréve, M. Lozada-Cassou, E. Sánchez, and E. González-Tovar, *J. Chem. Phys.*, **98**, 8905 (1993).

285. P. Sloth and T. S. Sørenson, *J. Chem. Phys.*, **96**, 548 (1992).

286. B. Svensson and Bo Jönsson, *Chem. Phys. Lett.*, **108**, 580 (1984).

287. D. Bratko and V. Vlachy, *Chem. Phys. Lett.*, **90**, 434 (1982).

288. C. S. Murthy, R. J. Bacquet, and P. J. Rossky, *J. Phys. Chem.*, **89**, 701 (1985).

289. L. Nilsson, L. Guldbrand, and L. Nordenskiöld, *Mol. Phys.*, **72**, 177 (1991).

290. P. Attard, D. Wei, and G. N. Patey, *J. Chem. Phys.*, **96**, 3767 (1992).

291. G. M. Torrie, *J. Chem. Phys.*, **96**, 3772 (1992).

292. L. B. Bhuiyan, C. W. Outhwaite, and S. Levine, *Mol. Phys.*, **42**, 1271 (1981).

293. J. Ennis, Ph.D. Thesis, Australian National University, Canberra, 1993.

294. P. Kékicheff, S. Marčelja, T. J. Senden, and V. E. Shubin, *J. Chem. Phys.*, **99**, 6098 (1993).

295. G. M. Torrie, P. G. Kusalik, and G. N. Patey, *J. Chem. Phys.*, **90**, 4513 (1989).

296. J. R. Henderson, *Mol. Phys.*, **59**, 89 (1986).

297. A. Luzar, D. Bratko, and L. Blum, *J. Chem. Phys.*, **86**, 2955 (1987).

298. F. Oosawa, *Biopolymers*, **6**, 1633 (1968).

299. F. Oosawa, *Polyelectrolytes*, Marcel Dekker, New York, 1971.

300. B. Davies and B. W. Ninham, *J. Chem. Phys.*, **56**, 5797 (1972).

301. V. N. Gorelkin and V. P. Smilga, *Sov. Phys. JETP*, **36**, 761 (1973).

302. D. J. Mitchell and P. Richmond, *J. Colloid Interface Sci.*, **46**, 128 (1974).

303. P. Richmond, *J. Chem. Soc. Faraday 2*, **70**, 1066 (1973).

304. A. P. Nelson and D. A. McQuarrie, *J. Theor. Biol.*, **55**, 13 (1975).

305. S. J. Miklavic, D. Y. C. Chan, L. R. White, and T. W. Healy, *J. Phys. Chem.*, **98**, 9022 (1994); S. J. Miklavic, *J. Colloid Interface Sci.*, **171**, 446 (1995).

306. P. Attard, *J. Phys. Chem.*, **93**, 6441 (1989).

307. P. Attard and G. N. Patey, *Phys. Rev. A*, **43**, 2953 (1991).

308. D. Grahame, *Chem. Revs.*, **41**, 441 (1947); *J. Chem. Phys.*, **21**, 1054 (1953).

309. B. W. Ninham and V. A. Parsegian, *J. Theor. Biol.*, **31**, 405 (1971).

310. P. G. Higgs and J.-F. Joanny, *J. Phys. (France)*, **51**, 2307 (1990).

311. P. Pincus, J.-F. Joanny, and D. Andelman, *Europhys. Lett.*, **11**, 763 (1990).

312. D. J. Mitchell and B. W. Ninham, *Langmuir*, **5**, 1121 (1989).

313. A. Fogden, D. J. Mitchell, and B. W. Ninham, *Langmuir*, **6**, 159 (1990).

314. A. Fogden and B. W. Ninham, *Langmuir*, **7**, 590 (1991).

315. J. Ennis, *J. Chem. Phys.*, **97**, 663 (1992).

316. R. E. Goldstein, A. I. Pesci, and V. Romero-Rochín, *Phys. Rev. A*, **41**, 5504 (1990).

317. S. R. Dungan and T. A. Hatton, *J. Colloid Interface Sci.*, **164**, 200 (1994).

318. S. Levine, J. Mingins, and G. M. Bell, *J. Electroanal. Chem.*, **13**, 280 (1967).

319. H. Ohshima and S. Ohki, *Biophys. J.*, **47**, 673 (1985).

320. Y.-C. Kuo and J.-P. Hsu, *J. Chem. Phys.*, **102**, 1806 (1995).

321. S. J. Miklavic and S. Marčelja, *J. Phys. Chem.*, **92**, 6718 (1988).

322. T. Åkesson, C. E. Woodward, and Bo Jönsson, *J. Chem. Phys.*, **91**, 2461 (1989).

323. T. Croxton, D. A. McQuarrie, G. N. Patey, and J. P. Valleau, *Can. J. Chem.*, **59**, 1998 (1981).

324. L. Onsager and N. Samaras, *J. Chem. Phys.*, **2**, 528 (1934).

325. J. S. Rushbrooke and M. Silbert, *Mol. Phys.*, **12**, 505 (1967).

326. G. Casanova, R. J. Dulla, D. A. Jonah, J. S. Rushbrooke and G. Saville, *Mol. Phys.*, **18**, 589 (1970).

327. Q. N. Usmani, S. Fantoni, and V. R. Pandharipande, *Phys. Rev. B*, **26**, 6123 (1982).

328. G. C. Aers and M. W. C. Dharma-wardana, *Phys. Rev. A*, **29**, 2734 (1984).

329. L. Reatto and M. Tau, *J. Chem. Phys.*, **86**, 6474 (1987).

330. P. Attard, *Phys. Rev. A*, **45**, 3659 (1992).

331. D. R. Bérard and G. N. Patey, *J. Chem. Phys.*, **97**, 4372 (1992).

332. M. Vertenstein and D. Ronis, *J. Chem. Phys.*, **87**, 4132 (1987).

333. B. Honig, K. Sharp, and A.-S. Yang, *J. Phys. Chem.*, **97**, 1101 (1993).

334. M. K. Gilson, M. E. Davis, B. A. Luty, and J. A. McCammon, *J. Phys. Chem.*, **97**, 3591 (1993).

335. K. E. Forsten, R. E. Kozack, D. A. Lauffenburger, and S. Subramaniam, *J. Phys. Chem.*, **98**, 5580 (1994).

336. A. A. Kornyshev, *Electrochim. Acta*, **26**, 1 (1981).

337. P. Attard, D. Wei, and G. N. Patey, *Chem. Phys. Lett.*, **172**, 69 (1990).

338. A. Chandra and B. Bagchi, *J. Chem. Phys.*, **90**, 1832 (1989).

339. A. Chandra and B. Bagchi, *J. Chem. Phys.*, **91**, 3056 (1989).

340. T. Fonseca and B. M. Ladanyi, *J. Chem. Phys.*, **93**, 8148 (1990).

341. S. L. Carnie and D. Y. C. Chan, *J. Chem. Phys.*, **73**, 2949 (1980).

342. D. Henderson and L. Blum, *J. Chem. Phys.*, **74**, 1902 (1981).

343. J. M. Eggebrecht, D. J. Isbister, and J. C. Rasaiah, *J. Chem. Phys.*, **73**, 3980 (1980).

344. J. C. Rasaiah, D. J. Isbister, and G. Stell, *J. Chem. Phys.*, **75**, 4707 (1981).

345. W. Dong, M. L. Rosinberg, A. Perera, and G. N. Patey, *J. Chem. Phys.*, **89**, 4994 (1988).

346. G. M. Torrie, A. Perera, and G. N. Patey, *Mol. Phys.*, **67**, 1337 (1989).

347. C. E. Woodward and S. Nordholm, *Mol. Phys.*, **60**, 415 (1987).

348. J. B. Sweeney, L. E. Scriven, and H. T. Davis, *J. Chem. Phys.*, **87**, 6120 (1987).

349. J. P. Badiali, *J. Chem. Phys.*, **90**, 4401 (1989).

350. Q. Zhang, J. P. Badiali, and W. H. Su, *J. Chem. Phys.*, **92**, 4609 (1990).

351. G. M. Torrie, P. G. Kusalik, and G. N. Patey, *J. Chem. Phys.*, **88**, 7826 (1988).

352. G. M. Torrie, P. G. Kusalik, and G. N. Patey, *J. Chem. Phys.*, **89**, 3285 (1988).

353. P. G. Kusalik and G. N. Patey, *J. Chem. Phys.*, **89**, 5843 (1988).

354. S. L. Carnie and G. N. Patey, *Mol. Phys.*, **47**, 1129 (1982).

355. J. M. Caillol, D. Levesque, J. J. Weis, P. G. Kusalik, and G. N. Patey, *Mol. Phys.*, **62**, 461 (1987).

356. B. M. Pettitt and P. J. Rossky, *J. Chem. Phys.*, **84**, 5836 (1986).

357. X. Dang and B. M. Pettitt, *J. Chem. Phys.*, **86**, 6560 (1987).

358. E. Guardia, A. Robinson, and J. A. Padro, *J. Chem. Phys.*, **99**, 4229 (1993).
359. R. A. Friedman and M. Mezei, *J. Chem. Phys.*, **102**, 419 (1995).
360. D. R. Bérard, M. Kinoshita, X. Ye, and G. N. Patey, *J. Chem. Phys.*, **101**, 6271 (1994).
361. D. R. Bérard, M. Kinoshita, X. Ye, and G. N. Patey, *J. Chem. Phys.*, **102**, 1024 (1995).
362. C. Y. Lee, J. A. McCammon, and P. J. Rossky, *J. Chem. Phys.*, **80**, 4448 (1984).
363. J. P. Valleau and A. A. Gardner, *J. Chem. Phys.*, **86**, 4162 (1987).
364. A. A. Gardner and J. P. Valleau, *J. Chem. Phys.*, **86**, 4171 (1987).
365. G. Rickayzen and M. J. Grimson, *J. Chem. Soc. Faraday Trans. 2*, **78**, 893 (1982).
366. A. T. Augousti and G. Rickayzen, *J. Chem. Soc. Faraday Trans. 2*, **80**, 141 (1984).
367. R. Kjellander, S. Marčelja, R. M. Pashley, and J. P. Quirk, *J. Phys. Chem.*, **92**, 6489 (1988).
368. R. Kjellander, S. Marčelja, R. M. Pashley, and J. P. Quirk, *J. Chem. Phys.*, **92**, 4399 (1990).
369. R. Kjellander, S. Marčelja, and J. P. Quirk, *J. Colloid Interface Sci.*, **126**, 194 (1988).

OSCILLATIONS AND COMPLEX DYNAMICAL BIFURCATIONS IN ELECTROCHEMICAL SYSTEMS

MARC T. M. KOPER*

Department of Electrochemistry, Debye Institute, Utrecht University, 3584 CH Utrecht, The Netherlands

CONTENTS

* Present address: Abteilung Elektrochemie, Universität Ulm, D-89069 Ulm, Germany.

Advances in Chemical Physics, Volume XCII, Edited by I. Prigogine and Stuart A. Rice.
ISBN 0-471-14320-0 © 1996 John Wiley & Sons, Inc.

I. INTRODUCTION

When maintained far from thermodynamic equilibrium, chemically reacting systems can exhibit a rich variety of spatio-temporal self-organization if they are governed by the appropriate nonlinear evolution laws. The most common way in which such a self-organization manifests itself, is through the occurrence of spontaneous reaction-rate oscillations. Although oscillations have been reported in many fields of chemical kinetics since the early nineteenth century, for a long time such phenomena were regarded as curiosities or anomalies with little more significance than that of an amusing friday-afternoon experiment. However, with the realization that such systems constitute some of the most clear-cut and intriguing examples of deterministic chaos, complex dynamical bifurcations, and other types of self-organization, the field of "chemical chaos" seems finally to have acquired the more respectable footing it deserves [1].

Although in electrochemical kinetics there are probably more examples of oscillating systems than in any other area of chemical kinetics, for many years systematic investigations of electrochemical oscillations have been less forthcoming than, for example, in homogeneous reacting systems, combustion, and heterogenous catalysis. Since the mid-1980s this situation has reversed dramatically, and at present it seems fair to say that electrochemistry has been exceptionally rewarding in providing us with interesting dynamical objects. Furthermore, some significant progress has been made in the theoretical understanding of electrochemical instabilities, oscillations and chaos.

The aim of this chapter is to give the reader an idea of what has been achieved in the study of electrochemical oscillations. For this purpose, I have refrained from being comprehensive or complete. Recently Hudson and Tsotsis [2] have done a tremendous job in writing a review article that covers practically all the relevant literature up to 1993. This review is an excellent follow-up to the much-quoted 1973 review of Wojtowicz [3] (for other reviews see [4]), which has become—with all the progress recently achieved—a bit outdated. It is clear that, given the comprehensive Hudson–Tsotsis review, the present article can only be warranted by taking a fundamentally different approach. This approach will essentially be an attempt to build a general framework within which the various types of oscillating electrochemical systems can be interpreted in a more or less coherent way. Such a framework should be one in which the electrical nature of the process is emphasized. Since in electrochemical systems there always exists a closed loop with respect to one of the reagents, namely, the electrons, it is clear that the design and the control of the electrochemical cell, or, more generally, the elements of the external circuit, will greatly influence the course of the events at the electrode under consideration. This is also the reason why, whenever the oscillatory behavior is being discussed, it is perhaps better to refer to electrochemical *systems* and not to electrochemical reactions. The result is—hopefully—a chapter that will present the reader with a representative view of electrochemical oscillations, stressing both their origin as well as the intriguing complexity of their various manifestations. However, although some of the most important concepts will be explained concisely at the appropriate places, a premise to this chapter is that the reader has some familiarity with nonlinear (chemical) dynamics. Practically all terminology associated with this field and used in the ensueing discussions is well explained at an elementary level in Scott's book [1].

The organization of this chapter can be summarized as follows: As the electrical nature of the cell plays such a prime role in its stability features, this chapter will start with an outline of how the stability problem is approached in electrical engineering. Various authors have already applied these techniques to electrochemistry, but in view of the renewed interest in bifurcations and oscillations, it seems worthwhile to readdress the so-called frequency domain stability in some detail, not in the least because of the beauty and the power of the method in generalizing and categorizing electrochemical instabilities. Section II therefore studies in some depth and generality the way in which a negative impedance element can influence a circuit's stability on the basis of a few standard methods from electrical control engineering. It will be shown that a measurement of the electrochemical system's frequency response (i.e. its complex electrical impedance) constitutes an elegant way of evaluating its stability and bifurcations, that will prove to be of use in subsequent sections when dealing with the dynamic behavior of explicit electrochemical systems. As in the reviews of Hudson and Tsotsis and Wojtowicz, examples are taken from different areas of electrochemistry, that is, cathodic reduction reactions at mercury (Section III), electrodissolution of metals (Section IV), electrocatalytic reactions (Section V), and semiconductor (photo-)electrochemical processes (Section VI). Attention will be restricted, however, to only a few systems for which the dynamics is well documented. Throughout, emphasis will be on the extent to which the electrochemical origin of the oscillations is understood in terms of the various models proposed and how they relate to the frequency response theory developed in Section II, and on the interpretation of the various dynamical bifurcation sequences and how these may relate to the various physical processes. This chapter will end with a few concluding remarks.

II. FREQUENCY RESPONSE METHODS AND STABILITY EVALUATION OF ELECTROCHEMICAL SYSTEMS

A. Introduction

Due to the experimental ability to make high-precision measurements, techniques based on *frequency response* or *alternating current (ac) impedance* are highly popular in electrochemical studies. By measuring an electrochemical system's response to a sinusoidally alternating input signal of small amplitude and comparing this response or output signal to that derived from a mathematical (phenomenological) model of the system, one can extract information about the electrode kinetics and its various complications. The interested reader is referred to [5–9] for detailed accounts of the various applications in this field.

A specific application of the impedance method that is of interest here is the ability of the impedance method to make predictions about the system's stability. Generally, the stability of a system is studied by linearizing the system's equations about the stationary reference state, and subsequently evaluating the local time evolution by solving the linear differential equations [10]. As is well known from standard calculus [10], solving coupled linear differential equations amounts to calculating the eigenvalues and the eigenvectors of the corresponding linearized Jacobian matrix. The stability is then uniquely determined by the sign of the eigenvalues of the Jacobian: all negative eigenvalues signify stability, one or more positive eigenvalues signifies instability. Since the eigenvalues normally depend on the system's parameters, the sign of one of the eigenvalues can change with a parameter, and the case where the real part of the eigenvalue vanishes, and where the linearized system is neutrally stable, heralds a bifurcation [11, 12]. Two of such (local) bifurcations are of particular importance:

- The saddle-node bifurcation, for which one of the real eigenvalues equals zero. At this bifurcation two stationary states merge, therefore indicating the existence of multiple steady states.
- The Hopf bifurcation for which the real part of a pair of complex conjugate eigenvalues equals zero. At this bifurcation, a (stable or unstable) periodic orbit is born [11, 12], signifying the emergence of sustained spontaneous oscillations. Although the existence of the Hopf bifurcation alone does not prove the existence of stable oscillations nor does the known existence of stable oscillations require a Hopf bifurcation, in a system with a sufficient number of parameters these two suppositions normally provide very reliable working hypotheses.

This section will discuss how the stability and bifurcations of an electrochemical system can be investigated by the impedance method. It is intuitively clear that if one determines the system's frequency response over a broad enough frequency range, one has all the linear information of the system and with it one should be able to make statements about the stability. A clear advantage of this technique is that it does not require any modelistic considerations, only the measurement of the system's frequency response under conditions where it is (conditionally) stable. The problem has been treated before in electrochemistry, notably by de Levie [13] and by Gabrielli and co-workers [14, 15], but without the explicit realization that it allows the recognition of Hopf and saddle-node bifurcations (of course, in the 1970s electrochemists were not aware of this terminology). This last point has important potential in studying a system's phase diagram, that is, a map in parameter space that indicates where the system will make transitions to oscillatory or multiple steady-state behavior. The analysis is based on a celebrated theorem from electrical control engineering known as Nyquist's stability criterion, already derived in 1932 [16–18]. The derivation of the Nyquist criterion, as outlined briefly in Section II.C, is based on rather abstract theorems from complex variable theory, but its most useful application for our purposes, namely, the identification of bifurcations, is easily understood on an intuitive basis.

It may be important to emphasize at the outset that the Nyquist criterion is restricted to an assessment of a system's *linear* stability, and in fact provides a very general tool in that respect. Unfortunately, it has no potential in the prediction of the nonlinear behavior of an unstable system.

B. Alternating Current Impedance Techniques in Electrochemistry

In this section, a short outline is given to the theory of ac impedance techniques as applied to electrochemistry [5–9]. Electrochemists familiar with impedance spectroscopy can safely skip this section.

Usually, an electrochemical cell is designed in such a way that only the working electrode has a significant impedance. The current through the working electrode–electrolyte interface is assumed to be divided between two pathways: A capacitive branch represented by the differential double-layer capacity C_d, and a faradaic branch representing the faradaic (electrochemical) reactions, designated by the unspecified impedance Z_F. Of course, the current must pass through an ohmic series resistance R_s, which includes the uncompensated cell resistance and—possibly—a series resistor deliberately connected in series with the working electrode, and therefore a general equivalent circuit of an electrochemical cell will look

166   MARC T. M. KOPER

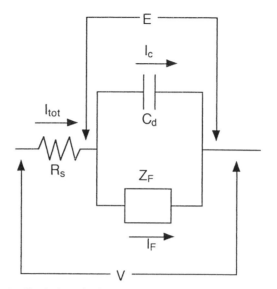

Figure 1. Equivalent-circuit representation of an electrochemical cell.

like the one shown in Fig. 1. Clearly, a strict separation between capacitive and faradaic currents is valid only on the condition that the two processes are not coupled. In the case of the adsorption of electroactive species, for instance, one should in principle account for a coupling between the charge on the electrode and the faradaic processes [5, 7].

In an ac impedance experiment one subjects the electrochemical cell to a small-amplitude sinusoidal perturbation ΔV superimposed on some applied bias potential V_s:

$$V = V_s + \Delta V = V_s + \Delta V^0 \exp(j\omega t) \tag{2.1}$$

As a consequence, the electrode potential E, the total current I, and total current density J will oscillate with the same frequency, only the latter usually with some (frequency-dependent) phase angle ϕ:

$$E = E_s + \Delta E = E_s + \Delta E^0 \exp(j\omega t),$$
$$I = I_s + \Delta I = I_s + \Delta I^0 \exp[j(\omega t + \phi)] \tag{2.22}$$

with $j = \sqrt{-1}$. The cell impedance Z_{cell} is defined as the ratio $\Delta V/\Delta I$, the interfacial impedance Z_{int} as $\Delta E/\Delta I$, and clearly the two are related by $Z_{int} = Z_{cell} - R_s$.

Since any impedance is defined by an amplitude ratio and a phase angle, it is convenient to treat them as complex numbers

$$Z = Z' + jZ'' \tag{2.3}$$

where $Z' = (\Delta E^0/\Delta I^0)\cos(\phi)$ and $Z'' = (\Delta E^0/\Delta I^0)\sin(\phi)$. Both R_s and C_d have simple and well-known impedances under sinusoidal excitation (R_s and $(j\omega C_d)^{-1} = -j/(\omega C_d)$, respectively), but the faradaic impedance $Z_F(\omega)$ is generally not simple and depends on the detailed kinetics of the electron transfer and the reactant's mass transfer.

Models for $Z_F(\omega)$ have been derived for a few simple—but representative—cases [5–9]. The simplest model conceivable is that of a resistor, usually referred to as the charge-transfer resistance R_{ct}. This model holds if there is no complication whatsoever to the electron-transfer process

$$\text{ox} + n\,e_M^- \leftrightarrow \text{red} \tag{2.4}$$

(the subscript M referring to "metal"), that is, mass transport is sufficiently fast, the electroactive species do not adsorb and do not react chemically. The quantity R_{ct} is then defined as $R_{ct}^{-1} = \partial I_F/\partial E = dI_F/dE$, where the first equality is the definition.[1] A common way of presenting the results of an ac impedance measurement is to plot Z'' against Z', in the so-called complex impedance plane. For the circuit of Fig. 1 with $Z_F = R_{ct}$ this plot is as shown in Fig. 2, where the various components can

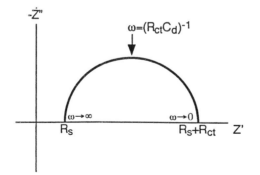

Figure 2. Complex-plane impedance diagram of Fig. 1 for the case that $Z_F = R_{ct}$.

---

[1] In most applications, except perhaps the one under consideration, it is clearly preferable to present the various definitions in terms of the current density, rather than current. Since the total current flowing through the cell has important implications for its stability, it is important to account for the area of the electrode.

be evaluated from the various characteristic points of the plot, as indicated in Fig. 2.

The most common complication that arises in a simple electrochemical reaction such as Eq. (2.4) is that of a finite mass-transfer rate. In this case one has to derive expressions for the ac response of the electrochemical system by including the mass-transport laws pertaining to the system under study. Generally, this gives rise to an additional impedance in the faradaic branch, Z_{mt}, as shown in Fig. 3(a). When diffusion is the only mass-transfer mechanism, this additional impedance is generally referred to as the Warburg impedance, having a characteristic frequency dependence. In the case of semiinfinite diffusion towards a spherical electrode, the mass-transfer impedance can be written as a parallel connection between a frequency-dependent and a frequency-independent part, as shown in Fig. 3(b), that we will refer to as Z_W and R_W^0, respectively, and

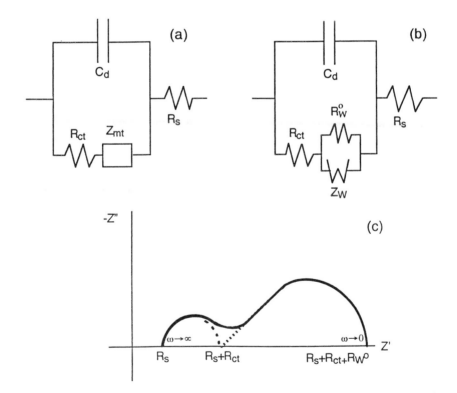

Figure 3. (a) Equivalent cell circuit including the role of mass transfer, as represented by the mass-transfer impedance Z_{mt}. (b) Randles equivalent circuit for diffusion towards a stationary sphere. (c) Complex-plane impedance diagram of Fig. 3(b).

for which the following expressions can be derived [7, 9]:

$$Z_W = -\frac{1}{(2D_{ox})^{1/2}} \frac{R_{ct}}{nF} \frac{\partial j_F}{\partial c_{ox}} \omega^{-1/2} (1-j) \tag{2.5a}$$

$$R_W^0 = -\frac{r_0}{D_{ox}} \frac{R_{ct}}{nF} \frac{\partial j_F}{\partial c_{ox}} \tag{2.5b}$$

where r_0 is the radius of the spherical electrode, D_{ox} is the diffusion coefficient of the ox species, and $j = \sqrt{-1}$. In these expressions we have assumed that $c_{red} = 0$, which is true for the experiments to be described in Section III. Also note that because of the coupling between charge transfer and mass transfer the charge-transfer resistance appears explicitly in the expressions for Z_{mt}, and that both Z_W and R_W^0 have the same sign as R_{ct} since cathodic currents are conventionally taken as negative. In most textbooks one will not encounter the term R_W^0 because one usually considers plane diffusion for which $r_0 \to \infty$ so that $R_W^0 \to \infty$. Plane diffusion is less interesting for our purposes because the only stationary-state solution it admits is the trivial one, namely, $c_{ox} = 0$ everywhere. Expressions for Z_{mt} are also available for the cases of so-called bounded (plane) diffusion and mass transport towards a rotating disk [5–9].

If both the electron-transfer and the mass-transfer impedance contribute within the measurable frequency range (mixed control), the impedance plot consists of two (partly overlapping) loops, as illustrated in Fig. 3(c) for a spherical electrode with diffusion as the only mass-transfer mechanism. The equivalent circuit for which $Z_F = R_{ct} + Z_W$ or $Z_F = R_{ct} + [1/Z_W + 1/R_W^0]^{-1}$, is generally referred to as Randles' equivalent circuit [5–9].

There is also important qualitative, diagnostic, information to be obtained from an impedance plot. As we saw in the last paragraph, the complication of an additional time-dependent variable normally gives rise to an additional semicircle or semicirclelike loop in the complex impedance plane. This allows an estimation of the number of important time-dependent variables one should incorporate into a phenomenological model from the number of loops in an experimentally obtained impedance plot. This can also be seen from the theoretical expression for the faradaic impedance in the case of N time-dependent variables x_i:

$$Z_F^{-1} = \left(\frac{\partial j_F}{\partial E}\right)_{x_i} + \sum_{i=1}^{N} \left(\frac{\partial j_F}{\partial x_i}\right)_{E,x_{j\neq i}} \frac{\Delta x_i}{\Delta E} \tag{2.6}$$

where each new variable x_i introduces a new typical time scale (in practice, time scales can be very similar and will result in overlapping or distorted loops in the impedance plot).

It is sometimes possible to rewrite Eq. (2.6) as an expression of an "equivalent" circuit consisting of well-known passive electrical elements such as resistors, capacitors, and inductors. For certain processes, one has to come up with typical electrochemical nonreducible elements, such as the Warburg impedance as we saw above. Such an equivalent circuit can be helpful in the interpretation of results, provided it does not become unduly complicated and has an unambiguous electrochemical meaning. The Randles circuit is a typical example; we will meet a few others in Sections V.A and VI.A.

C.   The Nyquist Stability Criterion

As before the working electrode has a potential E, resulting in a corresponding electric current I, which take the respective values E_s and I_s in the steady state. Note once more that at all times we distinguish between the electrode potential E and the applied (cell) potential V, and between the interfacial impedance Z_{int} and the cell impedance Z_{cell}. Both pairs are equal to one another only if $R_s = 0$. As we saw in Section II.B, in an impedance experiment one studies the relationship between small disturbances $\Delta E(t)$ and $\Delta I(t)$, that is,

$$E = E_s + \Delta E(t) \qquad I = I_s + \Delta I(t)$$

where the real-time interfacial impedance of the system is defined as the ratio $\Delta E(t)/\Delta I(t)$, provided that the disturbances are small enough so that their relation is linear. However, it turns out to be more convenient, and therefore more customary to evaluate the impedance in the complex frequency domain, denoted $Z(s)$, with $s = \sigma + j\omega$, defined as the ratio of the respective Laplace transforms of ΔE and ΔI:

$$Z(s) = \frac{\Delta E(s)}{\Delta I(s)} \tag{2.7}$$

where

$$\Delta E(s) = \int_0^\infty \Delta E(t) \exp(-st)\, dt$$

and similarly for $\Delta I(s)$.

To see how the properties of $Z(s)$ determine the stability of this linearized system, we can study the system's response to an arbitrarily chosen disturbance, such as the impulse disturbance $\Delta E(t) = \delta(t)$, so that

$$\Delta I(s) = \frac{1}{Z(s)} \tag{2.8}$$

Since we are considering a linear or linearized system, it follows from standard theory that $Z(s)$ can be expressed as a so-called rational function of s, that is, as the ratio of two polynomials in s that can be factorized in their respective roots [17, 18]

$$\Delta I(s) = \frac{1}{Z(s)} = Y(s) = \frac{K \prod\limits_{m=1}^{M} (s + \mu_m) \prod\limits_{n=1}^{N} (s^2 + 2\nu_n s + (\nu_n^2 + \beta_n^2))}{\prod\limits_{q=1}^{Q} (s + \lambda_q) \prod\limits_{r=1}^{R} (s^2 + 2\rho_r s + (\rho_r^2 + \omega_r^2))} \tag{2.9}$$

where M and N are the numbers of real and complex conjugate *poles* of the impedance, respectively, and Q and R are the number of real and complex conjugate *zeros* of the impedance, respectively. The quantity $Y(s)$ is the admittance.

The expression (2.9) can be simplified in a partial fraction expansion [10, 18]:

$$\frac{1}{Z(s)} = K \left[\sum_{q=1}^{Q} \frac{A_q}{s + \lambda_q} + \sum_{r=1}^{R} \frac{B_r}{s^2 + 2\rho_r s + (\rho_r^2 + \omega_r^2)} \right] \tag{2.10}$$

with A_q and B_r the so-called residues [10, 18].

For this expression the inverse Laplace transform is [10, 18]

$$\Delta I(t) = K \left[\sum_{q=1}^{Q} A_q \exp(-\lambda_q t) + \sum_{r=1}^{R} B_r \frac{1}{\omega_r} \exp(-\rho_r t) \sin(\omega_r t) \right] \tag{2.11}$$

The important point about this formula[2] is that we see that the zeros of the impedance correspond to the natural transients of the linearized system and are in fact the Jacobian eigenvalues that one would obtain

---

[2] This formula seems to exclude the possibility of nonexponential transients, such as occur for diffusion processes. However, such power-law decay modes can be constructed from an infinite series of exponential decay modes, which is intuitively clear if one replaces the partial differential equation by a large (infinite) number of ordinary differential equations, an approach one would follow in a numerical computer simulation. This is clearly equivalent to the fact that the Warburg impedance can be represented by a transmission line consisting of an infinite series of RC circuits [9].

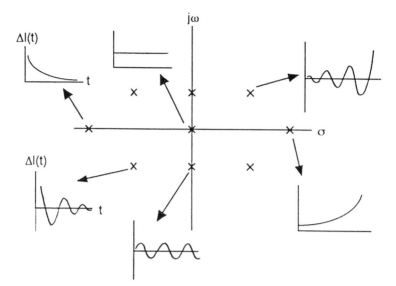

Figure 4. The various transient solutions of Eq. (2.11) in the Laplace s plane.

from a linear stability analysis. Therefore the sign of the zeros determines the stability in the linear regime, as illustrated in Fig. 4.

Physically, it is easy to see why a zero of the impedance should correspond to the system's natural transient response. If the system were driven with an input signal that is exactly equal to one of the natural transients of the system, there is resonance for which the output signal will equal the input signal with an infinite amplification factor. This means that the input signal does not experience any resistance or phase delay $(Z' = Z'' = 0)$.

From Fig. 4 it is seen that the (linear) stability of the system is ascertained when there are no zeros of the impedance in the right-hand part of the Laplace plane, inside the so-called Nyquist-contour [16–18] [Fig. 5(a)]. The number of zeros and poles of $Z(s)$ inside the Nyquist-contour can be derived from a mapping of the Nyquist-contour onto the complex impedance plane $Z = Z' + jZ''$. This follows from Cauchy's theorem from complex function theory, which can be stated as follows [18]: If a contour Γ_s in the s-plane encircles $\#Z$ zeros and $\#P$ poles of $Z(s)$ and does not pass through any poles or zeros of $Z(s)$ as the traversal is in the clockwise direction along the contour, the corresponding mapping Γ_Z in the $Z(s)$ plane encircles the origin of the $Z(s)$ plane $\#N = \#Z - \#P$ times in the clockwise direction (i.e., the number of

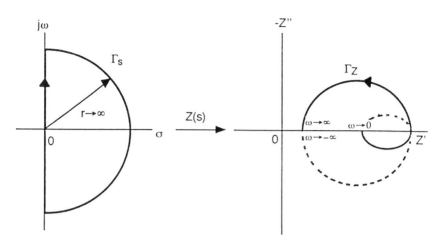

Figure 5. The s-plane Nyquist contour mapped onto the impedance plane by an impedance function $Z(s)$. The solid line is the impedance plot (positive frequency), the dashed line its mirror image (negative frequency).

clockwise encirclements minus the number of counterclockwise encirclements).

As it has become customary in electrochemistry to plot $-Z''$ against Z' [5–9], in such a case the number of clockwise encirclements (following Γ_Z in the direction from 0 to $+\infty$ frequency) is $\#N = \#P - \#Z$.

The usefulness of this theorem will become clear if one realizes that the mapped Nyquist contour Γ_Z is experimentally accessible by frequency response methods. This occurs because the mapping of the positive imaginary axis $s = j\omega$, $\omega > 0$ is simply the classical complex impedance plot $Z(j\omega)$, and the negative imaginary axis $s = j\omega$, $\omega < 0$ is simply its mirror image in the Z' axis [see Fig. 5(b)]. The portion from $\omega = -\infty$ to $\omega = +\infty$ with $r \to \infty$ is mapped onto the origin of the $Z(s)$ plane or onto the $Z'(\omega = +\infty)$ point, where in an electrochemical experiment $Z'(\omega = +\infty)$ is usually the ohmic loss in the external circuit R_s (comprising the solution resistance R_Ω). This can be seen if we write for the electrochemical system:

$$Z_{cell}(s) = R_s + Z_{int}(s)$$

so that

$$\lim_{r \to \infty} Z_{cell}(s)\big|_{s = re^{j\phi}} = R_s + \lim_{r \to \infty} Z_{int}(s)\big|_{s = re^{j\phi}}$$

which equals R_s as $Z_{int}(s)$ will be an expression of the type

$$Z_{int}(s) = \frac{1}{sC_d + Y_F(s)}$$

(where C_d is the double layer capacity and Y_F is the faradaic admittance), which normally vanishes in the case of infinite r ($=|s|$). The change in angle that ϕ makes from $+90°$ to $-90°$ will result in a corresponding change in angle of Γ_Z at the point $Z' = R_s$ [see Fig. 5(b)].

We are now in a position to state the Nyquist stability criterion [16, 18]: An electrochemical system is stable if and only if for the contour Γ_Z the number of counterclockwise encirclements ($=\#N$) of the origin of the $Z' - (-Z'')$ plane equals the number of poles ($=\#P$) in the right-hand s plane. Otherwise, the number number of zeros ($=\#Z$) in the right-hand s plane, and thus the number of instability directions of the steady state, equals $\#Z = \#P - \#N$.

In general, the Nyquist plot will give only the difference between the number of zeros and poles of the impedance or admittance function. As Bode points out (p. 164 of [17]), the determination of the number of zeros in the right-hand half-plane requires "that we must know that the structure is stable for some particular reference condition or if it is not stable what modes of instability it has." As will be illustrated in Section II.D, this particular reference condition may be the a priori knowledge of potentiostatic or galvanostatic stability. However, we do not need to know this reference condition to detect a change in stability, that is, if we want to study the possibility of bifurcations. This is easily seen as follows: The saddle-node bifurcation, for which one of the λ_q values equals zero, must occur if $Z(j\omega) = 0$ for $\omega = 0$. That is, if the complex impedance plot "ends in" the origin of the Z plane at $\omega = 0$, we know that we have a system undergoing a saddle-node bifurcation. Similarly, the Hopf bifurcation, for which one of the ρ_r values equals zero, must occur if $Z(j\omega) = 0$, with $\omega = \omega_r = \omega_H$. That is, if the complex impedance plot intersects the origin of the Z plane for some nonzero frequency ω_r, we know that we have a system undergoing a Hopf bifurcation with a critical frequency $\omega_H = \omega_r$ (Nyquist, in his original article [16], calls this the "singing point").

Again, these results have a simple physical interpretation: At the Hopf bifurcation, the linear system produces a natural undamped sine wave of frequency ω_H, so if the system is driven by a sine wave of exactly that frequency, due to the resonance the driving signal should not experience any resistance.[3] From the Hopf theorem [11, 12], we know that the Hopf bifurcation heralds the birth of a periodic orbit, that is, a sustained oscillatory solution. However, in practice, the periodic orbit can be either stable (physically observable) or unstable (physically unobservable), something that can only be predicted by a more involved nonlinear analysis. Therefore, the detection of a Hopf bifurcation does not guarantee sustained oscillatory behavior (there is an example in the literature of an electrochemical reaction model with a Hopf bifurcation but without stable periodic orbits [19]), but generally one can be quite confident that the presence of a Hopf bifurcation will give rise to stable periodic behavior somewhere in parameter space. Also, the impossibility of having a Hopf bifurcation is generally an excellent indication for the impossibility of having sustained oscillatory behavior.

D.   Application of the Nyquist Criterion to Electrochemical Cells

As various examples in Sections III–VI will illustrate, in the majority of cases an electrochemical cell loses stability due to ohmic losses in the external circuitry. Since the ohmic loss represents a simple shift of the impedance plot in the complex impedance plane, we can graphically test for the possibility of having instabilities from the impedance characteristics measured potentiostatically [13].

By choosing an external resistance R_s such that the $Z(j\omega)$ plot is made to pass through the origin of the complex impedance plane, we know that the system will undergo a bifurcation. If the intersection occurs for zero frequency, it is a saddle node, if the frequency is nonzero, it is a Hopf. The galvanostatic mode is approached for $R_s \rightarrow \infty$. A saddle node under galvanostatic conditions will be found at a potential E_s (and a corresponding current I_s) for which the impedance plot $Z(j\omega)_{\omega=0} \rightarrow -\infty$, and a Hopf if $Z(j\omega)_{\omega=\omega_H} \rightarrow -\infty$ ($\omega_H \neq 0$, but $Z(j\omega)_{\omega=0} =$ finite). This is evidently similar to requiring that $Y(j\omega) = 0$, so that we can formulate the following conditions for detecting bifurcations:

For a fixed applied potential, an electrochemical cell will exhibit:

- A saddle-node bifurcation if $Z(j\omega) = 0$, $\omega = 0$.
- A Hopf bifurcation if $Z(j\omega) = 0$, $\omega = \omega_H \neq 0$.

---

[3] Strictly speaking, a linearized system does not produce a neutrally stable sine wave at the bifurcation, since the nonlinear terms are no longer negligible in such a case. The true transient behavior at the exact bifurcation should be evaluated by a nonlinear technique known as center manifold theory [11, 12]. For ease of exposition, we will ignore this subtlety and will treat a linearized system as a linear system.

For a fixed applied current, an electrochemical cell will exhibit:

- A saddle-node bifurcation if $Y(j\omega) = 0$, $\omega = 0$.
- A Hopf bifurcation if $Y(j\omega) = 0$, $\omega = \omega_H \neq 0$.

For applying these conditions, knowledge about the number of impedance poles $\#P$ inside the Nyquist contour is not necessary. What we do know is that at these bifurcations, the number of impedance zeros $\#Z$ inside the contour will change, that is, $\Delta\#Z_{SN} = \pm1$ and $\Delta\#Z_H = \pm2$.

The theory described so far may appear somewhat arcane to the uninitiated reader. A few explicit examples should be helpful. In the first—and simplest—example the impedance response and its poles and zeros of a simple model system will be calculated explicitly, to get the flavor of the method. More details on the application of frequency response methods to electrochemical systems can be found in various textbooks and specialist reviews [5–9].

> *Example 1.* Since the electrochemical system's impedance (or cell impedance) is clearly the sum of the ohmic resistance and the interfacial impedance, one has
>
> $$Z_{cell}(s) = R_s + Z_{int}(s) = \frac{R_s Y_{int}(s) + 1}{Y_{int}(s)} \qquad (2.12)$$
>
> with $s = \sigma + j\omega$.
>
> An instability occurs if for one of the zeros $\sigma = 0$, that is, $Z(j\omega) = 0$. Assume that our interfacial process is a simple Randles-type system with negligible diffusion impedance (Fig. 2.2). Then the interfacial impedance can be expressed as
>
> $$Z_{int}(s) = \frac{1}{sC_d + R_{ct}^{-1}} = \frac{1}{Y_{int}(s)} \qquad (2.13)$$
>
> where C_d is the double-layer capacity and R_{ct} the charge-transfer resistance.
>
> The cell impedance $Z_{cell}(s)$ has a zero if $R_s Y_{int} + 1 = 0$, namely,
>
> $$s = -\frac{1}{C_d}\left(\frac{1}{R_s} + \frac{1}{R_{ct}}\right) \qquad (2.14)$$
>
> which is zero if
>
> $$R_{ct} = -R_s \qquad (2.15)$$

The cell impedance has a pole for $Y_{int}(s) = 0$, that is,

$$s = -\frac{1}{C_d R_{ct}} \qquad (2.16)$$

which clearly lies inside the Nyquist contour if $R_{ct} < 0$. If the system is studied at an electrode potential for which $R_{ct} < 0$, it will exhibit the impedance plots depicted in Fig. 6 when the value of R_s is changed from a low to a high value.

For the situation that $R_s \rightarrow 0$ [Fig. 6(a)], the impedance makes one clockwise encirclement of the origin, and has one pole [given by Eq. (2.16)] so that $\#Z = 0$ and the system is stable. A change in stability occurs at the critical value $R_s = R_s^{SN}$ [Fig. 6(c)], and for $R_s > R_s^{SN}$ the system no longer encircles the origin ($\#N = 0$) but $\#P$ still equals 1 so that $\#Z = 1$ and the system is unstable.

This simple example was already treated by de Levie [13], and the result that the system is unstable if $R_s > -R_{ct}$ is fully equivalent to the result obtained by a classical linear stability analysis in the time domain.

Example 2. It seems that if we want to evaluate the system's stability it is imperative to have an a priori knowledge of a number of poles $\#P$ inside the Nyquist contour. However, if the system is potentiostatically stable, we have a reference condition in the sense of Bode [17]. In that case, $\#Z = 0$ for $R_s \rightarrow 0$, and from the Cauchy theorem we have $\#P = \#N$, where the latter follows from an inspection of the experimental impedance plot. Since the poles of $Z_{cel}(s)$ are uniquely determined by the interfacial properties [$Y_{int}(s) = 0$, see

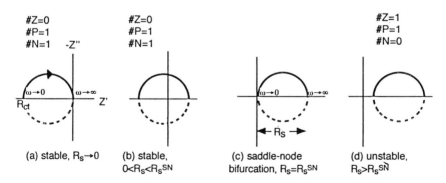

Figure 6. Destabilization in the impedance plane for a simple electrochemical process with a negative charge–transfer resistance R_{ct} by increasing the value of the external resistance R_s.

Eq. (2.12)], $\#P$ will not change as the value of R_s is increased. Therefore, we have an a priori knowledge of $\#P$ if the system is potentiostatically stable. Of course, with a standard equipment one would not be able to measure an impedance plot of a potentiostatically unstable system (although such a plot is well-defined mathematically), but it can sometimes be measured with an unconventional potentiostat with a negative output impedance, which stabilizes the steady state [20, 21].

Note that our knowledge of $\#P$ in the potentiostatic regime also no longer demands knowledge of an explicit expression for $Y_{int}(s)$, as in Example 1. So we can work directly with the impedance plot for $R_s \rightarrow 0$, without the need for a model of $Y_{int}(s)$.

Consider the impedance plot of Fig. 7(a). This is the kind of plot expected for a Randles-type system with a negative R_{ct} and concurrently negative Warburg diffusion impedances Z_W and R_W^0 [22–24] for spherical diffusion. An electrochemical reaction taking place via an adsorbed intermediate at a potential for which the charge-transfer resistance is negative, in the absence of mass-transport control, will also give a qualitatively similar-shaped impedance plot (see Section VI.A for an example). Figure 7(b–f) illustrate how

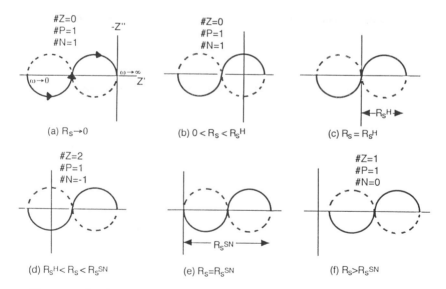

Figure 7. The destabilization process pictured in the impedance plane for an electrochemical process with an impedance plot consisting of two semicircles, with the possibility of both a Hopf and a saddle-node bifurcation.

this system is expected to undergo a Hopf and a saddle-node bifurcation, and that we can unambiguously determine $\#Z$ for every R_s, that is, we have full knowledge of the number of unstable directions of the steady state. Note also that the impedance plot is galvanostatically unstable with $\#Z = 1$, and is therefore unlikely to oscillate under galvanostatic conditions.

Example 3. We saw in Example 2 that the plot of Fig. 7 is galvanostatically unstable but not in an oscillatory fashion. Alternatively, consider the impedance plot of Fig. 8, a plot that has been obtained in a number of experimental systems (see Sections V and VI, Figs. 45 and 54). Since the plot makes two counterclockwise encirclements of the origin of $R_s \rightarrow 0$, we know that $\#P = 2$. For an external resistance $R_s > R_s^H$, the system experiences a Hopf bifurcation and becomes unstable with $\#Z = 2$, and this remains the case for $R_s \rightarrow \infty$ showing that the system is galvanostatically unstable with $\#Z = 2$.

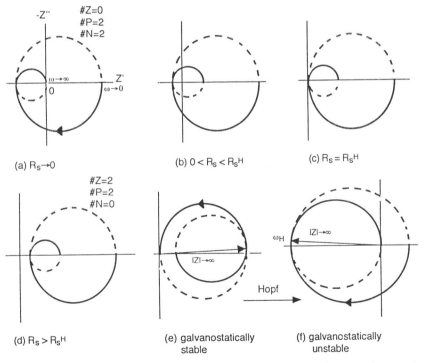

Figure 8. An example of a galvanostatically unstable impedance plot with $\#Z = 2$, illustrating [(e and f)] the process of a Hopf bifurcation in the galvanostatic mode.

In principle, $\#Z = 2$ could imply that the system is oscillatory, but in reality there may have occurred global bifurcations that the present analysis cannot detect. We already noted that a galvanostatic Hopf requires that $Y(j\omega)_{\omega=\omega_H} = 0$, or that $Z(j\omega)_{\omega=\omega_H}$ switches from $+\infty$ for $-\infty$ for some critical E_s. Therefore the transition from plot Fig. 8(e) to 8(f) must signify that a galvanostatic Hopf bifurcation has occurred. In Section V.A a simple equivalent cell circuit will be derived from electrochemical principles, which will exemplify this more concretely.

The plot of Fig. 8(a) is characteristic for galvanostatic oscillations. Note that although it exhibits a negative real impedance for a wide range of frequencies, its zero-frequency real impedance, which must equal the slope of the dc current–voltage curve, is positive. Therefore, the negative impedance cannot be read from the steady-state polarization curve, in contrast with the plots of Figs. 6 and 7: the negative impedance is "hidden." Electrochemically speaking, a plot such as Fig. 8(a) requires the coupling of at least two processes: a negative impedance process that is relatively fast, and a positive impedance process that is slower but that becomes dominant at the lowest frequencies. This already clearly shows that, mechanistically speaking, potentiostatic and galvanostatic oscillations must be different.

Example 4. As a last example we will treat a system for which it can be inferred from the fact that it is galvanostatically stable (or stable in the presence of a sufficiently large R_s) that it must be potentiostatically unstable. Figure 9 shows an impedance plot, the essential features of which are similar to an impedance plot obtained for the electrocrystallization of zinc in a Leclanché cell [25]. The entire polarization curve is galvanostatically stable, implying that $\#Z = 0$, from which it follows that for the plot of Fig. 9(b) $\#P = 0$. Therefore, in the potentiostatic control $\#Z = 1$. The latter feature is also clear from the fact that the plot has been obtained for the sandwiched branch of an S shaped I–E curve, and in fact the plot in Fig. 9(a) was extrapolated after due correction was made for the ohmic potential drop [25].

Gabrielli [14] pointed out another way of knowing that the plot of Fig. 9(a) is potentiostatically unstable. The plot is an example of an conformal mapping [10] since the direction around the mapped contour in the $Z'' - Z'$ plane [so not the $(-Z'')$–Z' plane] is the same as that around the Nyquist contour in the Laplace plane. This implies that the impedance function is analytic inside the Nyquist

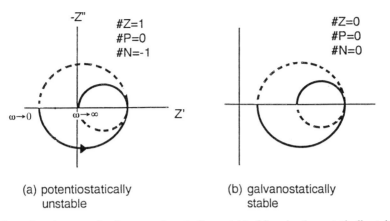

(a) potentiostatically
    unstable

(b) galvanostatically
    stable

Figure 9.   An example of a potentiostatically unstable (*a*) and galvanostatically stable (*b*) impedance plot. Note the possibility of a saddle-node bifurcation by inserting an external resistor.

contour, proving that it has no poles in the right-hand half-plane, $\#P = 0$. It then immediately follows that $\#Z = 1$.

Further examples to potentiostatically unstable impedance diagrams, and their relation to the existence of multiple steady states, can be found in Gabrielli's original papers [14].

E.   Virtues and Limitations of the Impedance Method in Stability Evaluation

This section has shown how the electrical impedance characteristics of an electrochemical cell can be related to its stability features by applying a few standard methods of electrical control engineering. In particular, a simple and unambiguous graphical test has been formulated for evaluating whether a system is going to exhibit a saddle node or a Hopf bifurcation, in both the fixed-potential and the fixed-current control. In underlining the power of this method, we can quote de Levie [13] that "this test is particularly useful since it does not require any detailed understanding of the electrode process, but is based exclusively on experimental impedance measurements."

From the point of view of electrode processes as dynamical systems, the method is useful for its ability to construct a linear stability diagram, that is, a map that draws the locations of the saddle node and Hopf bifurcations in control parameter space. It does not require the explicit search for oscillations, but merely the measurement of the system's frequency response under conditions where it is perfectly (but conditionally) stable. What could be especially interesting is that the method allows us to identify Hopfs that are difficult to identify by other methods, for instance, a subcritical Hopf the (unstable) periodic orbit of which rapidly terminates in a (global) homoclinic bifurcation. On the other hand, the impedance method remains an inherently linear technique that will never allow the prediction of the stability or the shape of the oscillations.

III. OSCILLATORY ELECTROCHEMICAL REDUCTION REACTIONS AT MERCURY

The first systematic study of current oscillations during reduction reactions at the dropping mercury electrode (DME) was carried out by Frumkin and co-workers [26, 27] in the early 1960s. During their studies of the influence of the structure of the electrical double layer on the electrode kinetics, they found that certain anion reductions ($S_2O_8^{2-}$, $PtCl_4^{2-}$, CrO_4^{2-}, $Fe(CN)_6^{3-}$, . . .) exhibit a region of negative polarization slope in their current–voltage characteristic if the concentration of supporting electrolyte is sufficiently low. Workers from Frumkin's group then established theoretically, by evaluating the charge balance equation for the electrochemical cell, the two most important conditions for observing oscillations [28, 29]: apart from a potential region with negative polarization impedance, the presence of a sufficiently large ohmic series resistor was found to be essential. About 10 years later, the subject was taken up again by de Levie [13], who became interested in the peculiar characteristics of the indium(III) reduction at the DME from thiocyanate or chloride solution, a process that was known to exhibit a negative polarization slope [30]. de Levie treated the problem of "the electrochemical oscillator" theoretically by the impedance plane method, his most important result being similar to Example 1 of Section II.C. This is however, as explained in the same section, not the most precise condition for observing spontaneous oscillations. The first real attempt to calculate explicitly the theoretical oscillation profiles by numerically solving the (simplified) kinetic differential equations was undertaken only 10 years later by Keizer and Scherson [31]. These authors based their model on a (unduly) detailed charge balance equation and a kinetic mechanism based on a scheme proposed by Pospisil and de Levie [32] for the In^{3+} reduction from SCN^- solution. Later, the Keizer–Scherson model was modified and improved to give a more transparent and realistic model [33], which allowed an extensive comparison with experiment, including an understanding of the complex and chaotic waveforms, which can be observed in these processes.

There are at least two reasons why reduction reactions at mercury could be advantageous in a systematic and fundamental study of oscillatory electrochemical cells:

1. These reactions usually feature relatively simple kinetics (at least compared to the kind of electrochemical reactions to be discussed in the remaining Sections IV–VI), which can be resolved using well-developed relaxation techniques, such as impedance voltammetry [5].

2. Mercury is an ideal electrode material with a perfectly smooth and reproducible surface. These features make the mercury electrode most suitable for developing theories of the electrode–electrolyte interface. Also, mass transfer to a stationary mercury sphere is well defined.

A. Indium(III) Reduction from Thiocyanate Solution

Current oscillations during the indium(III) reduction at the DME or HMDE (hanging mercury drop electrode) were studied by Tamamushi and co-worker [34, 35] and de Levie [13], and later in detail by Koper, Gaspard, and Sluyters [33, 36–41] in the context of dynamical systems theory. Figure 10 shows a polarogram and a true stationary-state current–voltage curve for this reaction at the HMDE. Current oscillations can occur in this system depending on the values of the applied potential V and the ohmic resistance of the external circuit R_s. Since the currents in this system are usually of the order of a few microamperes, this external resistance has to be large (several kiloohms) to induce oscillations, and should therefore be provided by an external ohmic resistor. Figure 11 depicts schematically the different stationary and periodic types of behavior of this system in the V–R_s parameter plane. One clearly recognizes the region of oscillations in this phase diagram. The region of steady-state multiplicity, bounded by the (unstable) saddle-node bifurcations, was determined by calculating the steady-state slope of the current–voltage curve, which is equivalent to determining the impedance at zero frequency (at a HMDE) as outlined in Section II. It is illustrative to show the difference in behavior at two values of R_s, one giving rise to oscillations and the other to bistability, by making (cyclic) voltammograms at low scan rate. Figure 12 shows two such curves (recorded under conditions different from those of Fig. 11). Figure 12(a) shows how for relatively small R_s the negative slope region is accompanied by current oscillations; for a higher value of R_s as in Fig. 12(b) a hysteresis between the two bistable steady states is observed. For the same conditions as the experiments of Fig. 12, the locus of the Hopf bifurcation was also

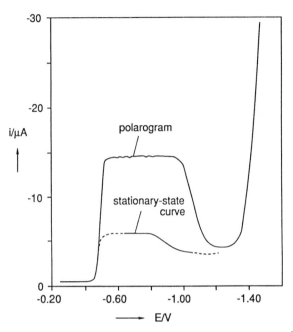

Figure 10. Current–voltage curve without external resistance for the In^{3+} reduction at the HMDE in the presence of thiocyanate; 11.3 mM In^{3+}, 5.0 M NaSCN, pH 3.0, $T = 5.0°C$. Drop time polarogram 10 s. (Reproduced from [41] with permission of the American Institute of Physics).

determined by the impedance method. Figure 13(a) shows a typical impedance plot from which the critical value R_s^H is extrapolated; in Fig. 13(b) the solid line represents the Hopf locus as determined from such plots at 15 different potentials. Clearly, the Hopf bifurcation determined in this way agrees well with the onset of oscillations when an external resistor is connected in series with the working electrode, as represented by the dots in Fig. 13(b). Note that, in contrast with Fig. 11, the conditions under which Fig. 13(b) was determined give rise to a crossing of the Hopf locus, whereas within the borders of Fig. 11 such a crossing is not observed. This "self-crossing" is indeed predicted by kinetic models to be explained below, and is a good example of the usefulness of the impedance method in determining the Hopf locus, because the models predict that the Hopf bifurcation near the crossing is subcritical and the oscillations are therefore unstable and thus unobservable.

As indicated in Fig. 11, there are large regions in the parameter plane where oscillations occur more complex than simple harmonic- or relaxation-type waveforms. These complex oscillations follow intricate but

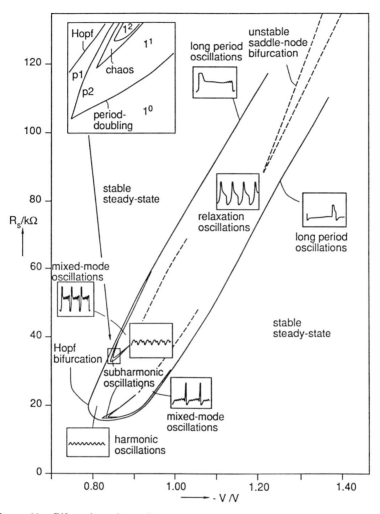

Figure 11. Bifurcation phase diagram for the In/SCN system where regions of stationary, oscillatory or complex behavior are mapped onto the V–R_s plane. Cell conditions as in Fig. 10. (Reproduced from [41] with permission of the American Institute of Physics.)

well-defined bifurcation sequences. In the largest part of the parameter plane, these complex oscillations appear as so-called mixed-mode oscillations (MMOs): waveforms consisting of alternating small and large amplitude excursions. To illustrate the typical appearance and evolution of these MMOs, Figs. 14 and 15 show the experimentally observed oscillatory responses of the system as the value of the external resistance

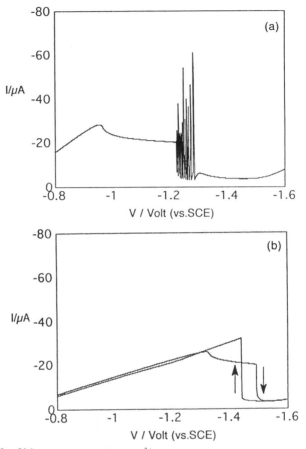

Figure 12. Voltammograms of the In^{3+} reduction on a HMDE from thiocyanate solution in the presence of the external resistor R_s. (a) $R_s = 10$ kΩ, (b) $R_s = 30$ kΩ. 9 mM In^{3+}, 5 M NaSCN, pH 2.9, $T = 25°C$.

R_s is changed from low to high value, at a fixed value of the applied potential V. At low R_s, the system is in a stable steady state of low current. At the critical value of R_s (in the case of Fig. 14 this was ~8.4 kΩ), the system exhibits a Hopf bifurcation and starts oscillating with small amplitude. At a slightly higher value of R_s (8.5 kΩ), the response changes into a high-period MMO, where a large number of growing oscillations are interspersed with strong relaxation spikes. These states are mainly periodic, and can therefore be designated with a symbolic notation L^S, where L signifies the number of large amplitudes and S is the number of small amplitudes in a periodic entity. The state in

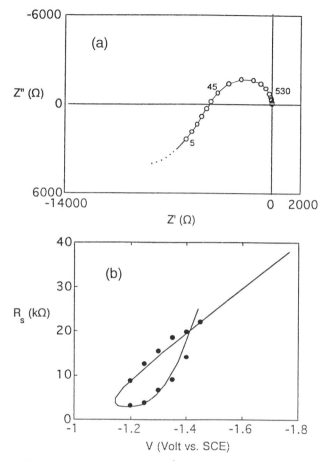

Figure 13. (a) Impedance diagram for the In^{3+} reduction on a HMDE from thiocyanate solution at $E = -1.075$ V vs. SCE. Note that this particular steady state exhibits a Hopf bifurcation for $R_s = 4.2$ kΩ and a corresponding $V = E + IR_s = -1.15$ V vs. SCE, where I is the steady-state current. (b) Line of Hopf bifurcations determined by the impedance method as explained under (a). Dots represent the onset of oscillations as observed by inserting an external resistor.

Fig. 14(c) is a 1^1 MMO, that in Fig. 14(d) is a 2^1. The number of small oscillations decreases on increasing the external resistance, until they finally disappear and the "phase" of purely monoperiodic relaxation oscillations is reached. However, there is a new sequence of MMOs for $R_s > 24.0$ kΩ, the 1^4 and 1^5 states of which are shown in Figs. 15 (a and b) (they are of course preceded by the 1^1, 1^2, and 1^3). Again, the sequence terminates by a sudden transition to smaller amplitude oscilla-

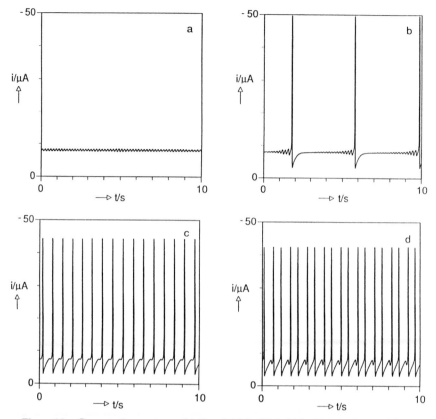

Figure 14. Current–time series at (*a*) $R_s = 8.4$ kΩ, (*b*) 8.5 kΩ, (*c*) 9.5 kΩ, and (*d*) 10.0 kΩ. Here $V = -0.906$ V, 7.3 mM In^{3+}, 5.0 M NaSCN, pH 3.0, $T = 5.0°$C. [Reproduced with permission from M. T. M. Koper, P. Gaspard, and J. H. Sluyters, *J. Phys. Chem.*, **96**, 5674 (1992). Copyright © 1992 American Chemical Society.]

tions, which become extinct for $R_s > 26.1$ kΩ, through another Hopf bifurcation.

Bifurcation scenarios of MMOs have received quite a lot of theoretical attention, because they have been observed in a wide variety of chemical systems. Further examples will be encountered at various places in this chapter. In general, there seem to be two distinguishable sequences: a periodic–chaotic sequence in which the transition from the 1^n to the 1^{n+1} state occurs via chaotic mixtures established through a period-doubling mechanism or a so-called tangent bifurcation, or a Farey sequence where the transition is accompanied by periodic mixtures of adjacent states, carrying symbolic notations such as $1^1 2$, a $1^2 (1^1)^2$, or 2^1, or 3^1, where the

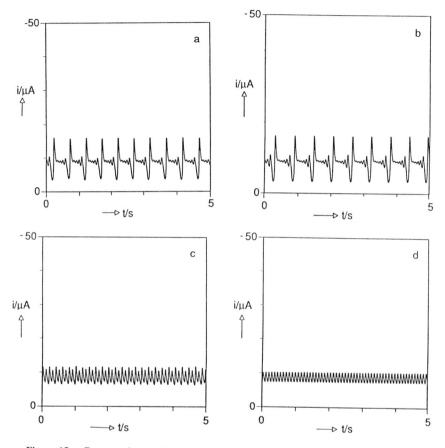

Figure 15. Current–time series at (a) $R_s = 25.2$ kΩ, (b) 25.5 kΩ, (c) 25.8 kΩ, and (d) 25.9 kΩ. Here $V = -0.906$ V, 7.3 mM In^{3+}, 5.0 M NaSCN, pH 3.0, $T = 5.0°$C. [Reproduced with permission from M. T. M. Koper, P. Gaspard, and J. H. Sluyters, *J. Phys. Chem.*, **96**, 5674 (1992). Copyright © 1992 American Chemical Society.]

latter two can be considered as periodic mixtures of a 1^1 and a 1^0. Such states as the $1^1 1^2$ and 3^1 have been observed in the indium–thiocyanate system [41], but more complete examples of these two bifurcation sequences have been reported for the electrodissolution of copper in phosphoric acid, to be discussed in Section IV.B. A more thorough discussion of the bifurcational origin of these MMOs will be deferred to the Sections III.B and III.D when we will discuss both an electrochemical as well as a geometrical model exhibiting these waveforms.

In general, the periodic–chaotic sequence becomes the dominant one for low values of V and R_s (as shown in the enlargement in Fig. 11), and therefore some clear-cut examples of deterministic chaos are best taken in this region. In Fig. 16 three typical time series from this region are shown: a period-two state that has bifurcated from the monoperiodic state via a period-doubling, exhibiting the period-doubling cascade to a deterministic chaotic state. A change in the control parameter R_s makes

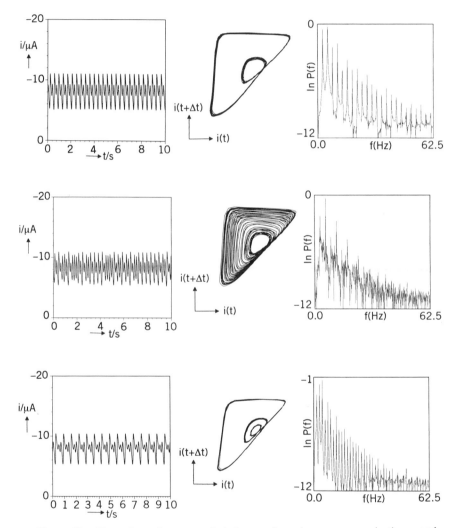

Figure 16. Illustration of some typical time series, phase space projections, and frequency spectra displaying a Feigenbaum-type bifurcation scenario. Here $V = -0.878$ V. From top to bottom: $R_s = 32.65$, 32.7, and 32.8 kΩ. Cell conditions as in Fig. 10. (Reproduced from [41] with permission of the American Institute of Physics.)

the chaotic state undergo a tangent bifurcation into a period-three state. Figure 16 also illustrates the phase–space projections and the frequency spectra of these three states. Evidence for the tangent bifurcation is given in Fig. 17. Figure 17(a) gives the next-minimum map of the chaotic state in Fig. 16, having approximately the familiar quadratic shape of the logistic mapping $x_{n+1} = ax_n(1 - x_n)$. When the third-next minimum is plotted against the minimum, Fig. 17(b) is obtained, which is clearly becoming tangent to the bisectrix, eventually getting trapped by the period-three state in Fig. 17(c).

B.   Model

As was already pointed out in Section II, the time evolution of an electrochemical system is governed not only by the mass balance of the different chemical species, but also by a charge balance equation. In the

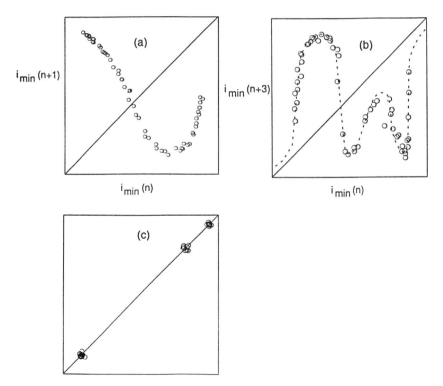

Figure 17.   (a) Next-minimum map of the chaotic time series of Fig. 16. (b) Third-next minimum map of the chaotic time series of Fig. 16, illustrating the tangent bifurcation to (c) third next minimum map of the period-three state from Fig. 16. (Reproduced by permission of the American Institute of Physics from [41].)

time domain, this charge balance is expressed as (see Fig. 1)

$$I = \frac{V - E}{R_s} = I_C + I_F = AC_d\frac{dE}{dt} + Aj_F \tag{3.1}$$

or

$$C_d\frac{dE}{dt} = \frac{V - E}{AR_s} - j_F(E) \tag{3.2}$$

where I denotes current, j current density, and A the electrode surface area.

A linear stability analysis of this equation will result in the same instability condition [38] that was established in Section II.C, namely,

$$Z_F < 0, |Z_F| < R_s \tag{3.3}$$

where Z_F is the steady-state polarization slope or the faradaic impedance (at zero frequency) $Z_F^{-1} = (dI_F/dE)_{E=E_{ss}}$. From the well-known expression for the faradaic current,

$$I_F = nFAk_f(E)c_0 , \tag{3.4}$$

(disobeying the usual sign convention for generality), where n and F have their usual meaning, $k_f(E)$ is the heterogeneous electron-transfer reduction rate constant and c_0 is the surface concentration of the electroactive species, it follows that for these simple reactions the negative polarization slope can have three different causes: (1) a negative dA/dE, (2) a negative $dk_f(E)/dE$, and (3) a negative dc_0/dE.

Cause (1). A negative impedance can develop when the effective electrode surface area decreases with increasing polarization. The potential-dependent formation of an ideally inhibiting film, such as in metal passivation, could be considered as an example of this type of negative impedance.

Cause (2): (a) Potential-Dependent Adsorption of an Inhibitor. An example of this category is the adsorption of an (organic) surface-active agent within the faradaic or diffusion-controlled potential region of the reduction reaction. One can either assume the adsorbed substance to totally block the surface, in which Case (1) as discussed above applies, or to increase the activation energy for electron transfer, so that Case (2) applies. The inhibitor is normally an organic substance that adsorbs in a certain potential region. For an uncharged organic, this region is often around the electrode's potential of zero charge. For a region of negative impedance to occur in this specific case, the half-wave potential of the reduction reaction should be more positive than the potential of zero charge.

Cause (2): (b) Potential-Dependent Adsorption of a Catalyst. The classical example of this type of negative impedance is the indium–thiocyanate oscillator. The idea is based on the fact that certain irreversible reduction reactions can be made reversible by the addition of halides (Cl^-, Br^-, or I^-) or halide-like ions (e.g., SCN^-). Their presence results in a substantial shift of the polarographic wave towards the reversible half-wave potential. Because these ions adsorb specifically at the mercury–water interface, it is assumed that this catalysis is furnished by a surface reaction [32]. A simple explanation would be a kind of chemical complexation that facilitates the electron transfer (see below). The catalyzing anions will be desorbed from the surface at sufficiently negative potentials, resulting in a loss of catalysis. According to the tables of Heyrovsky and Kuta [42], the shift in half-wave potential has been reported for In(III)/In in Cl^-, Br^-, and I^- solutions, and for Ni(II)/Ni in SCN^- solution. For the In case, the negative impedance is found to be most pronounced in the presence of Cl^- [22] or SCN^- [23].

Cause (3): Electrostatic Effect at Low Ionic Strength. According to a classical theory of Frumkin [43], Eq. (3.4) must take into account the effect of the electrical double layer. The effective driving force for the electrochemical reaction is not E, but $E - \phi_r$, where ϕ_r is the potential in the reaction plane. The surface concentration must also be corrected by a Boltzmann factor:

$$c_0 = c_0^* \exp(-ze\phi_r/kT) \qquad (3.5)$$

where c_0^* is the concentration just outside the electrical double layer and z is the valency of the electroactive species. As $\partial\phi_r/\partial E$ will almost always be positive, it follows that dc_0/dE can become negative if z is negative for a reduction reaction (i.e., anion reduction) or if z is positive for an oxidation reaction (i.e., cation oxidation). Because ϕ_r tends to zero with increasing ionic strength, the observation of this type of negative impedance requires a low concentration of supporting electrolyte. Examples of negative slopes in anion reduction are abundant, and oscillatory behavior was already reported by Frumkin et al. [26]. More quantitative conditions for observing oscillations in this type of systems have been worked out by Wolf et al. [44].

The occurrence of spontaneous current oscillations in reduction reactions at the HMDE with a negative polarization slope can be rationalized by accounting for the role of the mass transport, which in the presence of an excess of supporting electrolyte is predominantly taken care of by diffusion. We start with the exact equations and will then utilize the concept of the Nernst diffusion layer to obtain a mathematically tractable set of equations for which one can prove the existence of a limit cycle (i.e., spontaneous oscillatory solution). This simple model does not allow for a curvature of the concentration profile in the solution, so that as a next step a slightly refined three-variable model will be derived that will give an explanation for the more complex mixed-mode and chaotic oscillations.

For a spherical electrode, the concentration of the electroactive species as a function of the radial coordinate r and time t, denoted $c(r, t)$, is given by Fick's second law for semiinfinite spherical diffusion:

$$\frac{\partial c}{\partial t} = D \left(\frac{\partial^2 c}{\partial r^2} + \frac{2}{r} \frac{\partial c}{\partial r} \right) \qquad (3.6)$$

with D diffusion coefficient, subject to a constant boundary condition at infinity,

$$c(r = \infty, t) = c_{\text{bulk}} \qquad (3.7)$$

and Fick's first law as a boundary condition at the drop's surface $r = a$:

$$D \left(\frac{\partial c}{\partial r} \right)_{r=a} = \frac{j_F}{nF} \qquad (3.8)$$

(We do not worry too much about sign conventions.)

The solution for the stationary state of this diffusion problem reads as

$$c(r) = c_{\text{bulk}} - \frac{a}{r} \left(c_{\text{bulk}} - c_0 \right) \qquad (3.9)$$

with $c_0 = c_{r=a}$.

From Eq. (3.9) one obtains an expression for the Nernst diffusion layer thickness

$$\delta = \frac{c_{\text{bulk}} - c_0}{(\partial c / \partial r)_{r=a}} = a \qquad (3.10)$$

The diffusion-layer model tells us that, for the stationary state, the flux at the interface is proportional to the difference in the bulk and surface concentrations, with the drop radius as a proportionality factor. So the concentration profile can just as well be represented by a linear concentration profile connecting c_0 at $r = a$ and c_{bulk} at $r = 2a$, and a constant profile c_{bulk} from $r = 2a$ on.

For the sake of a simple explanation of oscillations at the HMDE, it is opportune to assume that the concentration profile in the diffusion layer can be represented at all times by Eq. (3.9) or its diffusion-layer equivalent as well when the system is not in a stationary state. An equation for the time evolution of c_0 can then be obtained by evaluating the various contributions to the mass balance, leading to [33]:

$$\frac{dc_0}{dt} = \frac{2}{a}\left\{-k_f(E)c_0 + \frac{D(c_{bulk} - c_0)}{a}\right\} \tag{3.11}$$

where δ has been put equal to a. The analysis is further simplified by introducing the dimensionless variables:

$$e = \frac{F}{RT}E \quad u = \frac{c_0}{c_{bulk}} \quad \frac{2D}{a^2}t \to t$$

(t denoting dimensionless time from now on), to obtain the dimensionless versions of Eqs. (3.1) and (3.11):

$$\varepsilon\frac{de}{dt} = \frac{v - e}{r} - k(e)u$$

$$\frac{du}{dt} = -k(e)u + 1 - u \tag{3.12}$$

where

$$v = \frac{F}{RT}V \quad \varepsilon = \frac{2RTC_d}{nF^2c_{bulk}a} \quad r = \frac{4\pi nF^2Dc_{bulk}a}{RT}R_s \quad k(e) = \frac{a}{D}k_f(E)$$

The dimensionless electrical current is $i = (v - e)/r$. Note that $0 < \varepsilon \ll 1$ because the potential is expected to relax on a much faster time scale than u. In this form $k(e) \gg 1$ indicates diffusion control, and $k(e) \ll 1$ kinetic control.

Equations (3.12) can be studied without any a priori assumption about the explicit form of $k(e)$ applying to a specific mechanism. The stationary state(s) of Eqs. (3.12) is (are) the intersection(s) of the e- and u-nullclines, that is, $de/dt = 0$ and $du/dt = 0$:

$$u = 1/(1 - k(e)) \quad \text{(the } u\text{-nullcline)}$$

$$u = (v - e)/rk(e) \quad \text{(the } e\text{-nullcline)}$$

These lines are conveniently studied in the $e-u$ phase plane (see Fig. 18).
The following result can be proved by straightforward application of
standard stability techniques: If, in the limit $\varepsilon \to 0$, the intersection occurs
on a part of the e-nullcline with negative (positive) slope, that is,
$de/du < 0$ $(de/du > 0)$, then the resulting steady state is always stable
(unstable). The potentially unstable situation $de/du > 0$ can only occur if
a region of negative $dk(e)/de$ exists. If the unstable steady state is the
only steady state, then a limit cycle (spontaneous oscillatory solution) will
develop around it.

The stability of the stationary states follows from the time evolution of

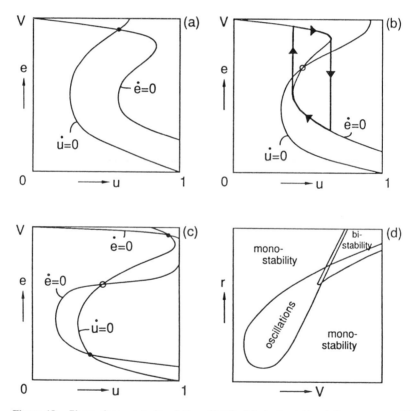

Figure 18. Phase-plane portraits of Eqs. (3.12): (*a*) single stable stationary state; (*b*)
unstable stationary state with corresponding limit-cycle oscillation; (c) two stable (filled
circles) and one stable (open circle) stationary states (bistability); (*d*) existence regions for
situations (*a*), (*b*),and (*c*) in the *v*–*r* parameter plane. (Reproduced from [45] with
permission of Elsevier Sequoia S.A.)

the small perturbations δe and δu, according to

$$\frac{d}{dt}\begin{pmatrix} \delta e \\ \delta u \end{pmatrix} = \begin{pmatrix} -\varepsilon^{-1}\left(\dfrac{1}{r} + u\dfrac{dk(e)}{de}\right) & -\varepsilon^{-1}k(e) \\ -u\dfrac{dk(e)}{de} & -1 - k(e) \end{pmatrix}\begin{pmatrix} \delta e \\ \delta u \end{pmatrix} \tag{3.13}$$

The eigenvalues of the Jacobian matrix in Eq. (3.13) are given by

$$\lambda^2 + T\lambda + \Delta = 0 \tag{3.14}$$

where

$$T = 1 + k(e) + \varepsilon^{-1}\left(\frac{1}{r} + \frac{1}{1 + k(e)}\frac{dk(e)}{de}\right)$$

$$\Delta = \varepsilon^{-1}\left(\frac{1 + k(e)}{r} + \frac{1}{1 + k(e)}\frac{dk(e)}{de}\right)$$

The stationary state undergoes a Hopf bifurcation if $T = 0$, $\Delta > 0$, and a saddle-node bifurcation if $\Delta = 0$. Consequently, stability of the stationary state is ensured when $T > 0$, unless $\Delta < 0$ giving a saddle fixed point. Instability is ensured when $T < 0$, and also when $T > 0$ and $\Delta < 0$ (simultaneously). The condition $T > 0$ gives

$$-\frac{dk(e)}{de} < \frac{1 + k(e)}{r} + \varepsilon(1 + k(e))^2 \tag{3.15}$$

and $\Delta < 0$ requires

$$-\frac{dk(e)}{de} > \frac{(1 + k(e))^2}{r} \tag{3.16}$$

In the limit $\varepsilon \to 0$ these conditions can never be satisfied simultaneously because $k(e) > 0$.

In the e–u phase plane, the e-nullcline (i.e., $de/dt = 0$) is given by

$$u = \frac{v - e}{rk(e)} \tag{3.17}$$

The slope du/de of the nullcline is negative when

$$-\frac{dk(e)}{de} < \frac{k(e)}{v - e} = \frac{1 + k(e)}{r} \tag{3.18}$$

which is equivalent to Eq. (3.15) for $\varepsilon = 0$, showing that, in this limit, a stationary state situated on a part of the e-nullcline with negative (positive) slope is always stable (unstable).

To prove the second part of our "theorem," that is, the existence of a limit cycle in the case of only a single unstable stationary state, we can use the Poincaré–Bendixson theorem [11, 12]. First, we construct a closed curve Σ_1 around the fixed point where the flow normal to the curve is always directed outwards (away from the fixed point) (see Fig. 19). Its construction is trivial as we can simply choose a circle of very small diameter around the fixed point. Next, we construct a second closed curve Σ_2 somewhere in the phase plane where the flow normal to the curve is always directed inwards. Then, if there is no fixed point in the region between Σ_1 and Σ_2 (this condition can be satisfied, for it is readily confirmed that it is possible to have only one unstable fixed point), the Poincaré–Bendixson theorem ensures the existence of at least one stable limit cycle in this region. An example of a closed-curve Σ_2 is shown in Fig. 19, bounded by $e = v$, $u = 1$, $u = 0$, and $e = e_{eq}$. A flow that is directed outwards on every part of this curve can be anticipated to be nonphysical by noting that it would bring us into conflict with the second law of thermodynamics on the various parts of the Σ_2 curve [i.e., negative diffusion coefficients, negative rate constant, negative (integral) resistance].

Three typical situations of the phase plane are illustrated in Fig. 18. Figure 18(a) shows a single stable stationary state, Fig. 18(b) shows a single unstable stationary state for which e and u will oscillate approximately following the arrowed curve, and Fig. 18(c) shows the existence of three stationary states, two of which are stable and the other unstable. The regions for the existence of these three situations can be drawn in the $v - r$ parameter plane, resulting in a kind of "fish-shaped" diagram

Figure 19. The two closed curves Σ_1 and Σ_2 in the $e-u$ phase plane. If a potential v is applied, which lies cathodic with respect to the equilibrium potential of the only redox couple present in the solution, then the normalized surface concentration u always lies between 0 and 1, and the electrode potential between v and e_{eq}; hence the phase plane flow at the border of Σ_2 should always be directed as the arrows indicate. Since it is possible to have a situation with only one unstable node or focus (inside Σ_1) [see Fig. 18(b)], there is no fixed point in the region between Σ_1 and Σ_2, and at least one stable limit cycle exists.

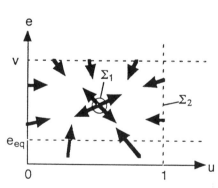

shown in Fig. 18(d), in agreement with experimental parameter planes of Figs. 11 and 13(b).

Although this simple model explains the occurrence of periodic oscillations, it cannot explain the more complex mixed-mode and chaotic oscillations that are often observed because this type of behavior requires at least three variables. In the next paragraphs a model will be built that removes this shortcoming.

To arrive at a three-variable model, an improvement of the linear diffusion-layer concept adopted in Eq. (3.11) has been suggested [37, 39]. As mentioned, this concept assumes that the concentration profile in the electrolyte always relaxes immediately to its corresponding (linear) stationary-state profile. Obviously, there must be a delay associated with this process when the boundary condition is oscillating, resulting in a time-dependent shape of the concentration profile. A simple way in which this can be accounted for qualitatively, without having to solve the exact partial differential equation (i.e., Fick's second law), is by the intro-duction of a second diffusion layer between the bulk and the electrode's surface, so that c_1 represents the concentration in the layer $a < r < 3a/2$, and c_2 the concentration in the layer $3a/2 < r < 2a$. A detailed balance of this geometry will lead to the following two equations [39]:

$$\frac{dc_1}{dt} = -\frac{24k_f(E)}{19a}c_1 + \frac{108D}{19a^2}(c_2 - c_1)$$

$$\frac{dc_2}{dt} = \frac{12D}{37a^2}(16c_{\text{bulk}} - 25c_2 + 9c_1)$$

$$(3.19)$$

Again, we define the dimensionless variables:

$$e = \frac{F}{RT}E \qquad u = \frac{c_1}{c_{\text{bulk}}} \qquad w = \frac{c_2}{c_{\text{bulk}}} \qquad \frac{108D}{19a^2}t \to t$$

to obtain the dimensionless vector field:

$$\frac{de}{dt} = \frac{v - e}{r} - mk(e)u$$

$$\frac{du}{dt} = -k(e)u + w - u \qquad (3.20)$$

$$\frac{dw}{dt} = \frac{19}{333}(16 - 25w + 9u)$$

where

$$v = \frac{F}{RT} V \qquad m = \frac{19nF^2 c_{\text{bulk}} a}{24RTC_{\text{d}}} \qquad k(e) = \frac{2a}{9D} k_{\text{f}}(E) \qquad r = \frac{432}{19} \pi D C_{\text{d}} R_{\text{s}}$$

Note that in order to arrive at these equations it was not necessary to introduce any new parameter with respect to the two-variable model Eqs. (3.12). The nondimensionalization of Eqs. (3.20) is slightly different from the somewhat more elegant scaling of Eqs. (3.12). The simulations to be described below, however, were originally carried out with the version given in Eqs. (3.20). (The peculiar coefficients in Eqs. (3.20) originate from the spherical geometry; had we chosen a linear geometry with a bounded diffusion layer, the coefficients for x, y, and z in the y-differential equation would have had the more familiar coefficients 1, -2, and 1, respectively.) It is important to remark that this "truncated" form of Fick's second law does not introduce any artifacts in the sense that qualitatively the same behavior as described below is obtained in a more realistic many-box description of the diffusion process [39].

In order to obtain an explicit expression for $k(e)$, a kinetic mechanism for the reduction of indium(III) in the presence of thiocyanate, which has been proposed by Pospisil and de Levie [32], can be used. Their mechanism consists of two successive heterogeneous steps:

$$\begin{aligned} \text{In}^{3+} + 2\text{SCN}^-_{\text{ads}} &\to \text{In(SCN)}^+_{2,\text{ads}} \qquad \text{(slow)} \\ \text{In(SCN)}^+_{2,\text{ads}} + 3e^- &\to \text{In}^0 + 2\text{SCN}^-_{\text{ads}} \qquad \text{(fast)} \end{aligned}$$

$$(3.21)$$

The rate-determining step of this mechanism is a purely chemical one, so that the potential dependence of its rate will only enter through the potential-dependent surface excess of the thiocyanate that is specifically adsorbed onto the mercury surface. If we assume the uncatalyzed, slow reduction of indium also to proceed, unaffected by the presence of SCN^-, an expression for the dimensionless rate constant $k(e)$ can be written:

$$k(e) = k_1 \theta^2 + k_2 \exp[n\alpha(e - e^\circ)] \qquad (3.22)$$

with e° denoting the dimensionless standard potential and α the transfer coefficient. The influence of all back reactions is neglected. The relative thiocyanate coverage θ will equal 1 at low e, due to its specific adsorption onto the electrode, but will tend to zero at higher e, as a result of Coulombic repulsion [46]. This may be modeled phenomenologically in a purely functional way by

$$\theta = \begin{cases} 1 & \text{for } e \le e_d \\ \exp[-b(e - e_d)^2] & \text{for } e > e_d \end{cases} \tag{3.23}$$

Any other sigmoidally shaped function would be equally appropriate, of course. Note that with Eqs. (3.22) and (3.23) it is assumed, for simplicity, that the uncatalyzed indium reduction is not affected by the prevailing thiocyanate coverage and that this coverage immediately relaxes to its potential-dictated value without any time delay (this is quite reasonable for the high concentration of the thiocyanate in the experiments).

The crucial property of Eqs. (3.22) and (3.23) lies in the fact that they give rise to a negative $dk(e)/de$ in some potential interval, that is, a negative impedance. In fact, one could regard Eqs. (3.22) and (3.23) as a purely functional expression of $k(e)$, without making any connection to the underlying chemistry (Eq. 3.21). Its shape is simply modeling a potentially unstable electrochemical reaction rate, irrespective of the molecular mechanism involved.

It turns out that Eqs. (3.20), (3.22), and (3.23) can explain the experimentally observed waveforms and their bifurcation scenarios in qualitative detail. The linear stability diagram is still of the fish-shaped (or cross-shaped) type, as in Fig. 18(d), although it now distinguishes between fixed points with only real eigenvalues (λ_1, λ_2, λ_3) and fixed points with one real and one pair of complex conjugate eigenvalues ($\rho \pm i\omega$, λ).

The typical solutions of the model are probably best illustrated by making a representative section of the parameter plane, similar to the experiment in Figs. 14 and 15. To this end, v is fixed and the value of the dimensionless resistance r is varied from low to high value. The exact parameter values can be found in the caption of Fig. 20.

At low r, the model predicts a stable stationary state, whereas for some critical value of r, the stationary state exhibits a Hopf bifurcation and the current will oscillate with small amplitude [Fig. 20(a)]. At a slightly higher value of r, the small-amplitude oscillation becomes a mixed-mode oscillation (MMO); some typical MMO states are depicted in Fig. 20(b–d). With increasing r, the following sequence of MMOs is observed, employing the previously introduced symbolic notation:

$$1^{21}, 1^{20}, \ldots, 1^3, 1^3(1^2)^n, 1^2, 1^2(1^1)^n, 1^1, 1^1(1^0)^n, 1^0$$

with n going from 10 to 12 in the numerical computation. In the right-hand side of the sequence, two neighboring states show inter-

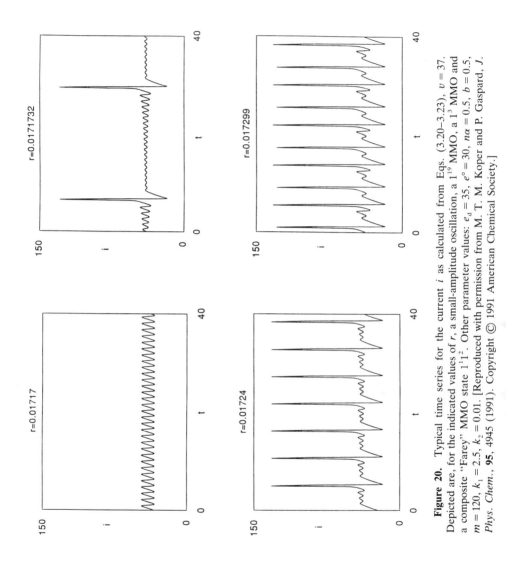

Figure 20. Typical time series for the current i as calculated from Eqs. (3.20–3.23), $v = 37$. Depicted are, for the indicated values of r, a small-amplitude oscillation, a 1^{19} MMO, a 1^3 MMO and a composite "Farey" MMO state 1^11^2. Other parameter values: $e_d = 35$, $e^o = 30$, $n\alpha = 0.5$, $b = 0.5$, $m = 120$, $k_1 = 2.5$, $k_2 = 0.01$. [Reproduced with permission from M. T. M. Koper and P. Gaspard, *J. Phys. Chem.*, **95**, 4945 (1991). Copyright © 1991 American Chemical Society.]

mediate periodic states that are concatenations of the two principal states. In the left-hand part, the windows of stability of these Farey states are so small that they are not observed. These MMO concatenations can be characterized by the so-called firing number F, which is defined as $F = S/(L + S)$. If the firing number of the concatenated states in between two neighboring principal states is plotted as a function of the control parameter r, Fig. 21 is obtained. The picture obtained is known as a devil's staircase, although here it is incomplete because only certain states are allowed, whereas for a complete devil's staircase, such as are associated with iterative maps on the circle, there is hierarchy of Farey states in between all other states, including nonprincipal ones, giving it a fractal structure.

For intermediate values of r, the model possesses a stable relaxation oscillation 1^0. This is the dominant waveform in the model. At higher r, another MMO sequence is observed, just as in the experiment, but now both chaotic and periodic states can be discerned. Apparently, the contractivity during reinjection is weaker here than in the MMO sequence at low r. Some typical time series are shown in Fig. 22. Generally, periodic states turn into chaotic ones through a cascade of period doublings, but for the chaotic states, both a series of period halvings and a tangent bifurcation can be the cause of their disappearance. Figure 22

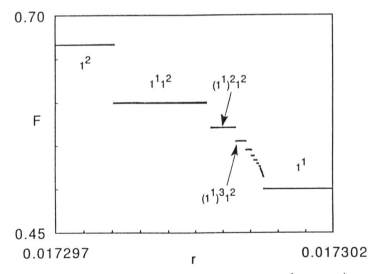

Figure 21. The firing number F of the MMO states between 1^2 and the 1^1 given as a function of the control parameter r. Other parameter values as in Fig. 20. (Reproduced from [39] with permission of the American Institute of Physics.)

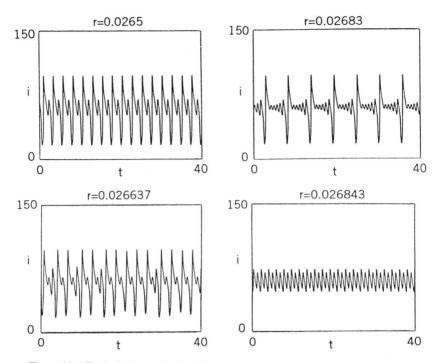

Figure 22. Typical time series for the current i as calculated from Eqs. (3.20–3.23), $v = 37$. Depicted are, for the indicated values of r, a 1^1 MMO, a 1^5 MMO, a chaotic MMO state in between the 1^1 and 1^2, and period-doubled small-amplitude oscillation. Other parameter values as in Fig. 20. [Reproduced with permission from M. T. M. Koper and P. Gaspard, *J. Phys. Chem.*, **95**, 4945 (1991). Copyright © 1991 American Chemical Society.]

shows a chaotic oscillation in between a 1^1 and a 1^2 state. The next-minimum maps and the phase-space trajectories of the chaotic states compare favorably with experiment [41]. The MMO sequence again suddenly terminates, giving way to smaller-amplitude states. A period-two small-amplitude oscillation is shown in Fig. 22. At high r, there is another Hopf bifurcation through which the stationary state is stabilized again. Figures 23(a and b) show the complicated parameter dependence of the various types of dynamical behavior, that is, steady-state, simple oscillations, MMOs, chaos and so on, in the v–r parameter plane.

The above scenario is clearly similar to that observed experimentally (cf. Figs. 14, 15–20, and 22). The same statement holds for the bifurcation diagram in the v–r plane.

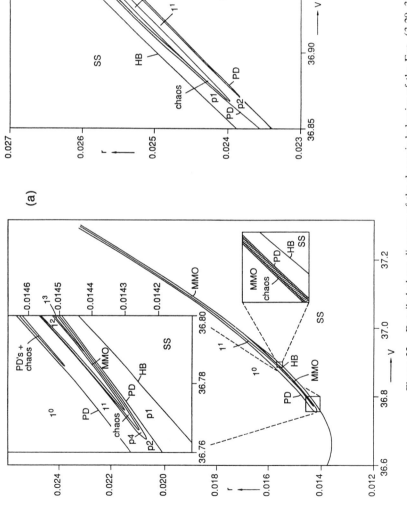
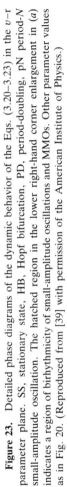

Figure 23. Detailed phase diagrams of the dynamic behavior of the Eqs. (3.20–3.23) in the v–r parameter plane. SS, stationary state; HB, Hopf bifurcation; PD, period-doubling; pN period-N small-amplitude oscillation. The hatched region in the lower right-hand corner enlargement in (a) indicates a region of birhythmicity of small-amplitude oscillations and MMOs. Other parameter values as in Fig. 20. (Reproduced from [39] with permission of the American Institute of Physics.)

205

C. Other Reduction Reactions

The model presented in Section III.B should in principle apply to all reduction reactions that exhibit a region of negative slope in their current–potential characteristic under mixed diffusion and activation control. Although the model was worked out for the explicit example of the reduction of In^{3+} from thiocyanate solution at a spherical electrode, it will be clear that any other reduction reaction having a similarly shaped $k(e)$ function should show an entirely similar behavior [provided there are good reasons to assume that $k(e)$ is time independent]. This concerns in particular the complex mixed-mode oscillations, since these result from a perfectly general extension of the mass-transport description that should be the same for all systems.

The prediction that all oscillatory reduction reactions at mercury should exhibit the same type of mixed-mode oscillations is indeed corroborated by experiment [45]. Although the stability of other systems varies from quite good to rather poor, all systems that have been studied show mixed-mode oscillations that never behave in contradiction with model predictions. These systems include the reduction of Cu^{2+} and Bi^{3+} in the presence of organic surface-active substances, the In^{3+} reduction from Cl^- medium and the Ni^{2+} reduction from SCN^- solution, and the reduction of the anions $PtCl_4^{2-}$ and $S_2O_6^{2-}$ from unsupported or poorly supported solutions. Figure 24 exemplifies some typical oscillatory states for the $PtCl_4^{2-}$ reduction.

D. The Mathematical Origin of Mixed-Mode Oscillations

The bifurcation sequences of MMOs described in Sections III.B and III.C are not only observed in simple electrochemical reduction reactions on mercury. There are several other electrochemical systems that exhibit very similar behavior, like the copper electrodissolution in phosphoric acid (see Section IV.B), the hydrogen oxidation on platinum (see Section V.A, although in that system the sequences are in fact slightly different), the zinc electrodeposition [47], and a few others [48]. Furthermore, they occur in other oscillating chemical reactions [1], such as the Belousov–Zhabotinskii reaction [49–51], gas-phase reactions [52], and heterogeneous catalysis [53, 54].

It is therefore not surprising that many authors have attempted to give a more general mathematical interpretation of these complex waveforms and their bifurcation sequences. Although a full discussion of the subject will not be given here, it is relevant to summarize some of the main ideas since these MMOs are such a recurrent phenomenon in the dynamics of nonlinear electrochemical kinetics. Almost all early interpretations of the

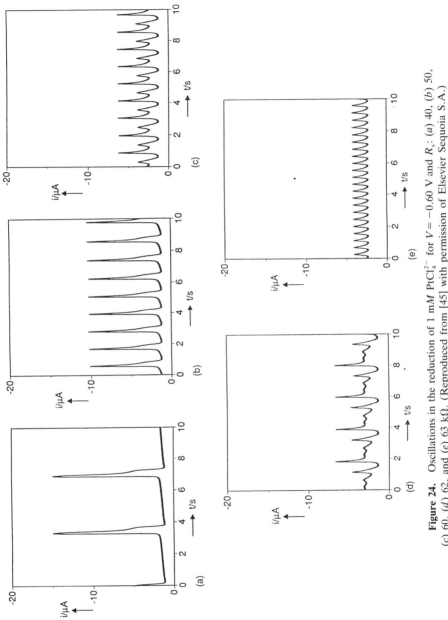

Figure 24. Oscillations in the reduction of 1 mM PtCl$_4^{2-}$ for $V = -0.60$ V and R_s: (*a*) 40, (*b*) 50, (*c*) 60, (*d*) 62, and (*e*) 63 kΩ. (Reproduced from [45] with permission of Elsevier Sequoia S.A.)

207

bifurcation sequences of MMOs were based on a celebrated theorem by
Shil'nikov [55] about the generation of chaos and highly ordered periodic–chaotic sequences in the neighborhood of a homoclinic orbit to a
saddle focus. A saddle-focus fixed point is a fixed point with Jacobian
eigenvalues ($\rho \pm i\omega$, λ) with $\rho/\lambda < 0$, and a homoclinic orbit to this saddle
focus is an invariant phase-space trajectory that tends to this fixed point
for both future and past (see Fig. 25). Shil'nikov proved that if $|\rho/\lambda| < 1$
then an infinite number of unstable periodic orbits and associated erratic
behavior will coexist with the homoclinic orbit, and the bifurcation
scenario expected in the neighborhood of this homoclinic orbit is shown

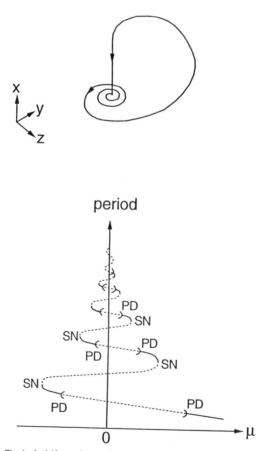

Figure 25. Typical bifurcation phenomena expected in the neighborhood of a
Shil'nikov homoclinic orbit to a saddle-focus fixed point with eigenvalues ($\rho \pm i\omega$, $-\lambda$), for
which $\frac{1}{2} < \rho/\lambda < 1$.

in Fig. 25 [12]. In this bifurcation diagram, there is a single bifurcation curve through which all periodic orbits are connected. The curve undergoes an infinite number of saddle-node bifurcations, and subsequent period doublings. Each saddle-node bifurcation creates another additional (small) oscillation around the saddle focus, whereas the homoclinic orbit itself would correspond to an infinite number of loops. If $\frac{1}{2} < |\rho/\lambda| < 1$, then the period orbit is stable between the saddle-node (tangent) bifurcation and the period-doubling bifurcation, so that one expects a stable MMO to become destabilized through a cascade of period doublings into chaos, which in turn gets trapped by the tangent bifurcation to result in the next stable MMO. This scenario is indeed very reminiscent of many experimental and numerical observations (see Sections III.A, III.B, and IV.B) and chaos in MMOs is therefore sometimes referred to as Shil'nikov chaos.

However, several authors have noted important discrepancies between the predictions of the Shil'nikov theorem and the MMO sequences [39, 41, 56–58]. First, the true homoclinic connection is never really found, and in many real or numerical experiments the sequence is often terminated before the scenario is completed, causing it sometimes to be referred to as an incomplete homoclinic scenario [39]. Second, but in fact related to the first point, the theorem predicts that it should be possible to visit the family of periodic orbits surrounding the homoclinic orbit by varying the relevant control parameter in both the negative as well as the positive direction with respect to the critical value. However, experimentally and numerically, the scenario is always one sided. The other side usually consists of some sequence of small-amplitude oscillations, as we saw in Section III.B.

An important clue to the origin of these deviations was provided by a numerical study on a simple hypothetical three-variable chemical reaction scheme carried out by Petrov et al. [59]. These authors showed numerically that the different stable and unstable MMO orbits are not connected but instead lie on isolated bifurcation curves. This property of MMO orbits was confirmed by Koper [60], who numerically studied the bifurcations of MMOs in the following three-variable scheme

$$\varepsilon_1 \frac{dx}{dt} = ky - x^3 + 3x - \lambda$$

$$\frac{dy}{dt} = x - 2y + z \qquad (3.24)$$

$$\frac{dz}{dt} = \varepsilon_2(y - z)$$

which is a three-variable version of a classical two-variable model first proposed by Boissonade and De Kepper (BDK) [61]. The BDK model is obtained in the limit $\varepsilon_2 \to \infty$, and was introduced by Boissonade and De Kepper to give a simple geometrical interpretation of a recurrent phase diagram in nonlinear chemical kinetics, the cross-shaped phase diagram. Equations (3.24) also possess a cross-shaped phase diagram (see Fig. 26), but in addition there are regions of MMOs (for $\varepsilon_2 = 1$) in an arrangement that is isomorphic to that observed in many phase diagrams of experiments or realistic models (such as those of the reduction reactions of the previous section; cf. Fig. 11).

The MMOs in Eqs. (3.24) are also isolated periodic orbits in a typical *one*-parameter cut. However, an interesting result is obtained when a detailed study is made in the parameter plane, that is, a *two*-parameter

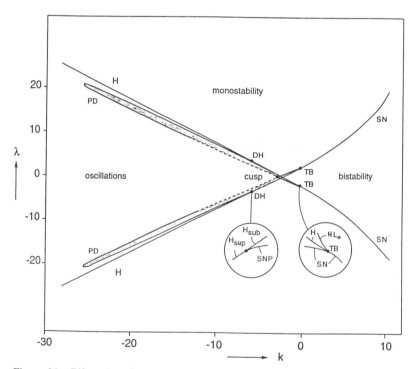

Figure 26. Bifurcation diagram of Eqs. (3.24) in the k plane. The shaded regions indicate the existence of complex and mixed-mode oscillations. H (sub- or supercritical) Hopf bifurcation; SN, saddle-node bifurcation; DH, degenerate Hopf bifurcation; TB, Takens–Bogdanov bifurcation; SL, saddle-loop bifurcation; SNP, saddle-node bifurcation of periodic orbits; PD, period-double bifurcation. $\varepsilon_1 = 0.1$, $\varepsilon_2 = 1$. (Reproduced from [60] with permission of Elsevier Science B.V.)

cut. Then, by an appropriate tuning of the parameters k and λ, all the unstable MMO orbits can be made to merge in a single codimension-two point (i.e., a kind of triple, or "multiple" point). This point lies on the locus of the homoclinic orbit born in the Takens–Bogdanov (TB) bifurcation, that is, the point where the steady state has two zero eigenvalues and the Hopf and saddle-node meet. This point is character-ized by a sudden change in twistedness of the orbit, that is, from nontwisted or twisted, as illustrated in Fig. 27(a–c). This bifurcation is known as the neutrally twisted homoclinic orbit or inclination switch. A recent mathematical analysis by Homburg et al. [62] has shown that at

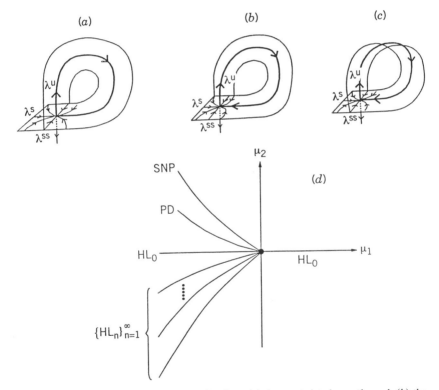

Figure 27. A homoclinic orbit bifurcating from (a) the nontwisted case through (b) the degenerate neutrally twisted case (inclination switch) to (c) the twisted case. Invariant manifolds indicated by their corresponding eigenvalues. (d) Conjectured unfolding of the codimension-2 inclination switch [62]. The parameter HL_0, homoclinic loop of a mono-periodic orbit, HL_n homoclinic loop of a period $(n + 1)$ orbit (i.e., the orbit makes n loops in phase space before the homoclinic connection is established). (Reproduced from [60] with permission of Elsevier Science B.V.)

certain conditions of the real eigenvalues of the saddle fixed point to which the homoclinic connection occurs, an infinite number of unstable periodic orbits (associated with chaos) is born in this bifurcation [Fig. 27(d)]. These unstable periodic orbits eventually deform and stabilize into the mixed-mode oscillations. This analysis clearly shows that on the one- and two-parameter level, mixed-mode oscillations have nothing to do with a Shil'nikov homoclinic orbit, but instead originate from another homoclinic structure known as the neutrally twisted homoclinic orbit.

IV. OSCILLATORY ELECTRODISSOLUTION OF METALS

A. Iron in Sulfuric Acid; Oscillations at the Active–Passive Transition

The classical example of an oscillatory electrochemical reaction is the anodic dissolution of metals in aqueous solution. Observations of current or potential oscillations during the chemically or electrically induced corrosion of various metals was a well-known phenomenon already in the nineteenth century [63], associated with names such as Joule [64] and Ostwald [65]. The most familiar and thoroughly studied system is probably the iron electrodissolution in sulfuric acid. Detailed dynamic studies on iron dissolution are not as rich and reproducible as for certain other anodic dissolution processes, such as the copper and nickel dissolution, to be treated in Sections IV.B and IV.C. However, it is an important and illustrative example that will allow us to appreciate some of the main principles of oscillatory electrodissolution.

Discussing the Fe dissolution in H_2SO_4 is started most easily by referring to two typical current–voltage curves found in the literature [20, 66–72], shown in Fig. 28(a and b). At potentials more anodic than −0.5 V (vs. SCE), iron dissolves electrochemically in sulfuric acid with an exponential characteristic (Tafel's law). As the applied potential is made progressively more positive, a current plateau is reached that is mass-transfer limited. It is commonly accepted that this regime is caused by the formation and precipitation of a $FeSO_4$ salt film [71–73], which puts an upper limit on the Fe^{2+} surface concentration, although the exact nature of the film is still disputed. The transition from this active dissolution to a passive state, where the electrode becomes covered with an essentially insulating oxide film, has been reported to occur in either one of two ways. Epelboin et al. [20] and Newman and Russell [70] reported a clear hysteresis due to a Z-shaped current–voltage curve, the sandwiched branch of which is unstable (although it can be stabilized with a special electronic device known as a Negative Impedance Converter [20]). They have shown that at low pH the hysteresis is caused by the high value of

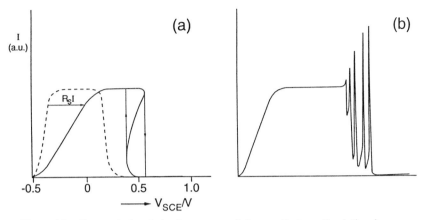

Figure 28. Two typical polarization curves of the anodic iron dissolution in concentrated sulfuric acid solution. (*a*) Z-shaped polarization curve, as caused by a nonnegligible ohmic electrolyte resistance; dashed curve represents the polarization curve corrected for the ohmic drop; see, e.g., [20, 67, 70]. (*b*) Current oscillations at the active–passive transition; see, e.g., [66, 68, 69, 72, 75, 86, 87, 88].

the ohmic drop in the electrolyte solution, resulting in a deformation of the current–voltage curve, as illustrated in Fig. 28(*a*) by the solid line. (In reality, things are more complicated than a simple IR correction since the dissolution does not occur homogeneously on the surface, giving rise to an inhomogeneous current density.) However, results are not fully unambiguous: Russell and Newman [70] showed that the Z-shaped polarization curve in 1 M H_2SO_4 can be ascribed entirely to ohmic drop, whereas Epelboin et al. [74] reported that dissolution of a ring electrode in 1 M H_2SO_4 leads to a Z-shaped curve even after correction for the ohmic drop.

On the other hand, Franck [67–69], Podesta et al. [75], and others observed that the active–passive transition is accompanied by strong current oscillations [Fig. 28(*b*)], in which the current repetitively switches from an active to a passive dissolution value. Very similar oscillations at the active–passive transition have been obtained for various other electrodissolving metals, for example, Au dissolution in HCl [76, 77], Zn dissolution in NaOH [78], Co in H_3PO_4 [79–81], and many others [2, 3]. Typical examples are given in Fig. 29. Note that, as a common feature of all these systems, the amplitude of the oscillations grows with increasing positive potential.

Just as the ohmic electrolyte resistance may cause the bistability and hysteresis in Fig. 28(*a*), it plays an important role in the origin of the

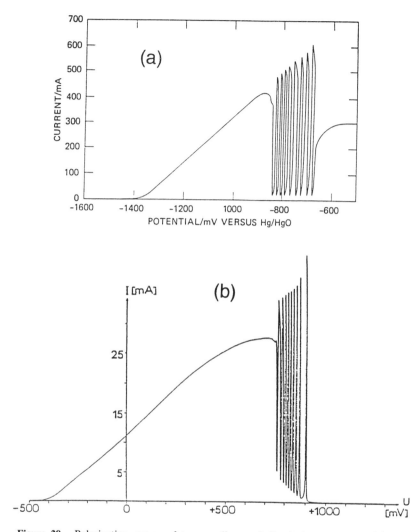

Figure 29. Polarization curves of two anodic metal dissolution processes. (*a*) Zn in NaOH, reproduced from [78] with permission of The Electrochemical Society. (*b*) Co in H$_3$PO$_4$. [Reproduced from [81] with permission of the author.]

oscillations in Fig. 28(*b*). Several authors have pointed this out, although there does not seem to exist a systematic study on iron dissolution in which the different dynamics in the V–R_s plane is studied (such as Fig. 11). Such studies have been carried out for the active–passive transition for the cobalt electrodissolution in H$_3$PO$_4$ by Sazou and Pagitsas [79, 80], confirming the expectation that at relatively low values of the external resistance oscillations are obtained, whereas at high external resistance a hysteresis is observed without oscillations.

Quite a few mathematical models have been published that attempt to explain the oscillations at the active–passive transition, the iron–sulfuric acid system being the classical example. The most-often quoted model is that by Franck and FitzHugh [69]. Although this model is important in being the first quantitative approach to electrochemical oscillations (1961), it has been criticized, mainly because it does not account for the role of the ohmic potential drop and because the kinetics of passivation in this model is treated in a rather nongeneric way. A modification of this model removing these two objections has been suggested by Koper and Sluyters [82]. This modified model essentially incorporates the same ingredients as an earlier model attempt by Russell and Newman [83, 84], with the main difference that the former is mathematically somewhat simpler, and therefore easier to integrate numerically and to study from a bifurcation point-of-view.

Instead of presenting the detailed equations of these two models, the most important phenomena that contribute to the oscillation cycle can be explained and visualized by means of an analysis in the phase plane spanned by the electrode potential E and by the acidity at the electrode surface pH_0 [84] [see Fig. 30(a)]. The two simplest possible reaction steps one can write down to model metal dissolution and passivation is the

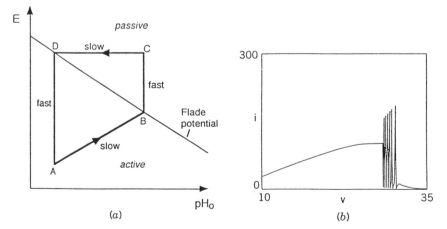

Figure 30. (a) Limit-cycle oscillation in the E–pH_0 phase plane. For detailed explanation, see text. (b) Linear sweep voltammogram in the model [82]. (Reproduced from [82] with permission of Elsevier Sequoia S.A.)

so-called dissolution–precipitation model [85]:

$$Me \rightarrow Me^{n+} + n\,e^- \qquad (4.1)$$

$$Me + mH_2O \leftrightarrow MeO_m + 2m\,e^- + 2mH^+ \qquad (4.2)$$

where Me denotes "metal" and where the insoluble oxide MeO_m covers the electrode as a passivating layer. The passivation reaction has a pH-dependent equilibrium potential (*the Flade potential* [67]) that is more anodic than the equilibrium potential of the dissolution reaction over the entire pH range of interest. The location of the Flade potential in the E–pH plane is drawn as the solid line in Fig. 30. In the solution there is also an anionic species A, with a finite solubility of the MeA salt (dissolution processes are often accompanied by the precipitation of salt films [71–73]). Let us assume that the electrode starts at point A in Fig. 30(*a*) in the E–pH_0 plane, at a potential that is not too anodic and at a relatively low pH_0 (the bulk pH, say) so that the dissolution reaction procedes rapidly and the passivation rate is negligible. Since H^+ should carry (part of) the current through the electrolyte, the surface region becomes depleted of H^+ and thus pH_0 increases. Also, there is a transient increase in the Me^{n+} concentration at the surface, which causes some precipitation of the MeA salt and a decrease in the rate of dissolution. In fact, one may assume that the MeA dissolution is the rate-determining step in total dissolution process. Therefore, the current I will drop, and consequently the electrode potential increases because $E = V - IR_s$, where V is the applied cell potential and R_s the uncompensated ohmic solution resistance. As E and pH_0 keep increasing, the pH-dependent Flade potential may be reached at point B (whether this really happens depends on the actual values of V and R_s and this can only be evaluated with a quantitative approach), that is, the potential-pH region where the passivation reaction becomes thermodynamically favorable. The electrode then passivates, I drops, as a result of which E increases leading to an even faster passivation; this is the crucial autocatalytic (or autoinhibition) process. If the passivation is a fast process, there is a sudden jump to point C (how fast it actually is depends also on the capacity of the double layer). The formation of a passive (mono-)layer does not consume too many H^+ ions so the pH_0 does not change appreciably in this short time interval. In the passivated state, there is only a small residual current density, and so the H^+ ions will diffuse back to the surface, whereas the electrode potential remains essentially constant. As the pH_0 recovers, the Flade potential is reached again and the passive layer will start dissolving autocatalytically, since the increasing current forces the electrode potential into the cathodic direction. Now the system returns to the initial state with low E and low pH_0 and the cycle can repeat. The entire cycle constitutes a relaxation oscillation.

This simple model can explain some qualitative features of the experimental metal electrodissolution exhibiting current oscillations at the active–passive transition [82]. Figure 30(b) shows a linear sweep voltammogram, resembling the experimental curves in Figs. 28(b) and 29. However, the model is certainly too simple to warrant any detailed (quantitative) comparison to experiments. Although the model supports complex mixed-mode and chaotic oscillations for certain parameter values [82], this feature is due to the mere fact that it is built on three dynamical variables the choice of which is not particularly unique. The situation should be contrasted with the model for the oscillatory reduction reactions on mercury considered in Section III, where one can be fairly confident that the model incorporates all the important features of the experimental systems. Things are much more complicated in the case of metal electrodissolution. First of all, the dissolution–precipitation scheme is a caricature of the passivation kinetics, and only provides the simplest possible interpretation of the negative polarization slope associated with the passivation phenomenon. Also, the model does not include a diffusion-layer relaxation, an effect that was shown to be important in the generation of complex oscillations for the simple reduction reactions on mercury. However, the most important simplification is probably the neglect of spatial variations in the variables in the tangential directions. Not only is the activity of the electrode unlikely to be perfectly homogeneous, but the geometry of the working electrode will also induce nonuniform current and potential distributions that have important consequences for both the local and global dynamics. An exciting and revealing example of this effect was reported recently by Hudson et al. [86] who observed that what appears as a period doubling in the current–time series of the oscillatory dissolution of an iron disk, originates from a spatio–temporal bifurcation from a ring of activity into two half-rings of alternating activity.

The current oscillations occurring at the active–passive transition of the iron electrodissolution in H_2SO_4 are usually simple relaxation oscillations. However, Russell and Newman [84] and Diem and Hudson [87] observed high-frequency, irregular, small-amplitude oscillations in the electropolishing region just before the active–passive transition. Diem and Hudson calculated correlation dimensions ranging from 2.4 up to about 6, indicating that the time series may be deterministic chaotic instead of noisy, although no clear scenarios to chaotic behavior were found. More complex oscillations can also be obtained if halides such as chloride or bromide are added to the sulfuric acid base electrolyte [88]. These halides cause pitting corrosion by chemically attacking the passive film formed at potentials more anodic than the Flade potential. Sazou et al. [88] observed that this may induce bursting oscillations, that is, complex waveforms in which trains of relaxation oscillations are alternated by periods of steady-state passivity (Fig. 31). Heuristically, such oscillations can be explained by assuming that the pitting corrosion is a relatively slow process that sweeps the system back and forth between the passive and the oscillatory state. Bursting oscillations are quite interesting from a bifurcation point-of-view and have been studied in some detail in biophysical context [89].

B.   Copper in Phosphoric Acid

The electrodissolution of copper in various electrolytes has been an active area of research in nonlinear chemical dynamics. The oscillatory copper dissolution in acidic chloride has been studied in great detail by Hudson and Bassett, who have reviewed their results in [4(c)]. In this chapter attention will be restricted to the electropolishing of copper in phosphoric acid, since this is a nice example of an electrochemical system for which a complex but well-defined bifurcation structure has been reported. Oscillations in this system were first reported by Jacquet in 1936 [90]. Detailed studies of this oscillatory system have since been carried out, where from the dynamic point of view the series of papers by Albahadily and Schell [91–93], to be discussed in some detail below, is most notable.

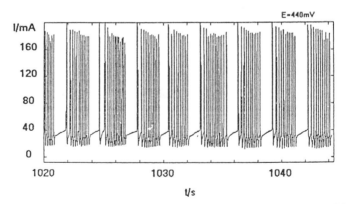

Figure 31. Bursting oscillations during the dissolution of a Fe disk in 2 M H$_2$SO$_4$ + 0.04 M Cl$^-$; rotation rate 1000 rpm, $V = 440$ mV vs. SCE. (Reproduced from [88] with permission of Elsevier Science Ltd.)

A typical current–voltage curve, reproduced from [94], is given in Fig. 32. Note that the curve does not exhibit a clear transition to a passive state. Glarum and Marshall [94] reported that the current plateau follows a square-root dependence on the disk rotation rate, but is dependent only on solution viscosity, and not on the Cu^{2+} bulk concentration or the nature of the phosphate species. They suggest an important role of water

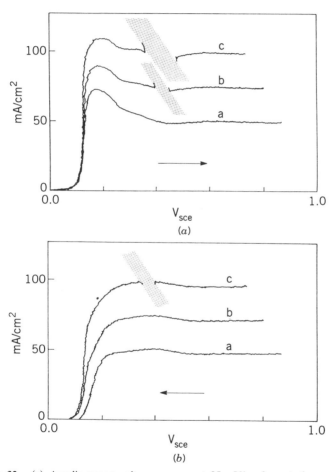

Figure 32. (*a*) Anodic sweep voltammograms at 25 mV/s of a rotating copper disk electrode in 85% phosphoric acid. For plots *a–c*, the disk rotation speeds are 100, 225, 400 rad/s. The shaded regions indicate areas swept by oscillations. (*b*) cathodic sweeps complementing the anodic sweeps. (Reproduced from [94] with permission of The Electrochemical Society.)

depletion in the boundary layer, as water is consumed by the hydration demands of the Cu^{2+} ions.

Although there is not a very distinct negative slope in the current–voltage curve, Glarum and Marshall [95] clearly established the role of the negative impedance characteristics in generating the oscillations. Their frequency response measurements exhibit a distinct frequency range for which the real impedance (or real admittance, in terms of which Glarum and Marshall plot their results) is negative, with the real admittance at zero frequency tending to the origin of the admittance plane, in agreement with the current plateau in the stationary current–voltage curve. It is these negative impedance characteristics, in conjunction with the ohmic electrolyte resistance, that cause the current oscillations. The crucial role of the ohmic potential drop was also clearly established by Glarum and Marschall [94]. They noted that oscillations were generally not observed for copper disks of small diameter or low-rotation speed (resulting in a small current, and hence a small ohmic potential drop), but, as Glarum and Marschall stress, "under conditions where oscillations were never otherwise seen, they could be induced with a sufficiently large (series) resistance."

The origin of the negative impedance characteristics is most likely related to the formation of surface films. An important contribution regarding the nature of these surface films was made by Tsitsopoulos et al. [96, 97]. By in situ ellipsometric measurements [96], they showed that the thickness and/or composition of these films oscillate with the same period as the current. By ex situ X-ray photoelectron spectroscopy (XPS) and scanning electron microscopy (SEM) measurements [97], they demonstrated that during these oscillations the surface is covered by a porous Cu_2O film (the porosity explaining the absence of complete passivation), the pores being filled with $Cu(OH)_2$. The actual blocking should occur by the $Cu(OH)_2$, as in the passive (low-current) state of the oscillatory cycle the presence of $Cu(OH)_2$ is more profound (compared to Cu_2O) than in the active state. Tsitsopoulos et al. [97] describe the oscillatory cycle by the local pH changes, due to changes in the migration current, which trigger the alternating formation and dissolution of the $Cu(OH)_2$ film. However, they do not make the essential link with the ohmic cell resistance in the generation of the oscillations.

Given the shear complexity of the dissolution process and the variety of surface films involved, one may be somewhat skeptical about the possibility of such a system being sufficiently regular and deterministic to hold to the ordered sequences of bifurcations, as discussed in Section III for the mechanistically much simpler reduction reactions on mercury. Surprisingly, the remarkable papers by Albahadily and Schell [91–93] on the dynamics of the electropolishing of a rotating copper disk in 85% phosphoric acid constitute one of the most complete studies of bifurcation scenarios of mixed-mode oscillations in nonlinear chemical kinetics.

Albahadily and Schell [91] presented a phase diagram of the system in a parameter plane spanned by the applied potential V (note that, in view of Glarum and Marschall's results, this *cannot* be called the working electrode potential, as Albahadily and Schell sometimes refer to it) and the disk rotation rate, which is essentially a way to control the thickness of the diffusion layer (δ in Section III). Experiments were carried out at $T = -20°C$, for which Albahadily and Schell reported a much improved reproducibility with respect to room temperature. The boundary between steady-state and oscillatory phases is shown in Fig. 33. The bifurcation along the solid line was identified as a Hopf bifurcation. Close to the transition to the oscillatory state, one observes sequences of mixed-mode oscillations.

Albahadily et al. [92] note that the sequences can be assigned to one of two classes: (1) "periodic–chaotic sequences," in which intervals of periodicity are separated by an interval of chaotic states that resemble random mixtures of adjacent patterns, and (2) "Farey sequences," which can be defined as periodic sequences for which a 1:1 correspondence with ordered sequences of rational numbers can be established. Farey se-

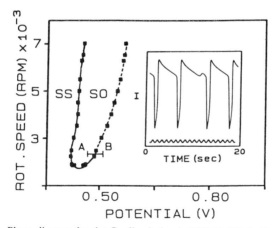

Figure 33. Phase diagram for the Cu dissolution in 85% H_3PO_4 in the parameter plane spanned by the disk rotation speed and the applied potential. SS, stable stationary states, SO sustained oscillations. Solid lines consist of points at which Hopf bifurcations occurred; dashed lines consist of points where MMOs appeared. $T = -17.5°C$. Small amplitude oscillation in the insert occurred for rotational speed = 2200 rpm and $V = 427$ mV (vs. SCE); large amplitude oscillation for 7000 rpm and 567 mV. (Reproduced from [91] with permission of the American Institute of Physics.)

quences typically occur close to the Hopf bifurcation with relatively low disk rotation rate [92]. Measured waveforms of a few typical mixed-mode states are shown in Figs. 34 and 35. The authors note that each MMO state possesses either only one small amplitude oscillation per set of large oscillations (Fig. 34, this page), or only one large amplitude oscillation per set of small oscillations (Fig. 34, facing page). The MMOs exhibit concatenations similar to those described in the Sections III.B and III.C; some examples are shown in Fig. 35. They allow the construction of a Farey tree, in which a (concatenated) MMO state is constructed from its nearest neighbors [92]. Albahadily et al. carefully constructed a parame-

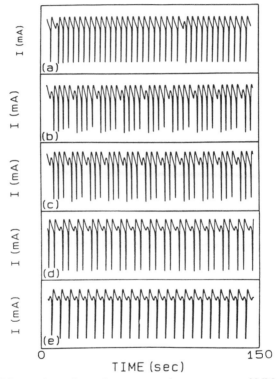

Figure 34. Measured waveforms that correspond to a sequence of MMOs. The current (ordinates) runs from -2.60 to -0.55 mA for all waveforms. Rotation rate = 1600 rpm. This page: (*a*) MMO state close to the "limiting 1^0 state", $V = 426.70$ mV; (*b*) 4^1 state, 427 mV; (*c*) 3^1 state, 427.05 mV; (*d*) 2^1 state, 427.75 mV; (*e*) 1^1 state, 428.75 mV. Facing page: A continuation of the sequence shown on this page: (*a*) 1^2 state, 430.15 mV; (*b*) 1^3 state, 430.30 mV; (*c*) 1^4 state, 430.40 mV; (*d*) 1^5 state, 430.45 mV; (*e*) MMO state near the end of the sequence, 430.49 mV. Cu in 85% H_3PO_4, $T = -20°C$. (Reproduced from [92] with permission of the American Institute of Physics.)

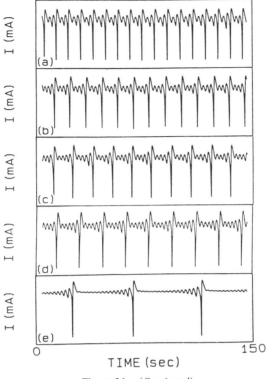

Figure 34. (*Continued*)

ter plane that depicts the regions where the various complex oscillations were found. This phase diagram is given in Fig. 36, an explanation of the various symbols can be found in the figure's caption. A surprising and in fact rather puzzling thing about this phase diagram in view of the results described in Section III.D, is the occurrence of small-amplitude torus oscillations. These are born from the primary small-amplitude orbit in a secondary Hopf. Torus oscillations can also be found in the three-variable Boissonade–De Kepper model [Eqs. (3.24)] (by decreasing the parameter ε_2) [60], but only in combination with torus-like mixed-mode oscillations in which the occurrence of sets of consecutive small and large oscillations in a periodic entity becomes more likely. However, as noted above, Albahadily and Schell stress the absence of such MMO states, and did not find any evidence for an underlying torus behavior of the MMOs, despite the observation of the small-amplitude torus oscillations.

The periodic–chaotic sequence of MMOs occurs close to the Hopf bifurcation at relatively high-rotation speed [93]. A phase diagram in the

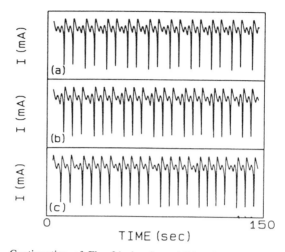

Figure 35. Continuation of Fig. 34 showing additional Farey states. (*a*) $1^1 1^2$ state, 429.70 mV; (*b*) $(1^1)^2 1^2$, 429.57 mV; (*c*) $(1^1)^3 1^1$, 429.47 mV. (Reproduced from [92] with permission of the American Institute of Physics.)

parameter plane illustrates the relative sizes of the regions for which different MMOs were found (Fig. 37). This diagram resembles the diagrams of Figs. 11 and 23(*b*) of the indium–thiocyanate oscillator and its model, and certain parts of the phase diagram of the BDK-model [60]. As is clear from this diagram, the transition between two consecutive states of Fig. 37 is accompanied by chaotic oscillations. Two examples are shown in Fig. 38, which are chaotic waveforms appearing as a mixture of the adjacent 1^1 and 1^2 state. The chaotic character of the time series in Fig. 38(*b*) is clearly demonstrated by the next-amplitude map [Fig. 38(*c*)]. The experiments of Schell and Albahadily [93] show that the typical bifurcation route through which a periodic MMO is transformed into its nearest neighbor, starts with a period-doubling bifurcation to a $(1^n)^2$ state, the subharmonic of the 1^n parent state. A cascade of such period-doubling leads to chaos. The chaotic oscillation disappeared through a tangent bifurcation, whereby the system collapsed onto the newly born 1^{n+1} state. This process is clearly evidenced by the third-next-maximum maps of Fig. 39.

The bifurcations of this intriguing system are remarkably similar to those observed in the indium–thiocyanate oscillator and the three-variable BDK model, as described in Section III. A region of steady-state bistability has not been reported, but this is not so surprising if one realizes that the steady-state polarization curve does not show a distinct

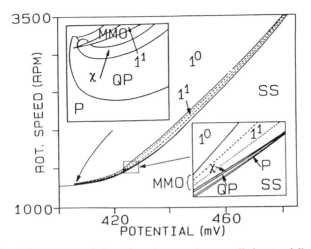

Figure 36. The parameter plane (rotation speed vs. applied potential) divided into regions where different dynamical behavior was observed. The largest variety of behavior was found in the thin banana-shaped region, whose various parts are labeled in the lower right inset, which is an enlargement of the rectangle indicated by the arrow. Parameter ranges in this rectangle: 424.0–429.5 mV, 1450–1600 rpm. P = periodic oscillations with small amplitudes. QP = quasiperiodic oscillations. χ = chaotic oscillations. MMO = mixed-mode oscillations. Note that at lower potentials the small-amplitude "P" was observed to deform smoothly into a 1^0 relaxation oscillation (this also occurs in the In/SCN system, see Fig. 11). The inset in the upper left corner is an enlargement of the parameter ranges 405.5–408 mV, 1280–1310 mV. (Reproduced from [92] with permission of the American Institute of Physics.)

region of negative slope, which determines the possibility of having saddle-node bifurcations, as we saw in Section II.

C. Nickel in Sulfuric Acid

As a third and last example of complex bifurcations in metal electrodissolution, we discuss the Ni dissolution in sulfuric acid in the so-called transpassive region. This is an interesting system, as both the dynamics as well as the impedance characteristics are well documented, and kinetic models have been proposed to explain these experimental results. A current–voltage curve for nickel dissolution in sulfuric acid is shown in Fig. 40. With progressively more anodic potentials, initially, in region A, nickel dissolves actively until the formation of an oxide film [presumably similar to reactions (4.1–4.2)] becomes thermodynamically favorable (region B), and the electrode passivates (region C). A transpassive dissolution occurs in region D (the chemical origin of which will be discussed below), whereas at very anodic potentials oxygen evolution

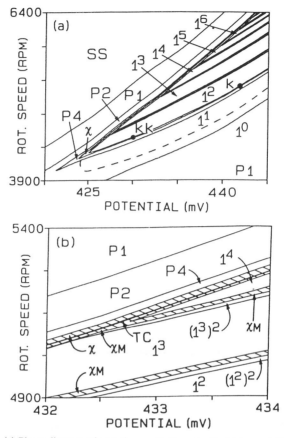

Figure 37. (*a*) Phase diagram of complex oscillations for the higher rotation speeds. (*b*) An enlargement of a region in (*a*). L^s periodic MMO state, $(L^s)^2$ = subharmonic of MMO state; χ_M chaotic MMO state, SS = stationary state, P1 = small-amplitude periodic oscillation; P2 = subharmonic of P1; P4 = second subharmonic of P1; χ chaotic small-amplitude oscillation. TC = approximate boundaries of the curve at which a transition from small-amplitude chaos to chaotic MMOs occurred. k and kk are the points at which the waveforms shown in Fig. 38 were measured. Boundaries should be considered to represent general trends. (Reproduced from [93] with permission of the American Institute of Physics.)

becomes dominant (region E). Also note the region of "secondary passivity" in between regions D and E.

The earliest account of oscillatory behavior in the transpassive dissolution of Ni is due to Hoar and Mowat [98] (1950). Two remarkable papers (considering the years 1960 and 1962 in which they were published) were

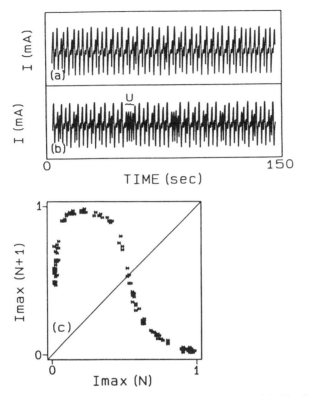

Figure 38. (*a*) A waveform of chaotic MMO measured at point k in Fig. 37, which lies between the 1^1 and 1^2 states. $V = 442.0$ mV, 5503 rpm. Current scale runs from -2.50 to -0.70 mA. (*b*) The same as in (*a*) but measured at point kk in Fig. 37, 430.0 mV, 4540 rpm. (*c*) A one-dimensional map constructed from the time series in (*b*). The $(n + 1)$th maximum of the time series is plotted against the nth maximum. (Reproduced from [93] with permission of the American Institute of Physics.)

written by Osterwald [99, 100]. Figure 41 is redrawn from Osterwald's original paper [99], with an extension based on a more recent experimental study by Lev et al. [101], as explained below and in the figure's caption. Osterwald studied the nickel transpassive dissolution using a parallel connection of a potentiostat and a galvanostat, allowing a rapid switching from one mode of operation to another. By means of the potentiostat, the electrode was brought to a state represented by some point on the left-hand, ascending branch of the current–voltage curve, and the control was switched from potentiostatic to galvanostatic, the latter being previously adjusted so that the value of the current did not change. Osterwald observed that when the electrode potential was chosen

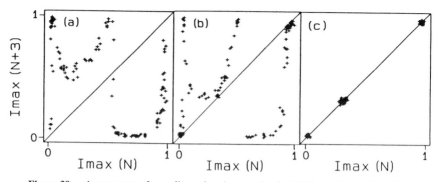

Figure 39. A sequence of one-dimensional maps; the $(n + 3)$th maximum from the time series is plotted against the nth maximum. (a) The third power of the map shown in Fig. 38(c). (b) A map constructed from data collected for a value of the rotation speed that is slightly less than at which the 1^2 state becomes stable. The map function outlined by the iterates of the map is almost tangent to the bisectrix, $V = 430.0$ mV, 4547 rpm. (c) A plot of iterates constructed from data collected for a value of the rotation that is slightly larger than that used in (b). The underlying map function now crosses the bisectrix at six points, three stable and three unstable fixed points; the iterates are dispersed around the three stable fixed points which correspond to the maxima of the 1^2 state, $V = 430.0$ mV, 4556 rpm. (Reproduced from [93] with permission of the American Institute of Physics.)

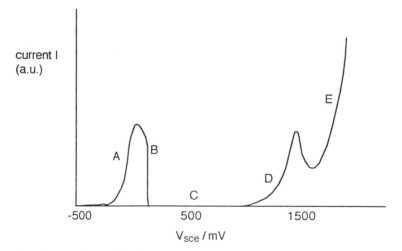

Figure 40. Typical polarization curve of the anodic Ni dissolution in H_2SO_4. Region A: potential region of active dissolution; Region B: primary passivation; Region C: passivated electrode; Region D: transpassive dissolution ; Region E: oxygen evolution.

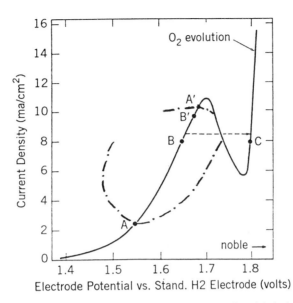

Figure 41. Polarization curve obtained potentiostatically of a nickel electrode in 1 N H_2SO_4 in the transpassive region. Points A, B, A', and B' are explained in the text. The dashed–dot curve represents the amplitude of the potential oscillations measured galvanostatically (see [101]). (Reproduced from [99] with permission of The Electrochemical Society.)

between A and B or between A' and B', in the galvanostatic mode the potential started oscillating as illustrated in Fig. 42(a). When the potential was lower than A or higher than A', damped potential oscillations were observed tending to the original potential value, as in Fig. 42(b) (clearly, the descending branch in Fig. 41 is galvanostatically unstable). Interestingly, when in the potentiostatic mode the electrode was brought to a point in between B and B', in the galvanostatic mode the potential did not give sustained oscillations but instead escaped to the right-hand ascending branch. We now interpret this as evidence that homoclinic bifurcations have occurred at B and B', in which the limit-cycle oscillation collides with the descending saddle-type branch. This is illustrated by drawing the potential amplitudes of the sustained oscillations, as found by the more recent experiments of Lev et al. [101]. To the best of the author's knowledge, Osterwald's experiment is the first in which a chemical oscillator exhibits so unambiguously a homoclinic bifurcation (1960!), although Osterwald himself was not aware of this, of course. Osterwald [100] attempted a theoretical description of the system that does not seem to be fully correct, but upon which he drew a few

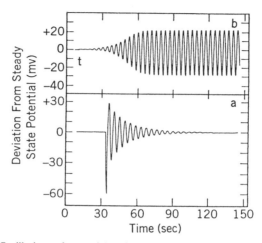

Figure 42. Oscillations of potential under galvanostatic conditions of a nickel electrode in 1 N H_2SO_4. (a) damped oscillations for a current density of 2.26 mA cm^{-2}; (b) sustained oscillations for a current density of 2.87 mA cm^{-2}. (Reproduced from [99] with permission of The Electrochemical Society.)

conclusions that are surprisingly to-the-point. His interpretation will be mentioned briefly below, in the treatment of the more recent studies of this system.

As mentioned, Lev et al. [101, 102] carried out a comprehensive bifurcation study of the nickel dissolution in sulfuric acid as a function of the applied current and the solution acidity. The resulting phase diagram is shown in Fig. 43. Lev et al. also observed more complex periodic and chaotic oscillations appearing through period doubling and torus bifurcations, and a region of birhythmicity. The existence regions of these phenomena are illustrated in Fig. 43.

The same group of researchers recently proposed a kinetic model for the anodic nickel dissolution, which reproduces the qualitative features of their experimental phase diagram. Haim et al. [103] claim their model to be similar to a mechanism suggested by Keddam et al. [104], who based their model on impedance measurements. An inspection of the paper by Keddam et al. [104] reveals, however, that the chemistry of the two models is substantially different, and even contradictory at some points. Both models will be discussed below.

According to Haim et al. [103], the nickel electrode is initially covered by oxide film due to the primary passivity (region C of Fig. 40). Nickel ions migrate through the oxide layer and can dissolve only through the surface fraction that is not covered by adsorbed species. There are three of such surface species in the Haim model, oxide (in the form of NiO),

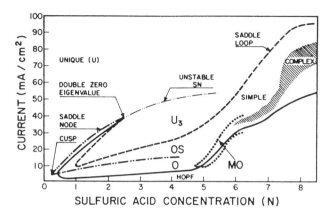

Figure 43.   Grand bifurcation map of the Ni dissolution: typical phase planes are unique stationary state (U_3) (this occurs in Fig. 42 for applied currents below A and above the current maximum); three stationary states (U_3), one stable and two unstable ones (this occurs in Fig. 42 for applied current between B and B′); only oscillatory states (O) (this occurs in Fig. 42 for applied currents between A and the current minimum); bistability of an oscillatory and a stationary state (OS) (this occurs in Fig. 42 for applied currents between A′ and B′ and between B and the current minimum); bistability (two simultaneously stable states) near the cusp (this occurs in Fig. 42 for applied currents between A′ and the current maximum); and birhythmicity (MO) (this does not occur in Fig. 42). The Hopf, saddle-node and saddle-loop bifurcations are identified in the figure; the dotted line borders the region of birhythmicity by saddle-node bifurcations of periodic orbits. [Reproduced with permission from O. Lev, A. Wolffberg, L. M. Pismen, and M. Sheintuch, *J. Phys. Chem.*, **93**, 1661 (1989). Copyright © 1989 American Chemical Society.]

hydroxide (NiOH), and adsorbed bisulfate ($NiHSO_4^-$). The nickel ion dissolution is modeled by a first-order, acid-catalyzed, Tafel-like reaction:

$$Ni \rightarrow Ni^{2+} + 2e^- \qquad (4.3)$$

with a rate

$$r_1 = k_b C_h \left(1 - \sum_i \theta_i\right) \exp\left(\frac{bFE}{RT}\right) \qquad (4.4)$$

where C_h is the proton concentration in the solution, bF/RT the Tafel slope, and E is the electrode potential. Strictly speaking, Reaction (4.3) is the active electrodissolution that one would expect to take place in region A of the current–voltage curve. It is not really clear from the original Haim paper how Reaction (4.3) relates to their earlier statement that the electrode is fully covered with an oxide film. Perhaps the rate law

[Eq. (4.4)] should be regarded as a more formal expression for the small leakage current through the oxide layer. Formation of surface hydroxyl groups and an oxide layer is also taken into account through the reactions:

$$Ni + H_2O \rightarrow NiOH + H^+ + e^- \qquad (4.5)$$

$$NiOH \rightarrow NiO + H^+ + e^- \qquad (4.6)$$

which dissolves chemically through

$$NiO + 2H^+ \rightarrow Ni^{2+} + H_2O \qquad (4.7)$$

Reaction (4.7), through purely chemical, is given an electrochemical rate law (Tafel law) and thus becomes strongly potential dependent. Haim et al. [103] do not discuss why a priori this should be so, but this assumption is crucial in the generation of the oscillations, as will be shown below.

The actual secondary passivation is, according to Haim et al., caused by bisulfate adsorption (ads):

$$Ni + HSO_4 \leftrightarrow NiHSO_{4,ads}^- \qquad (4.8)$$

At higher sulfuric acid concentration the bisulfate concentration of the solution becomes significant, and since bisulfate has a greater tendency (than sulfate) to adsorb onto the electrode surface it competes for surface sites and blocks the nickel ion path to the solution. Indeed, as Keddam et al. [104] mention, the secondary passivity is observed only in sulfuric acid electrolyte, indicating an essential role played by the anion. The bisulfate coverage θ_h is assumed to follow a potential-dependent isotherm:

$$\theta_h = \frac{K_a C_h}{1 + K_a C_h}(1 - \theta) \qquad (4.9)$$

with the equilibrium constant $K_a = K_a^\circ \exp(\alpha E)$, and where θ signifies the sum of the oxide and hydroxide coverage. This is reasonable since the negatively charged HSO_4^- will adsorb more strongly with more positive (anodic) potential. Finally, the Haim model takes into account oxidation of the solvent at very anodic potential:

$$H_2O \rightarrow 2H^+ + \tfrac{1}{2}O_2 + 2e^- \qquad (4.10)$$

which is assumed to occur with negligible rate on the oxide-covered part of the electrode.

The final model is reduced to two variables, the electrode potential E and the total oxide–hydroxide coverage θ_h. After the parameter values are estimated in a rather lengthy procedure, the model bifurcation diagram is calculated numerically in the $J - C_h$ parameter plane [103]. The result is shown in Fig. 44 and is seen to reproduce qualitatively the experimental diagram of Fig. 43.

As mentioned, Haim et al. based their model on papers by Keddam et al. [104], who studied the transpassive nickel dissolution by means of impedance measurements. Before discussing the kinetic Keddam model, it is interesting first to have a look at their impedance data. Figure 45 shows the steady-state polarization curve and the impedance diagrams obtained at the various potentials A–F. Note that the transition from plot C to D must be accompanied by a plot for which the real impedance Z' becomes infinitely negative for some finite (i.e., nonzero) frequency. According to the theory outlined in Section II, such a transition represents a Hopf bifurcation under galvanostatic conditions, and this is clearly in agreement with the results obtained by Osterwald and Lev, who observed the development of sustained potential oscillations in this region (Fig. 42). Keddam et al. noted that at potentials close to the current maximum, the impedance diagram again takes a shape similar to B or C. This is also in quite remarkable agreement with the original Osterwald experiment, who observed a second Hopf bifurcation at the point A' in his experiment, close to the current maximum. Keddam et al., who were well aware of Osterwald's work, noted that their impedance data were an experimental verification of Osterwald's deduction [100] that oscillations require a second-order dynamical system [i.e., a system with (at least) two dependent variables], and that therefore the impedance should reveal

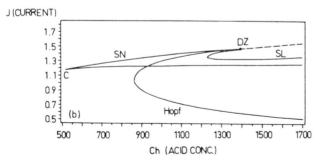

Figure 44. Theoretical bifurcation diagram predicted by the model proposed by Haim et al. [103], to be compared with Fig. 43. [Reproduced with permission from D. Haim, O. Lev. L. M. Pismen, M. Sheintuch, *J. Phys. Chem.*, **96**, 2676 (1992). Copyright © 1992 American Chemical Society.]

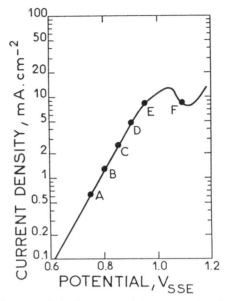

Figure 45. Steady-state polarization curve of the transpassive dissolution of (111) Ni single crystal in $1M$ H_2SO_4 at 25°C. A to F correspond to the impedance diagrams on the facing page. Indicated frequencies in hertz. (Reproduced from [104] with permission of The Electrochemical Society.)

a negative real part for some nonzero frequency, in spite of the positive zero-frequency slope. In this respect, Epelboin and Keddam [105] remarked that impedance plots D and E correspond to an oscillation circuit under constant current regulation. Strictly speaking, this is not an entirely accurate statement, as in fact all we can say about plots D and E is that under constant current load they will give rise to the unstable steady state with two unstable directions ($\#Z = 2$). Whether this runaway condition for the linearized system will result in sustained and stable oscillations for the real nonlinear system cannot be assessed, as is clearly illustrated by the region B–B′ in Fig. 41, where the impedance diagram has a similar shape, but no oscillations occur. On the other hand, since we know that the destabilization has occurred through a Hopf bifurcation, the existence of sustained oscillations is quite likely. As a final point, note that the impedance data indicate that a Hopf bifurcation will also occur in the fixed applied potential mode if the electrolyte (or external) resistance exceeds about 30 Ω cm^{-2}/A (A electrode area). Indeed, oscillations have been observed in this "almost potentiostatic mode" [106].

The kinetic model that Keddam et al. [104] presented in their most

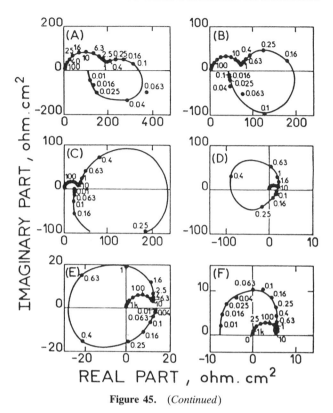

Figure 45. (*Continued*)

recent paper, is quite different from that of Haim et al. [103]. Keddam et al. also invoke the role of the HSO_4^- anion, but instead of a blocking of the surface they suggest a catalytic effect of the anion on the dissolution of the passive layer. This suggestion is based on their observation that an increase in the sulfate concentration (at constant pH) leads to a higher dissolution current (this could never be explained by the Haim model). In the Keddam-model, initially, the electrode is completely covered by a divalent passive layer, as the result of reactions such as Eqs. (4.5) and (4.6). At sufficiently anodic potentials, SO_4^{2-} and HSO_4^- attack the passive film and transform it into a more soluble species. The secondary passivity is explained by another type of passive film such as $Ni(OH)_3$. This explanation does not really address the question of why the secondary passivity only occurs in sulfate media. Keddam et al. do not give an explicit reaction chemistry, but only a simplified skeletal model that specifies the valency of the intermediate nickel species. We will not reproduce the details of the Keddam model here, but simply suffice with the remark that the model qualitatively reproduces most of the impedance data, showing that the linear stability diagram should be similar to the experiment and the Haim model (determination of global bifurcations would require explicit numerical integration of the full nonlinear model), although the underlying chemistry is quite different from the Haim model.

In addition to their experiments on the time-dependent behavior of the nickel dissolution, Lev et al. [106] also carried out experiments in which they measure the local current and potential distribution by equally spreading an array of 15 microelectrodes along a 20-cm nickel wire. In the galvanostatic mode, they observed that locally the oscillations are accompanied by traveling waves or nonmoving antiphase oscillations, in which there is a 180° phase shift between the current oscillations at the wire edges, so that the space-averaged current is constant. Of course, the "global" electrode potential, as measured sufficiently far from the electrode, still oscillates. The Haim model as outlined above was extended in order to account for these observations [107].

We can conclude from this section that the nickel dissolution in sulfuric acid is a prototype example of a "hidden" negative impedance oscillator, as clearly illustrated by the impedance plots of Fig. 45. It is fair to note that this fact was already mentioned by Osterwald [100], although maybe not in a particularly lucid fashion, and later by others [105, 108, 109]. As already explained in Section II.C, such a hidden negative impedance requires the coupling of a negative impedance process and a positive impedance process. In the Haim model the negative impedance process is the bisulfate adsorption that blocks the electrodissolution. It is not difficult to show that the positive impedance process leading to the galvanostatic potential oscillations in the Haim model, is the chemical but potential-dependent dissolution of the oxide layer, reaction (4.7). To show this, the Haim model can be studied in a form that is different though fully equivalent to the form suggested in the original paper. The model has two variables, electrode potential E and the sum of the oxide and hydroxide coverage, denoted by θ. It can be shown that a model consisting of coverage-like variables (in addition to the electrode potential) can be scaled to the following form [109]:

$$\varepsilon \frac{dE}{dt} = J - r_{nd} \qquad \frac{d\theta}{dt} = r_{of} - r_{od} \qquad (4.11)$$

where J is the applied current density, r_{nd} is the rate of nickel dissolution, r_{of} is the rate of oxide formation, r_{od} is the rate of oxide dissolution, and $\varepsilon = [C_d \Gamma_m / n F] \approx 0.1$, where Γ_m is the total number of surface sites. For simplicity, it is assumed that the oxygen evolution reaction (4.10) can be neglected in the region of the potential oscillations. Although the rate of oxide formation r_{of} is assumed to take place through Reactions (4.5) and (4.6), Haim et al. do not include it in the total faradaic current. The stability of the stationary states of Eqs. (4.11) can be studied by calculating the eigenvalues of the corresponding Jacobian matrix, for which the characteristic Eq. (3.14) can be derived with

$$T = \varepsilon^{-1}\left(\frac{\partial r_{nd}}{\partial E}\right) - \frac{\partial r_{of}}{\partial \theta} + \frac{\partial r_{od}}{\partial \theta}$$

$$\Delta = \varepsilon^{-1}\left[\left(\frac{\partial r_{nd}}{\partial E}\right)\left(\frac{\partial r_{od}}{\partial \theta} - \frac{\partial r_{of}}{\partial \theta}\right) - \left(\frac{\partial r_{nd}}{\partial \theta}\right)\left(\frac{\partial r_{od}}{\partial E} - \frac{\partial r_{of}}{\partial E}\right)\right]$$

where the partial derivatives are evaluated in the steady state. The conditions for a Hopf bifurcation are that $T = 0$ and $\Delta > 0$,

$$T = 0 \rightarrow \frac{\partial r_{nd}}{\partial E} = \varepsilon\left(\frac{\partial r_{of}}{\partial \theta} - \frac{\partial r_{od}}{\partial \theta}\right) < 0$$

$$\Delta = \varepsilon^{-1}\left[-\varepsilon\left(\frac{\partial r_{od}}{\partial \theta} - \frac{\partial r_{of}}{\partial \theta}\right)^2 - \left(\frac{\partial r_{nd}}{\partial \theta}\right)\left(\frac{\partial r_{od}}{\partial E}\right) + \left(\frac{\partial r_{nd}}{\partial \theta}\right)\left(\frac{\partial r_{of}}{\partial E}\right)\right] > 0$$

so that it clearly follows that $\partial r_{nd}/\partial E$ must be negative to have a zero trace, since $\partial r_{of}/\partial \theta < 0$ and $\partial r_{od}/\partial \theta > 0$. Since $\partial r_{nd}/\partial \theta < 0$, and $\partial r_{od}/\partial E > 0$ as well as $\partial r_{of}/\partial E > 0$, it follows that only a sufficiently positive $\partial r_{od}/\partial E$ can make the determinant positive. This clearly shows that if the chemical dissolution of the oxide layer by Reaction (4.7) were not assumed potential dependent, the Haim model would not be able to explain the galvanostatic potential oscillations.

V. OSCILLATORY ELECTROCATALYTIC OXIDATION PROCESSES

The study of oxidation reactions on electrocatalytic materials is a research field of substantial interest in electrochemical engineering for its importance in the design of more efficient fuel cells [110]. Kinetic investigations of the anodic oxidation of small organic molecules, hydrogen, hydrazine, and carbon monoxide are often frustrated by the occurrence of current and potential oscillations. For a long time, these oscillations were considered as rather pathological peculiarities, and their explanation was frequently assumed to be fairly trivial. In this respect, it is illustrative to quote an apparently quite plausible explanation for voltage oscillations under galvanostatic conditions, as given by Liebhavsky and Cairns [110] in their 1968 monograph on fuel cells and batteries: "A plausible general explanation for these phenomena is easy to give. During anodic reaction, inert compounds form either chemically or electrochemically and remain on the anode surface to block the reactions that yield electrons. When this blocking has gone far enough on an anode at which the overvoltage can increase (as it can in a galvanostatic experiment or in a fuel cell under resistive load), the overvoltage does increase to the point at which the species responsible for the blocking are oxidized. This purging action clears the anode surface so that the reactions yielding electrons can again proceed at reduced overvoltage. ... A quantitative explanation of voltage oscillations cannot be given."

It can be considered as the main aim of this section—and of this entire chapter for that matter—to show that caution should be exercised with assuming that things are that simple, and that the answer to the important question of why the system does not settle down in a stable steady state imperatively requires a (semi-) quantitative approach.

There are a number of mathematical models that attempt to explain oscillations in electrocatalytic reactions, but many of these are based on assumptions that lack a sound experimental basis. It is only quite recently that an acceptable qualitative model has been suggested for potential oscillations during hydrogen oxidation on platinum, the main features of which are supported by voltammetric and impedance measurements. It therefore seems opportune to start our discussion with this system, also because it serves as another noteworthy example of an electrochemical oscillator with a hidden negative impedance. As a second example, the oxidation of formic acid and formaldehyde on Pt-group metals is discussed, and perhaps this system is the most widely studied electrochemical oscillator. In spite of this activity, which has yielded some intriguing dynamical behavior, there is still quite a lot of conflicting information in the literature. Hence, there is, as yet, no fully satisfactory explanation of the oscillations in this system.

A. Hydrogen Oxidation on Platinum

The hydrogen redox reaction $2H^+ + 2e^- \leftrightarrow H_2$ is perhaps the most intensively studied electron reaction in electrochemistry. Not only has it been an important reaction in the elaboration of various important concepts of electrochemical kinetics [68], but the oxidation of molecular hydrogen also remains a prominent candidate for the anode reaction in electrochemical energy conversion. Oscillations during hydrogen oxidation, mainly on Pt, have been reported by many authors, but it seems that Horányi and Visy [111] were the first to undertake a systematic study of the influence of the electrolyte composition on the oscillations, in particular of electrosorbing cations and anions. This subject was taken up again by Kodera, Yamazaki and colleagues [112], who also developed a mathematical model in which the oscillations are caused by the attractive forces between the underpotential deposited metal atoms [113]. Krischer, Wolf, and co-workers [114–116] recently rejected this explanation and suggested a more realistic model that seems to include all the important qualitative features of the oscillatory H_2 oxidation on Pt in the presence of cations (Cu^{2+}, Bi^{3+}, Cd^{2+}, or Ag^+) and anions (Cl^- or Br^-). Both the experimental system as well as the model serve as a good example of a "hidden negative impedance oscillator" so that a detailed discussion of this system is worthwhile.

Figure 46(a) shows the hydrogen oxidation current in the absence and presence of a small amount of Cu^{2+} in the electrolyte [114]. In sulfuric

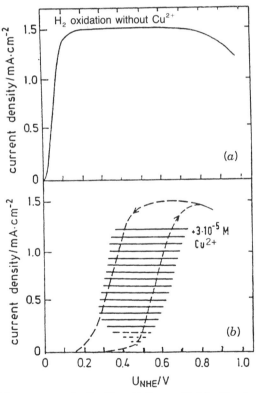

Figure 46. Voltammograms for the H_2 oxidation on Pt (a) without and (b) with Cu^{2+}. The hatched region in (b) represents the oscillation amplitudes observed in the galvanostatic control. Electrolyte 0.5 M H_2SO_4, 10^{-3} M Cl^-, saturated with H_2. (Reproduced from [114] with permission of the author.)

acid electrolyte, the hydrogen oxidation on platinum is known to be a fast process [67], leading to a diffusion-limited current for relatively low overpotential [Fig. 46(a)]. In the presence of Cu^{2+}, copper exhibits an underpotential deposition (upd), and the isotherm determined by Krischer et al. [115] suggest that the relative copper coverage decreases from 1 (monolayer) to 0 from about 0.3 to 0.6 V (vs. NHE). This is clearly the region where the hydrogen oxidation current is seen to increase in Fig. 46(b), suggesting that the catalytic action of the copper surface is negligible, and in fact inhibiting the hydrogen oxidation. In the galvanostatic operation, the presence of Cu^{2+} and Cl^- leads to strong potential oscillations around the positively sloped branch in the polarization curve [see Fig. 46(b)].

In addition to the influence of cations in the electrolyte (Ag and Bi have an effect very similar to Cu), there is also a strong effect of the anions Cl^- and Br^-. These halides tend to adsorb onto the platinum surface (more strongly with more anodic potential) and to reduce the oxidation current. Although it is tempting to ascribe the decrease in oxidation current at potentials more anodic than 0.8 V to the oxidation of the Pt surface (OH adsorption), cyclic voltammetry shows that the onset of oxide formation is shifted to about 1.0 V in the presence of Cl^- [114, 115]. Therefore it is suggested [114, 115] that the negative slope is caused by the combined (specific and electrostatic) adsorption of SO_4^{2-}, HSO_4^-, and—in particular—Cl^- (or Br^-). Ring–disk voltammetry at different scan rates further supplies the important information that Cu upd is a relatively slow and Cl^- adsorption a relatively fast process, which also explains why in the negative slope region (0.6–0.9 V) there is negligible hysteresis, whereas the positive slope region (0.3–0.6 V) displays a substantial hysteresis [114, 115].

The occurrence of the galvanostatic potential oscillations has been studied in the parameter space spanned by the applied current I, bulk copper concentration $c_{Cu^{2+}}$, and bulk chloride (or bromide) concentration c_{Cl^-} [56, 114, 116, 117]. The existence regions of simple or more complex (mixed-mode) oscillations are illustrated in Figs. 47 and 48. From these figures it can be seen that the critical current value for the onset of potential oscillations increases with decreasing halide concentration and increasing copper concentration. Another meaningful observation is that oscillations are not observed below a certain c_{Cl^-}, whereas no such lower limit seems to exist for $c_{Cu^{2+}}$ (although in the complete absence of copper cations no oscillations are to be expected, of course [see Fig. 46(a)]).

As can be seen from Figs. 47 and 48, the experiments yield a regular structure of complex oscillations. An example of a typical sequence of MMO oscillations (on Pt in the presence of Cu^{2+}) is shown in Fig. 49.

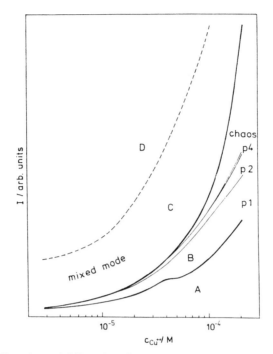

Figure 47. Experimental bifurcation diagram of the existence regions of different oscillation forms during the galvanostatic H_2 oxidation on Pt as a function of the Cu^{2+} concentration; pN period-N oscillations. [Reproduced with permission from K. Krischer, M. Lübke, W. Wolf, M. Eiswirth, J. L. Hudson, and G. Ertl, *Physica D* **62**, 123 (1993).]

Note that these MMOs are different from those observed for the In/SCN and the Cu/H_3PO_4 systems (Sections III and IV.B, respectively). In the present case, the small oscillations do not always grow continuously before giving way to a relaxation peak, but have a more complex structure. The middle sequence of Fig. 49 illustrates this, where it is seen that the small oscillations are chaotic and the system escapes from this small-amplitude chaotic attractor every once in a while. Krischer et al. [56] made a detailed study of these MMOs for the case of copper cations and chloride anions, and argued that their appearance and bifurcations are consistent with the existence of a so-called interior crisis [56]. "Interior crisis" is a general name for a global bifurcation in which a small-amplitude chaotic attractor becomes tangent to the separatrix associated with a saddle-type fixed point or orbit. As soon as the chaotic attractor reaches the other side of the separatrix, this will lead to a large excursion (i.e., large amplitude oscillation). Some reinjection process will throw the system back onto the chaotic attractor, until it reaches the

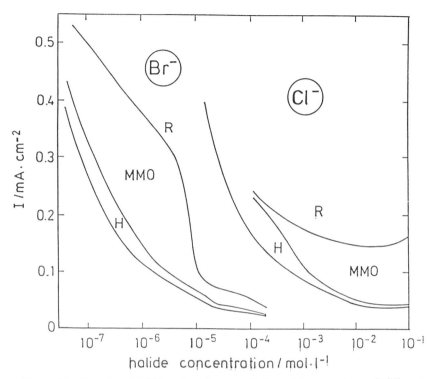

Figure 48. Experimental bifurcation diagram of the existence regions of different oscillation forms during the galvanostatic H_2 oxidation on Pt as a function of the chloride and bromide concentration. H, harmonic oscillations, MMO mixed-mode oscillations, R relaxation oscillations. (Reproduced from [118] with permission of Elsevier Sequoia S.A.)

other side of the separatrix again to make another large excursion. With increasing distance from the interior crisis bifurcation, the average length of intervals on the original small attractor should decrease continuously, in agreement with what has been found experimentally [56].

Wolf et al. [118] proposed a simplified model to reproduce the basic qualitative phenomena of this system. The model consists of three variables, the electrode potential E and the electrode surface coverages of Cl^- (denoted by θ_a) and Cu (denoted θ_m). The hydrogen oxidation current is assumed to be potential dependent, but only through the potential-dependent values of θ_a and θ_m. The time dependence of the respective coverages is modeled with the phenomenological rate laws:

$$\dot{\theta}_a = v^a_{ads} - v^a_{des} \qquad (5.1)$$

$$\dot{\theta}_m = v^m_{ads} - v^m_{des} \qquad (5.2)$$

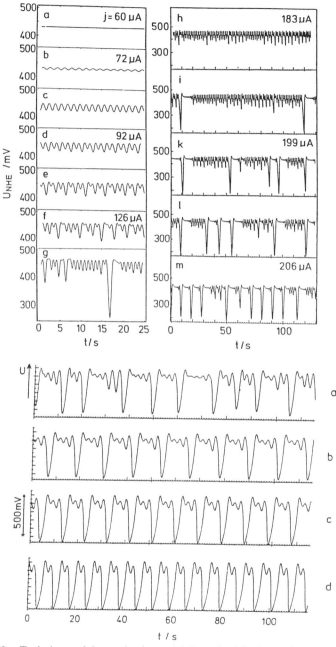

Figure 49. Typical sets of time series (potential E vs. time) for increasing values of the applied current density for three different electrolytes. Top left (a–g): 1 M HClO$_4$, 1.5×10^{-4} M Cu^{2+}, 5×10^{-4} M Cl$^-$, H$_2$ saturated. Top right (h–m): 1 M HClO$_4$, 3×10^{-5} M Cu^{2+}, 3×10^{-5} M Cl$^-$, H$_2$ saturated. Bottom (a–d): time series showing a transition from aperiodic (a) to periodic (b–d) MMOs; 1 M HClO$_4$, 2×10^{-5} M Cu^{2+}, 3×10^{-5} M Cl$^-$, H$_2$ saturated. (a) $J = 99.7$, (b) 101.5, (c) 109, and (d) 123.5 μA cm^{-2}. (Reproduced from [116] with permission of the VCH Verlagsgesellschaft mbH.)

with

$$v_{ads}^a = k_a c_a (1 - \theta_a) \exp[a_a(E - E_0^a)]$$
$$v_{des}^a = k_a \theta_a \exp[-a_a(E - E_0^a)]$$
$$v_{ads}^m = k_m c_m (1 - \theta_m) \exp[-a_m(E - E_0^m)]$$
$$v_{des}^m = k_m \theta_m \exp[a_m(E - E_0^m)]$$

(The dimensionless bulk concentrations c_a and c_m are fractions of the standard concentration 1 M.) Setting these derivatives equal to zero, one obtains the potential-dependent steady-state coverages and the potential-dependent isotherms shown in Fig. 50. Due to the high anodic overpotential, the hydrogen oxidation current is set equal to the rate-determining (dissociative) adsorption of hydrogen onto the available free sites, namely,

$$j_{H_2} = k_{ads}^{H_2} c_{H_2}^{surf} (1 - \theta_a - \theta_m)^2 \tag{5.3}$$

where $k_{ads}^{H_2}$ is the adsorption rate constant in amps per square centimeter. A steady-state approximation is now made for the surface concentration $c_{H_2}^{surf}$, which is replenished by diffusion:

$$j_{H_2} = k_{dif}^{H_2}(c_{H_2}^{bulk} - c_{H_2}^{surf}) \tag{5.4}$$

Equating these two expressions, one gets

$$j_{H_2} = \frac{k_{ads}^{H_2} c_{H_2}^{bulk} (1 - \theta_a - \theta_m)^2}{1 + \dfrac{k_{ads}^{H_2}}{k_{dif}^{H_2}} (1 - \theta_a - \theta_m)^2} \tag{5.5}$$

The differential equation for the electrode potential E follows from the charge balance equation:

$$C_d \dot{E} = j_{appl} - j_{H_2} - j_m \tag{5.6}$$

where j_m is the current density associated with the underpotential deposition of a monolayer of metal:

$$j_m = -q_{mono} \dot{\theta}_m \tag{5.7}$$

and j_{appl} is the applied current density.

Two bifurcation diagrams calculated by Wolf et al. [118], giving the

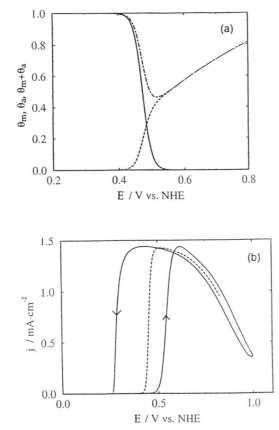

Figure 50. (*a*) Calculated coverages as a function of E in a solution with $c_m = 10^{-4}$ and $c_a = 10^{-3}$. Solid line: θ_m, dashed line θ_a, dot–dashed line $\theta_a + \theta_m$. (*b*) Calculated hydrogen oxidation current density as a function of potential in a solution with $c_m = 10^{-4}$ and $c_a = 10^{-3}$. Dashed line: stationary curve, solid line: calculated cyclic voltammogram at a scan rate of 20 mV/s. (Reproduced from [118] with permission of Elsevier Sequoia S.A.)

locations of the most important bifurcations in the $j_{appl} - c_m$ and $j_{appl} - c_a$ planes, are shown in Fig. 51. Note that the locus of the Hopf bifurcation, implying the onset of sustained oscillation, is in the qualitative agreement with the experimental diagrams of Figs. 47 and 48. That is, the model correctly predicts that for increasing c_m the minimum j_{appl} needed to induce potential oscillations is shifted to higher values, whereas the opposite occurs for increasing c_a.

As outlined in Sections II and IV.C, galvanostatic potential oscillations are expected if a negative impedance process (Cl$^-$ adsorption in the present case) has a higher rate than the positive impedance process (Cu

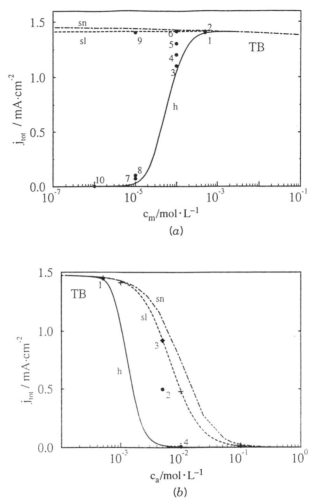

Figure 51. Theoretical bifurcation diagrams, Hopf (h), saddle-loop (sl), saddle-node (sn) and Takens–Bogdanov (TB) bifurcation as a function of (a) c_m and j_{appl} at $c_a = 10^{-3}$ and (b) as a function of c_a and j_{appl} at $c_m = 10^{-4}$. (Reproduced from [118] with permission of Elsevier Sequoia S.A.)

upd), so that roughly one would expect $k_a > k_m$. The model confirms this condition since Wolf et al. present a bifurcation diagram in the $j_{appl} - k_a$ plane showing that the Hopf bifurcation disappears in a Takens–Bogdanov bifurcation (two zero eigenvalues) for $k_a = 1.126 \ M \ s^{-1} = 1.126 \ k_m$.

Recently, an impedance analysis of the hydrogen oxidation on platinum has been carried out in order to test the presence of a "hidden" negative impedance on the positively sloped branch in Fig. 46(b) [119]. To get an idea of the kind of impedance plots expected for this system, first the above model may be further simplified by adiabatically eliminating the variable θ_a, since this has to be a "fast variable" in order to induce oscillations. So θ_a now depends parametrically on E and θ_m:

$$\theta_a^{ss} = \frac{1 - \theta_m}{1 + \exp[-2a_a(E - E_0^a)]/c_a} \tag{5.8}$$

If a sine wave excitation ΔE is superimposed on the dc electrode potential E_s, θ_m will give rise to a delayed response:

$$E = E_s + \Delta E = E_s + \Delta E^\circ \exp(j\omega t) \rightarrow \theta_m = \theta_{m,s} + \Delta\theta$$

$$= \theta_{m,s} + \Delta\theta_m^\circ \exp[j(\omega t + \phi)] \tag{5.9}$$

so that the faradaic impedance is defined by [see Eq. (1.6)]

$$Z_F^{-1} = \left(\frac{\partial j_F}{\partial E}\right)_{\theta_m} + \left(\frac{\partial j_F}{\partial \theta_m}\right)_E \frac{\Delta\theta_m}{\Delta E} \tag{5.10}$$

where

$$\left(\frac{\partial j_F}{\partial E}\right)_{\theta_m} = \left(\frac{\partial j_F}{\partial \theta_a}\right)_{E,\theta_m} \left(\frac{\partial \theta_a^{ss}}{\partial E}\right)_{\theta_m} + \left(\frac{\partial j_F}{\partial E}\right)_{\theta_a,\theta_m}$$

$$\left(\frac{\partial j_F}{\partial \theta_m}\right)_E = \left(\frac{\partial j_F}{\partial \theta_a}\right)_{E,\theta_m} \left(\frac{\partial \theta_a^{ss}}{\partial \theta_m}\right)_E + \left(\frac{\partial j_F}{\partial \theta_m}\right)_{\theta_a,E}$$

$$\frac{\Delta\theta_m}{\Delta E} = \frac{\partial \dot{\theta}_m}{\partial E} \bigg/ \left(j\omega - \frac{\partial \dot{\theta}_m}{\partial \theta_m}\right)$$

The model makes definitive predictions about the signs of three of the four partial derivatives, namely,

$$\left(\frac{\partial j_F}{\partial \theta_m}\right)_E < 0 \qquad \frac{\partial \dot{\theta}_m}{\partial E} < 0 \qquad \frac{\partial \dot{\theta}_m}{\partial \theta_m} < 0 \tag{5.11}$$

whereas $(\partial j_F/\partial E)_{\theta_m}$ can have either the positive or the negative sign. This allows us to interpret the faradaic impedance as the impedance of an equivalent electrical circuit of nonreducible elements, as given by

$$Z_F^{-1} = \left(\frac{\partial j_F}{\partial E}\right)_{\theta_m} + 1 \bigg/ \left[\frac{j\omega}{\left(\frac{\partial j_F}{\partial \theta_m}\right)_E \frac{\partial \dot\theta_m}{\partial E}} - \frac{(\partial \dot\theta_m / \partial \theta_m)}{\left(\frac{\partial j_F}{\partial \theta_m}\right)_E \frac{\partial \dot\theta_m}{\partial E}}\right]$$

$$= \frac{1}{R_{ct}} + \frac{1}{j\omega L + R_0} \tag{5.12}$$

hence

$$R_{ct} = \left(\frac{\partial j_F}{\partial E}\right)_{\theta_m}^{-1} \qquad L = \frac{1}{(\partial j_F / \partial \theta_m)_E (\partial \dot\theta_m / \partial E)} (>0)$$

$$R_0 = -\frac{\dfrac{\partial \dot\theta_m}{\partial \theta_m}}{(\partial j_F / \partial \theta_m)_E (\partial \dot\theta_m / \partial E)} (>0)$$

Taking into account that $Z_{tot} = [1/Z_F + j\omega C_d]^{-1}$, the total interfacial impedance can be represented by the circuit of Fig. 52. It may be noted that the same equivalent circuit may be derived from other basic assumptions regarding the electrochemical kinetics [120, 121], and it is therefore by no means unique to this particular system. (Neither is the fact that this circuit can explain the experimental impedance rather well an irrefutable proof that the model is correct or that no physical effects of importance have been left out.)

This circuit predicts a number of qualitatively different impedance plots. Let us assume that as we change the dc potential E_s, first the charge-transfer resistance R_{ct} goes through zero, and for higher E_s the

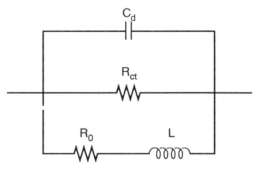

Figure 52. Equivalent circuit representation of Eq. (5.12). Symbols are explained in the text.

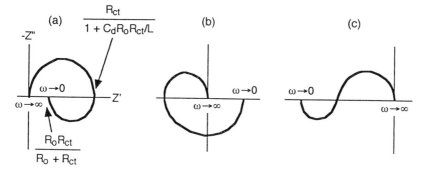

Figure 53. Sequence of three typical impedance diagrams of the circuit in Fig. 53 illustrating the transition from (a) a galvanostatically stable steady-state, exhibiting a Hopf bifurcation to (b) a galvanostatically unstable steady state with $\#Z = 2$, exhibiting a saddle-node bifurcation to (c) a galvanostatically unstable steady state with $\#Z = 1$.

polarization slope $1/R_{ct} + 1/R_0$ becomes negative. This is what one would one expect to be a typical behavior on the basis of Fig. 46(b). The sequence of qualitatively different impedance plots is depicted in Fig. 53. For (relatively) low E_s, the plot in Fig. 53(a) is obtained. An important frequency here is that at which the real impedance axis is intersected, which occurs for $Z' = R_{ct}/(1 + C_d R_0 R_{ct}/L)$. [Note that this is equal to R_{ct} if the time constant for the $R_{ct}C_d$ part of the circuit ($\tau_{RC} = R_{ct}C_d$) is much shorter than that of the $R_0 L$ part ($\tau_{RL} = L/R_0$)]. From the theory of Section II, we can expect a galvanostatic Hopf bifurcation if Z' suddenly switches from $+\infty$ to $-\infty$, or more precisely $Y(\omega_c) = 0$, provided that the polarization slope remains positive. Hence, $1 + C_d R_0 R_{ct}/L = 0$ and $1/R_{ct} + 1/R_0 > 0$, or

$$R_{ct} = -\frac{L}{C_d R_0} \wedge \frac{C_d R_0^2}{L} < 1 \text{ (galvanostatic Hopf)} \qquad (5.13)$$

where the latter inequality is normally satisfied if the electrical double-layer capacity is small enough. The new plot after this transition is shown in Fig. 53(b). This plot is transformed to the plot in Fig. 53(c) if the polarization slope changes sign, which corresponds to a maximum in the I–E curve, that is, for $1/R_{ct} + 1/R_0 = 0$, or

$$R_{ct} = -R_0 \text{ (galvanostatic saddle-node)} . \qquad (5.14)$$

This typical sequence of impedance plots has indeed been found experimentally in the hydrogen oxidation on platinum. Figure 54(a) illustrates the transition through a galvanostatic Hopf bifurcation in the

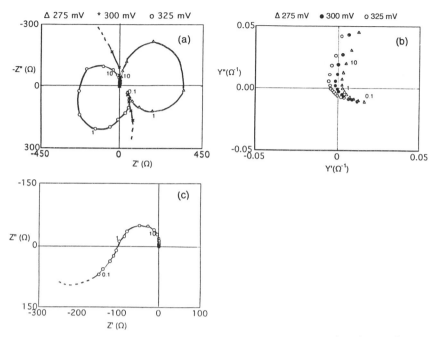

Figure 54. (*a*) Impedance diagrams of the H_2 oxidation on Pt undergoing a galvanostatic Hopf bifurcation. Electrolyte 0.5 M H_2SO_4, 10^{-4} M Cu^{2+}, 10^{-4} Cl^-, saturated with H_2. All potentials refer to Ag/AgCl/sat. KCl. Frequencies range from 10^5 to 10^{-1} with five frequencies per decade on a log-scale. (*b*), as in (*a*) but now plotted in the complex admittance plane. (*c*) Impedance diagram on a branch in the $I–V$ curve with negative slope. Electrolyte 0.5 M H_2SO_4, 10^{-3} M Bi^{3+}, 10^{-2} M Cl^-, H_2 saturated. $V = 800$ mV.

presence of copper cations (a similar transition was observed by Keddam and co-workers for the Ni dissolution in H_2SO_4, as has already been discussed in Section IV.C). Figure 54(*b*) plots the same data in the admittance plane, where it is more easily seen that the Hopf bifurcation is indeed characterized by a zero admittance. Figure 54(*c*) shows an example of the impedance plot in Fig. 53(*c*), which was taken on the negative slope branch of the hydrogen oxidation in the presence of bismuth cations, although the lowest frequencies that should converge to real axis were too unreliable to be measured. Incidentally, in most cases that the impedance plot was measured for an E_s on the negatively sloped branch, the lowest frequencies normally clustered around the negative real axis without unambiguously crossing it, yielding approximately a semicircle in the second quadrant. In principle, this must involve a transition from the plot in Fig. 53(*c*) to a single semicircle, occurring for $R_{ct} = -R_0$ and $C_d/L = 1/R_0^2$. This event is known as a Takens–Bogdanov

(TB) bifurcation in dynamical system theory [11, 12], which is a single point in a two-parameter bifurcation diagram where the Hopf and saddle-node lines meet (note the presence of such a TB bifurcation in the phase diagrams of Fig. 51).

Finally, two points may be noticed from this analysis. First, the experimental plots are in good agreement with the plots predicted by the simple analysis in which the anion fractional coverage was eliminated adiabatically, at least as far as the qualitative shape is concerned, and this can be interpreted as evidence for the idea that the limit of high k_a is a realistic approximation. For the original model simulations in [116], k_a was taken equal to 10, for which impedance plots can be simulated, which are quantitatively different from those of Fig. 54(a), in particular at the highest frequencies. A higher k_a seems quantitatively more realistic, although the qualitative story remains the same, of course. Second, the impedance behavior clearly predicts the possibility of observing current oscillations under potentiostatic conditions, provided a sufficiently large external resistance is added to the circuit [for the plot of Fig. 54(a), one needs about 250 Ω]. These current oscillations can indeed be observed experimentally [119].

B. Formic Acid and Formaldehyde Oxidation on Platinum and Platinum-Group Metals

There is an extensive literature on the oxidation of small organic molecules on noble platinum- or platinum-group metals, primarily because such substances could serve as cheap and safe anolytes for fuel cells. A good deal of the research literature has been devoted to the occurrence of spontaneous current and potential oscillations, especially during formic acid (HCOOH) and formaldehyde (HCHO) oxidation, as these two are among the simplest organic molecules. Such periodicity was first observed by Mueller [122] and has been studied ever since by many authors. Although important progress has been made, concerning both dynamic as well as electrochemical aspects of the oscillatory behavior, there is as yet no consensus about the fundamental feedback mechanism leading to the instabilities and oscillations. It is remarkable that most authors invoke an autocatalytic surface chemistry as the cause of the oscillations, and the role of the external circuit remains underexposed in practically all studies. Only the recent paper by Raspel and Eiswirth [123] explicitly mentions this important point, and represents a justification for the assumption of a negligible influence of the external circuit (i.e., the ohmic potential drop).

Since the various experimental findings are only occasionally in good qualitative accord, presumably owing to the large number of control parameters that may vary from one laboratory to the other (trace impurities, electrode pretreatment, electrode size, cell construction, cell temperature, transport conditions, external circuit, and control...), it is neither feasible nor useful to give an exhaustive overview. The discussion will therefore start with some commonly accepted qualitative aspects of the mechanism of HCOOH oxidation on platinum [124–127], and will subsequently treat some of the more relevant studies pertaining to the oscillatory behavior, and discuss how these papers have employed (or modified) the mechanism in order to explain some of the experimental observations. It may be good to note that, in general, the oxidation of a small one-carbon organic molecule to carbon dioxide requires a change in the oxidation state of the carbon atom of 4 (formic acid), 5 (formaldehyde), or 6 (methanol) and, in addition, the breaking and formation of chemical bonds, so that the mechanism may be expected to be very complex and to involve several intermediates. Figure 34 from a recent review by Beden et al. [124] on the mechanistic aspects of the oxidation of small organics illustrates this nicely.

In spite of these unfavorable complications, a consensus has now grown in the literature about the idea, originally suggested by Capon and Parsons [125], that the oxidation of formic acid (and of other small organics, for that matter) on platinum takes place via a dual path mechanism:

$$\text{HCOOH} \nearrow \text{reactive surface intermediate} \rightarrow CO_2 \atop \searrow \text{unreactive surface intermediate (poison)} \rightarrow CO_2 \quad (5.15)$$

The two pathways will be referred to as the direct and indirect path, respectively. Spectroelectrochemical measurements have identified the poisoning intermediate as adsorbed carbon monoxide, which can be bonded in more than one configuration to the platinum surface. For the reactive intermediate, an adsorbed radical ·COOH species has generally been proposed.

Therefore, the reactive, direct path can in its simplest and possibly naive form be written as consisting of two consecutive steps:

$$HCOOH + Pt \rightarrow \cdot COOH_{ad} + H^+ + e^- \quad (5.16)$$

$$\cdot COOH_{ad} \rightarrow Pt + H^+ + CO_2 + e^- \quad (5.17)$$

where Pt denotes an unoccupied surface site. The indirect path involves the heterogeneously catalyzed dissociative adsorption of formic acid leading to the production of adsorbed CO:

$$HCOOH + Pt \rightarrow CO_{ad} + H_2O \qquad (5.18)$$

Although other proposals for the formation of the poison have been made, it is very likely a nonelectrochemical process, that is, not involving electron transfer, because the oxidation state of the C atom in HCOOH and CO is the same. Removal of the poisoning adsorbed CO occurs electrochemically by combining with adsorbed OH, which is formed if the electrode potential is sufficiently anodic to induce oxidation of the Pt surface:

$$H_2O + Pt \leftrightarrow OH_{ad} + H^+ + e^- \qquad (5.19)$$

$$CO_{ad} + OH_{ad} \rightarrow 2Pt + H^+ + CO_2 + e^- \qquad (5.20)$$

Reactions (5.18)–(5.20) provide the simplest possible version of the indirect path. We can use the above reactions to interpret some features of a typical cyclic voltammogram of the formic acid oxidation on Pt, as shown in Fig. 55 (2 M NaCOOH, 0.5 M H$_2$SO$_4$, 50°C) [128].

During (electrochemical) pretreatment and the subsequent relaxation at open circuit, the platinum-surface gets poisoned with adsorbed CO$_{ad}$ according to Reaction (5.18) or some other nonelectrochemical mechanism. Therefore, on the forward scan, in the anodic direction, the oxidation current is very low owing to the poisoned surface [126, 145]. Only after the platinum surface becomes oxidized around 0.4–0.5 V, can the poison be electrochemically oxidized by Reaction (5.20), at potentials more anodic than about 0.6 V. At even more anodic potentials, a further electrochemical oxidation of the surface may occur, for which one generally writes a reaction such as [124, 126, 129]:

$$OH_{ad} \rightarrow O_{ad} + H^+ + e^- \qquad (5.21)$$

which can result in a secondary passivation involving some higher valency surface oxides. In the reverse scan, in the cathodic direction, the oxide layer is reduced around 0.4 V (the initial stages of oxide-layer formation are usually found to be reversible), and since the surface is now free from adsorbed CO$_{ad}$, a strongly enhanced oxidation current is observed as compared with the forward scan, as there are now sufficient sites for the direct path. In this interpretation, the negative slope observed in the reverse scan, which clearly might be important for the explanation of the

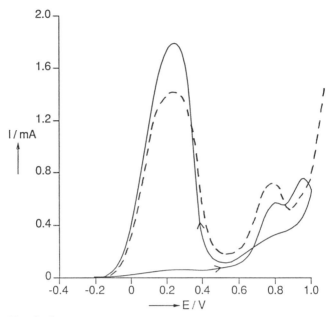

Figure 55.   Cyclic voltammogram of 2.0 CHOONa in 0.5 H_2SO_4 at a platinum rotating disk (3000 rpm) at 40°C, scan rate 50 mV s^{-1}. The dashed line represents the presumed steady polarization curve of the formate oxidation in absence of the indirect path (see [131–133]).

oscillations, is due to (hydr)oxide formation on the platinum surface. The reverse scan can therefore be interpreted as a reasonable approximation to the reaction's stationary current–voltage curve in the absence of the indirect path, as illustrated by the dashed curve in Fig. 55. When the cathodic potential limit in the cyclic voltammogram is chosen in the potential region of oxide formation, Lamy et al. [130] observed a strongly reduced hysteresis in the sweep, indicating improved reversibility of the poison removal. It may be noted that these authors interpret the low current in the anodic scan differently by invoking a surface blockage by adsorbed hydrogen at sufficiently cathodic potentials. Note that in the dashed curve in Fig. 55, we have drawn two regions of negative slope, due to the primary and secondary passivation, respectively. Several authors pointed out the existence of two such passivating regions [131–133]. It will become clear below why this may be meaningful.

Current oscillations have indeed been observed under the same conditions for which the cyclic voltammogram was recorded, provided a sufficiently large external resistor (1–4 kΩ) was connected in series with the working electrode [128]. A few examples of the more complex oscillations are shown in Fig. 56, where the electrode potential E is plotted instead of the current I (of course, $E = V - IR_s$). The most typical sequence of states observed with increasing applied potential V can be summarized as follows:

1. At low V the steady-state bifurcates in a Hopf.
2. A transition to MMO occurs [Figs. 56(a and b)], where the number of small amplitudes becomes fewer with increasing V, finally resulting in the following reference.
3. A simple relaxation oscillation.

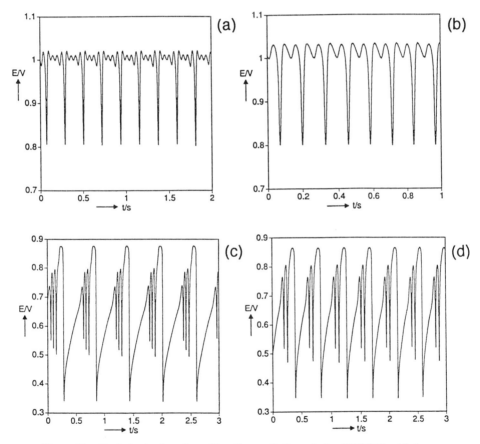

Figure 56. Typical mixed-mode and bursting oscillations during HCOOH oxidation on a Pt disk. (a) 1^3 MMO, $V = 2.00$ V, $R_s = 4000$ Ω, 3000 rpm; (b) 1^1 MMO, $V = 2.00$ V, $R_s = 4000$ Ω, 3000 rpm; (c) 1^1 burst, $V = 2.48$ V, $R_s = 4000$ Ω, 3000 rpm; (d) 1^2 burst, $V = 2.29$ V, $R_s = 4000$ Ω, 3000 rpm.

4. The simple relaxation oscillation gets interspersed with an even larger amplitude, and thus transforms to a 1^n "bursting" oscillation, with n from 1 to many, the 1^3 and 1^2 states shown in Fig. 56(c and d). The number of smaller peaks becomes fewer, finally resulting in a simple 1^0 relaxation peak.

5. Transition to steady state.

These observations are remarkably similar to earlier results by Albahadily and Schell [134] who carried out a detailed study of the characterization of (complex) oscillatory states in the galvanostatic oxidation of formic acid on platinum (2 M NaCOOH, 1 M H$_2$SO$_4$, 50°C). Figure 57 shows the measured (approximately stationary) potential responses at different values of the applied current. This figure demonstrates that the system supports an oscillatory response over a significant range of values of the applied current and that the system exhibits bistability, that is, one of two different forms of asymptotic behavior can be realized under the same conditions. Albahadily and Schell give an explanation of this bistability, which can be understood on the basis of the dashed curve in Fig. 55, that is, the stationary I–E curve with neglect

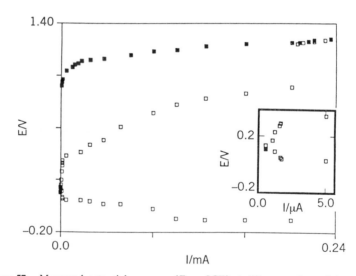

Figure 57. Measured potential response (E vs. SCE) at different values of the applied current I for the oxidation of 2.0 COOHNa in 1.0 M H$_2$SO$_4$ on a Pt rotating disk at 50°C. Open squares represent data recorded after changing the current in the positive direction. Two open squares for one value of the current represent maximal and minimal values of an oscillating potential. Closed squares represent data recorded after changing the current in the negative direction. Inset: an enlargement of the low current region. [Reproduced from F. N. Albahadity and M. Schell, *J. Electroanal. Chem.*, **308**, 151 (1991).]

of the indirect path. By disregarding the second passivation region, this curve clearly illustrates that a range of I values should exist for which the electrode potential can choose between three values, one of which is always unstable. This galvanostatically unstable branch is due to oxide-layer formation, or what Albahadily and Schell call the oxide-feedback mechanism [134]. In this simple interpretation, however, the low-potential direct oxidation path should be galvanostatically stable, as Albahadily and Schell remark.

According to Albahadily and Schell, destabilization of the low-potential oxidation branch occurs through the coupling with the indirect path. Formation of the poison on the electrode will cause substantial increases in E, until a potential is reached where the oxidation of the platinum surface starts. Oxygenated species such as OH then react autocatalytically with the poison, cleaning the electrode surface. Albahadily and Schell propose, following a suggestion by Wojtowicz et al. [135], that Reaction (5.20) may need another free site, namely,

$$OH_{ad} + CO_{ad} + Pt \rightarrow 3Pt + CO_2 + H^+ + e^- \qquad (5.22)$$

which is autocatalytic in the generation of free sites. In fact, Wojtowicz [135] originally suggested that the rate law for Reaction (5.20) involves a factor of $\exp(\gamma\theta_{OH})$, which also has the properties of an explosive rate law if $\gamma < 0$. The autocatalytic character of the cleaning of the electrode is essential to obtain oscillations, as Albahadily and Schell stress, and they present a detailed kinetic model based on these ideas that indeed allows for oscillations, although they do not mention whether the bifurcation diagram of Fig. 57 is qualitatively reproduced by this model (this would clearly provide an important test, more important than the explanation of oscillations alone).

It may be noticed that the schematic bifurcation features illustrated in Fig. 57, are actually quite similar to those of the hydrogen oxidation on platinum as depicted in Fig. 46(b). One may therefore argue against the idea of the autocatalytic "cleaning" reaction by noting that it is not imperative to invoke such an additional feedback mechanism in order to explain galvanostatic potential oscillations. The oxide feedback mechanism is in itself sufficient to explain potential oscillations, provided that it is coupled to a positive impedance process that hides the negative slope associated with the oxide formation. For instance, a simple "toy model" has been considered recently [109] in which a potential-dependent but nonelectrochemical adsorption precedes the charge transfer. Combined with a negative impedance, for example, due to oxide passivation, this simple model can qualitatively reproduce the bifurcation diagram of Fig.

57. Although there is no claim that this model is the correct kinetic interpretation of the oscillations, it demonstrates that an additional autocatalysis is in principle redundant for explaining the experimental findings of Albahadily and Schell.

As mentioned above, Albahadily and Schell carried out a detailed study of intriguing oscillation patterns that occur in the oscillatory region of Fig. 57. Sequences of waveforms were found on increasing the applied current in small increments; this is similar to the sequence described above in which the applied potential V is increased for a fixed value of R_s. At low current, from 1.2 to 1.75×10^{-3} mA (electrode surface ~0.8 cm^2), a sequence of MMO states occurs, with an increasing number of small oscillations on increasing I. These transformed into a family of large-amplitude oscillations with (very) long period (7 h) presumably involving a homoclinic orbit at $I \approx 4.15 \times 10^{-3}$ mA. After this critical value the period shortens again giving way to another sequence of MMOs in which the number of small oscillations decreased with increasing I. For 5.30×10^{-3} mA a simple monoperiodic waveform is observed, which for $I > 6.00 \times 10^{-2}$ mA becomes mixed with an even larger amplitude spike, resulting in the complex "bursting-type" oscillation pattern of Fig. 58. Upon further increasing I, the ratio of small over large relaxation oscillations becomes smaller, as illustrated with the various waveforms in

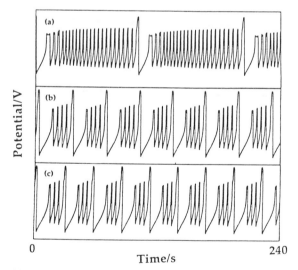

Figure 58. Measured waveforms of mixed "bursting" oscillations. Conditions as in Figs. 5.12. Potential ranges from −200 to 900 mV. (*a*) $I = 60.0$ μA. (*b*) 1^4 state, 64.0 μA (*c*) 1^3 state, 65.0 μA. [Reproduced from F.N. Albahadity and M. Schell, *J. Electroanal. Chem.*, **308**, 151 (1991).]

Fig. 59. These complex waveforms can be designated a symbolic notation L^S, L and S denoting the number of large and small amplitudes in one periodic entity, respectively. Note from Fig. 59 the existence of complex Farey states, that is, periodic mixtures of adjacent principal states. A largely similar behavior—including the bistability of Fig. 57—has been found in the formaldehyde HCHO oxidation on Pt [136, 137].

In all waveforms presented by Albahadily and Schell, the maximum

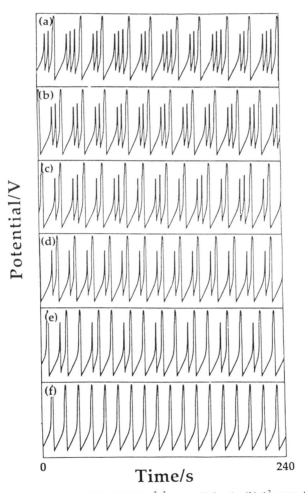

Figure 59. Continuation of Fig. 58. (a) $1^2 1^3$ state, 68.0 μA. (b) 1^2 state, 74.0 μA. (c) $1^1 1^2$ state, 82.0 μA. (d) 1^1 state, 88.0 μA. (e) 2^1 state, 100.0 μA. (f) 1^0 state, 103.0 μA. [Reproduced from F. N. Albahadity and M. Schell, *J. Electroanal. Chem.*, **308**, 151 (1991).]

potential reaches values where oxide layers must be formed on the
electrode. For the peak potentials of Figs. 58 and 59, the potential is
sufficiently anodic to assume that there must be a coupling to later stages
of the Pt-surface oxidation, for example, through Reaction (5.21). It
therefore seems that the potential oscillates in a potential region where
both the primary as well as the secondary passivation in the polarization
curve of Fig. 55 must be considered. Based on this idea, a simple
geometrical model may be suggested to explain these complex bursting
patterns, similar to the three-variable BDK model discussed briefly in
Section III.D.

To this end, consider the following three-dimensional vector field:

$$\varepsilon_1 \dot{x} = f(x, y, z) \qquad \dot{y} = g(x, y, z) \qquad \varepsilon_2^{-1} \dot{z} = h(x, y, z) \qquad (5.23)$$

where the characteristic time constants ε_1 and ε_2 can take any value
between 0 and 1, and express the idea that the variables x, y, and z may
evolve on (widely) different time scales. In a paper devoted to various
types of bursting patterns observed in a biochemical–biophysical context,
Rinzel [89] described how bursting oscillations can arise due to a slowly
changing variable z, which modulates the faster spike generating dynamics of the x–y subsystem. Assume that the x–y subsystem is given by
a BDK-type expression such as

$$\varepsilon_1 \dot{x} = y - F(x) \qquad \dot{y} = b_1 x + b_0 - y \qquad (5.24)$$

where $F(x)$ is a polynomial in x up to some order m, that is, $F(x) = \sum_{n=0}^{m} a^n x^n$. In the classical theory of relaxation oscillations, as considered by
Boissonade and De Kepper [61] and many others, $m = 3$. In Section
III.D, we saw how the linear coupling of such an $m = 3$ subsystem to a
third variable z can lead to mixed-mode oscillations. The qualitative form
of the bursting oscillations in Figs. 56, 58, and 59 can only be reproduced
if we assume that $m = 5$. The choice $m = 5$ is not surprising if one realizes
that the stationary-state curve in Fig. 55 has approximately the shape of a
fifth-order function. In Fig. 60, it is shown in four typical phase portraits
how by a small change in the parameter b_0, the effect of which is to shift
upwards the y-nullcline with respect to the x-nullcline the solution of the
x–y subsystem can change from a small relaxation-type periodic solution
to a large relaxation-type periodic solution. This happens since above the
critical value of b_0, the rate of flow at the left-hand maximum of the
x-nullcline is high enough to pass the intermediate maximum and to reach
the right-hand branch. (There does not seem to be a bifurcation
associated with this sudden jump in amplitude; rather the phenomenon is
canard-like.) Bursting oscillations can be obtained if we allow for a

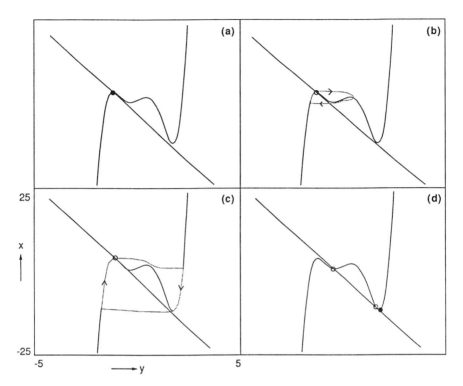

Figure 60. Existence of small- and large-amplitude limit cycles and multiple steady states in Eqs. (5.24). (a) $b_0 = -4.5$, (b), $b_0 = -4.0$, (c) $b_0 = -3.8$, (d) $b_0 = -3.0$. Other parameters: $a_5 = 1$, $a_4 = -0.92$, $a_3 = 5.15$, $a_2 = 1.0$, $a_1 = 3.15$, $a_0 = 0.5$, $b_1 = -5$, $\varepsilon_1 = 0.2$. ● stable steady state, ○ unstable steady state.

modulation of b_0 by the z variable, for example,

$$\varepsilon_1 \dot{x} = y - F(x) \qquad \dot{y} = b_1 x + b_0 - y + z \qquad \varepsilon_2^{-1}\dot{z} = c_1 y + c_0 - z \quad (5.25)$$

(The coupling of z may also be to the x-differential equation, as can be seen by putting $\bar{y} = y - z/\varepsilon_2$ and writing out the new $(x, \bar{y}, z)$ flow.) If the parameters are appropriately tuned, one can obtain complex periodic solutions as depicted in Fig. 61, where a few time series for the variable x are shown, obtained for different values of the control parameter a_0, with a_1 changing according to $a_1 = k_0 + k_1 a_0$. Both the shape of the oscillation pattern as well as the bifurcation sequence are in remarkable agreement with the experimental data in Figs. 58 and 59. Although Eq. (5.25) is clearly not acceptable as an electrochemical explanation of the bursting oscillations, this simple geometrical analysis strongly suggests that in any

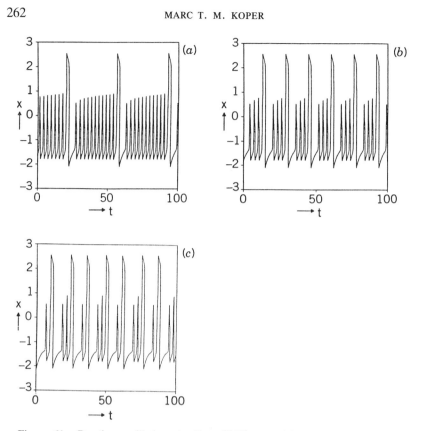

Figure 61. Bursting oscillations in Eqs. (5.25). $a_1 = 3.3-0.3a_0$, $c_1 = -2$, $\varepsilon_2 = 0.2$, $c_0 = -1$ (*a*) 1^{11} burst for $a_0 = 0.34$, (*b*) 1^3 burst for $a_0 = 0.4$, (*c*) $1^2 1^1$ burst for $a_0 = 0.501$. Other parameter values as in Fig. 60.

electrochemical model of these complex waveforms, one should take into account both the primary and the secondary passivation, which give rise to the doubly humped shape of the polarization curve. It may be noted that the bursting waveforms of Figs. 58 and 59 have also been observed in other processes, that is, the chloride–thiosulfate reaction [138], and that the above model may likewise provide a geometrical (*not* a physical) interpretation.

In an attempt to identify spectroscopically the nature of the surface species that participate in the oscillation process, Hachkar and co-workers [139–141] investigated in some detail the formaldehyde oxidation on rhodium. This system was found to give rise to current oscillations only in alkaline medium. Mass transfer has a small and rather unsignificant effect in the production of the oscillations. The most important result from their work is shown in Fig. 62. By measuring in situ during the oscillations the ultraviolet–visible (UV–vis) reflectivity changes [140], which are due to changes in the surface oxide, an in-phase oscillation of the degree of surface oxidation is observed. By calculating the charge associated with the surface oxidation, the spectral peaks at $\lambda = 400$ and 620 nm were ascribed to Rh(III) and Rh(I) surface species, the latter presumably arising from the reaction

$$Rh + OH^- \rightarrow RhOH + e^- \tag{5.26}$$

Note also from Fig. 62 that there is no significant phase delay between

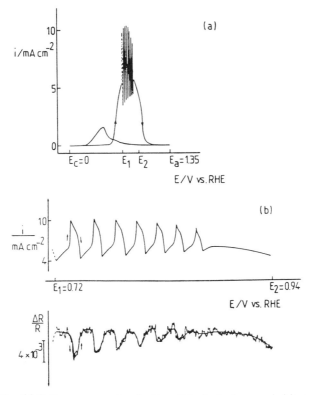

Figure 62. (*a*) Voltammogram of a rhodium disk electrode recorded in a spectroelectrochemical cell at $v = 20$ mV/s, 5°C in 0.1 M NaOH + 0.1 M HCHO. The first oscillation (in dashed line) is used as a triggering pulse for data acquisition. (*b*) The simultaneous recording of the oscillating part of the voltammogram ($i - E$) and the corresponding reflectogram ($\Delta R/R - E$; λ) in the potential domain of oscillations: 0.72–0.9 V vs. RHE. (Reproduced from [141] with permission of Elsevier Sequoia S.A.)

current oscillations and reflectivity oscillations, indicating that on the time scale of the oscillations the surface oxidation is a relatively fast (reversible) process. It is also clear that a high degree of surface oxidation corresponds to a low current (passivation) and vice versa. Calculations of the charge involved in one oscillation cycle reveal that about 10–20% of the surface is involved in the oscillation process. Attempts were also made by these authors to identify oscillations of organic surface species (like adsorbed CO) with electromodulated infrared (IR) spectroscopy, but unfortunately without success [141].

In an earlier paper, Hachkar et al. [139] reported that current oscillations during HCHO oxidation on Rh were strongly favored when the electrode roughness or the cell temperature was increased, or in general, when a minimum current was reached. Recent results [142] have shown that this minimum current should in fact be interpreted as a minimum ohmic drop. Current oscillations can always be surpressed by lowering the concentration of HCHO or by lowering the disk rotation rate, and can then be induced by a sufficiently large external resistor in series with the working electrode. This is illustrated in the voltammograms of Fig. 63. This figure also shows the occurrence of galvanostatic potential oscillations around the positively sloped HCHO oxidation branch. A sudden transition to very anodic potentials takes place when the oscillatory potential collides in a homoclinic bifurcation with the negatively sloped $I–E$ branch, a phenomenon similar to that illustrated in Fig. 41 for the Ni dissolution in sulfuric acid. On the basis of this figure, it is evident to suspect a hidden negative impedance on the positively sloped oxidation branch. The impedance diagrams indeed confirm this expectation, as also illustrated in Fig. 63. Note that the large radius of impedance plot for $V = -0.45$ V (vs. SCE) implies that it must be very close to a galvanostatic Hopf bifurcation, in good agreement with the onset of the galvanostatic potential oscillations in the galvanostatic $I–E$ curve. These results are important as the first explicit evidence for the relevance of the external cell control in explaining electrocatalytic oscillations and the "hidden negative impedance" concept as the origin of both current oscillations in potentiostatic (or almost potentiostatic) and potential oscillations in galvanostatic control.

At this point, it may be worthwhile to comment briefly on the kind of information that can be extracted from in situ (spectroscopic) measurements UV–vis [140], ellipsometry [96, 178], luminescence [159], IR [176], electrochemical quartz microbalance (ECQM) [143], etc. on the simultaneous oscillation of some nonelectrical surface property. In general, if the interfacial potential or current of the electrochemical system oscillates (as it does in practically all electrochemical oscillators), it is quite obvious

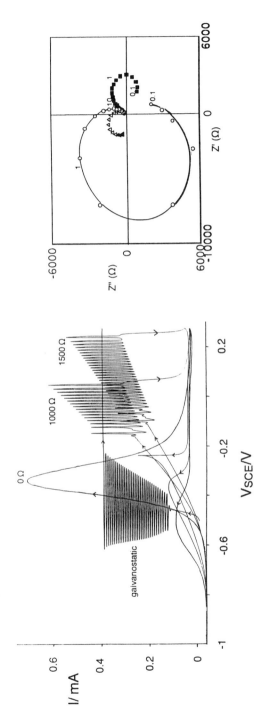

Figure 63. Voltammograms of 0.1 M HCHO in 0.1 M NaOH on a Rh rotating disk electrode for various values of the external resistance, 0, 1000, and 1500 Ω (internal cell resistance 95 Ω), $v = 10$ mV/s, 3000 rpm; amperogram recorded at 0.01 mA/s. Impedance diagrams taken at ■ −0.50 V, ○ −0.45 V, △ −0.35 V (vs. SCE).

265

that *all* other interfacial properties will oscillate as well, at least to some extent. One should therefore be careful with assuming that the observation of the simultaneous oscillation of some other nonelectrical surface feature is meaningful to the elucidation of the oscillation mechanism. In general, in a mechanism of a chemical oscillator, Eiswirth et al. [144] have shown the usefulness of distinguishing between essential and nonessential species, that is, species that play a crucial role in the generation of the oscillation and species that simply oscillate along without really affecting the basic instability mechanism, respectively. From the earlier discussions in this section, it seems quite reasonable to assume that (hydr)oxide layers play a crucial role in the oscillation mechanism so that the UV–vis measurement by Hachkar et al. [141] probably detects an essential species.

In concluding our discussion of the oscillations during the electro-catalytic oxidation of small organic molecules, we mention that several authors studied the structural effects of the electrode on the oscillations during formic acid oxidation on various low-index platinum single-crystal planes [123, 145–147]. Although the results from the three different groups show some important discrepancies regarding the role of mass transport, pH, and surface structure, all authors noted a significant difference between the three low-index planes Pt(100), Pt(110), and Pt(111). Current oscillations in the cyclic voltammogram on Pt(100) are easily observed (Fig. 64), but observed only with great difficulty on Pt(110). A full analysis of these effects is extremely difficult and can at the present stage not be undertaken without a substantial deal of speculation. However, both Markovic and Ross [147], as well as Raspel and Eiswirth [123], seem to agree about the main destabilizing process in the mechanism, namely, the autocatalytic removal of CO_{ad} by OH_{ad}. Markovic and Ross claim that the extremely sharp current peak in the CO oxidation on Pt(100) is evidence for its autocatalytic oxidation. Raspel and Eiswirth have built a relatively simple model that includes Reactions (5.19) and (5.20) as the basic autocatalytic mechanism. They show the possibility of a Hopf bifurcation by using stoichiometric network analysis (SNA) [144]. This autocatalysis is furnished by a reversible OH^- adsorption Reaction (5.19), which can then be included in Reaction (5.20) to give

$$CO_{ad} + OH^- + Pt \rightarrow H^+ + CO_2 + 2e^- + 2Pt \qquad (5.27)$$

which is autocatalytic in the number of free sites. Note that this interpretation of the autocatalysis is different from that invoked by

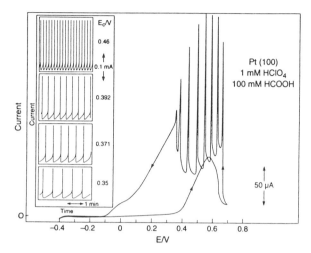

Figure 64. Potentiodynamic curves and potentiostatic transients (inserts) in 1 mM NaClO$_4$ + 100 mM HCOOH on Pt(100). [Reproduced with permission from N. Markovic and P. N. Ross, *J. Phys. Chem.*, **97**, 9771 (1993). Copyright © 1993 American Chemical Society.]

Wojtowicz [135] and Schell and co-workers [134], who assume that Reaction (5.20) itself requires an additional free site [Reaction (5.22)].

In advocating this mechanism, Raspel and Eiswirth explicitly mention that the effect of ohmic drop is negligible in their measurements. Although the current in most of their experiments is indeed quite low, and the Luggin–Haber capillary will certainly surpress a good deal of the ohmic drop, they do not give a value for the ohmic cell resistance so that it remains difficult to analyze the effect quantitatively. It is still tempting to relate some of the effects reported by Markovic and Ross on the influence of the solution composition to a change in the electrolyte conductivity rather than to a change in any particular surface chemistry as the authors do.

In closing this section, let us go through the main points once again.

- The oxidation of small organic molecules may proceed through two different pathways, one involving only reactive surface intermediates and the other involving a strongly adsorbed unreactive intermediate that "poisons" the electrode and should probably be identified with adsorbed carbon monoxide.

- In most of the experiments discussed, the conditions of mass transport play no or no really significant role, indicating that the oxidation current is mainly kinetically controlled.

- The formation of surface oxides inhibits the direct oxidation pathway, and in terms of the polarization curve this will give rise to a branch of negative slope. There is evidence for two such negative-slope passivation regions, the first presumably connected with surface hydroxide formation [Pt(I)OH, Rh(I)OH], the second to some higher valency oxide. An analysis of complex bursting oscillations, as observed by two different groups, also indicates the importance of the secondary passivity.

- Oscillations have been observed under all conceivable electrical conditions of the cell: (presumed) potentiostatic, potentiostatic with external resistor, galvanostatic, and open-circuit conditions. The latter can be achieved by the addition of a redox couple that is a good oxidizing agent for formic acid or for another organic molecule [148], and should in fact be considered as a special case of galvanostatic conditions where the (approximately) constant "external" current is supplied by the other redox couple. Instabilities have also been reported during cyclic voltammetry of the oxidation of organic molecules [149]. In these experiments, one observes, for instance, a period-doubled voltammogram for which the current traces back a previous path only every other potential cycle, and even further period doublings to a chaotic voltammogram for which the current never traces back a previous cycle although the potential cycle remains the same [149].

- Most authors seem to agree about the basic instability mechanism leading to the oscillatory instability, namely, the autocatalytic removal of the poison CO_{ad} by the adsorbed hydroxide OH_{ad}, Reactions (5.22) or (5.27). It seems important to settle the autocatalytic nature of this process more convincingly, as it would lend further credit to Raspel and Eiswirth's claim [123] that the formic acid oxidation can oscillate under truly potentiostatic conditions. However, recent experiments [142] have shown that the current oscillations observed in the voltammograms of the formaldehyde oxidation on Rh and Pt, are due to a nonnegligible influence of the ohmic electrolyte resistance, although they were previously believed to be purely potentiostatic. Figure 63 clearly illustrates this fact. To explain the oscillations in these I–E curves, it is not necessary to invoke an autocatalytic reaction such as Eqs. (5.22) or (5.27). It is rather the coupling of the negative interfacial impedance, most likely caused by an (hydr)oxide formation on the surface by Reactions (5.19) or (5.26), with the ohmic electrolyte resistance, that would provide the simplest explanation of the current oscillations [109]. To explain the galvanostatic potential oscillations would require additional assumptions regarding the interfacial (electro)chemistry, in order to account for the hidden negative impedance [109].

VI. OSCILLATORY SEMICONDUCTOR ELECTRODE REACTIONS

The interest in the electrochemical properties of semiconductor materials (e.g., for the application in device processing and electrochemical solar energy conversion) has initiated the study of many electrode reactions on commercially important semiconductors. Two such reactions have attracted attention with respect to the occurrence of spontaneous oscillations: the reduction of H_2O_2 on various semiconductors and the etching of silicon in fluoride media.

There are interesting and new features about semiconductor electrode oscillations that warrant this separate section. First of all, semiconductors are sensitive to light and can therefore give rise to new phenomena such as photocurrent and photopotential oscillations and oscillatory light emission. Second, the potential distribution across the semiconductor–electrolyte interface is markedly different from the metal–electrolyte interface [150], and it is worthwhile to study how this may affect the possibility of instabilities.

A.   Hydrogen Peroxide Reduction on GaAs and
Copper-Containing Semiconductors

Spontaneous current and/or potential oscillations during H_2O_2 reduction have been observed on several (noble) metal electrodes, such as platinum [151, 152], silver [153] and gold [154], and semiconductor electrodes [155–161]. On platinum, two rather detailed studies of the complex bifurcation sequences of bursting-type oscillations have been carried out by Fetner and Hudson [151] and by van Venrooij and Koper [152]. Here, we will be confined to semiconductor electrodes, although there are some interesting analogies with platinum that we will mention at the end of this section. We will start with a discussion of oscillations at GaAs [159], as in this system the semiconductor properties of the electrode seem to play an unambiguous role, and will then refer to very similar behavior observed for Cu-containing semiconductors [155–158, 160] where it seems that the semiconducting nature of the electrode is less important.

Figure 65 depicts a cyclic voltammogram of hydrogen peroxide reduction on n-GaAs that illustrates a typical behavior of the system [159]. In the cathodic scan, a steady increase in cathodic current is observed from -0.2 V (vs. SCE) on, which is associated with a decrease in the semiconductor band bending and a resulting increase of conduction

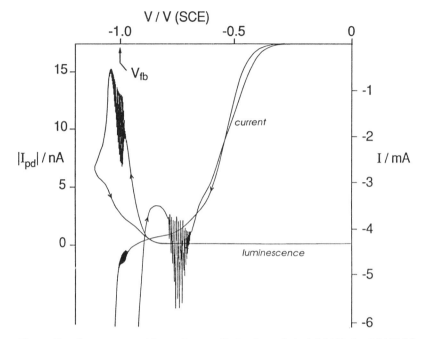

Figure 65. Current–potential scan for an n-GaAs electrode in 1.0 M H_2O_2, 1 M H_2SO_4 at 40°C. Scan rate 2 mV/s. Also shown is the electrode's light emission detected as a photodiode current. [Reprinted with permission from M. T. M. Koper, E. A. Meulenkamp, and D. Vanmaekelbergh, *J. Phys. Chem.*, **97**, 7337 (1993). Copyright © 1993 American Chemical Society.]

band electrons at the surface. From -0.6 to -1.0 V a current plateau is observed that is proportional to the hydrogen peroxide bulk concentration but that does not depend on the rotation rate of the GaAs disk, and must therefore be of a kinetic nature [161]. On the return scan, it is in this region that spontaneous current oscillations are observed; the oscillations can also be observed in the cathodic scan if the scan rate is sufficiently low. The oscillations are sustained if a fixed potential is applied, and are accompanied by oscillations in the light emission of the electrode [159], which indicates the injection of holes during the reduction process (see below). The oscillations were reported to be set in through a Hopf bifurcation, and were observed to exhibit period-doubling bifurcations to chaotic oscillations, followed by a sudden transition to larger amplitude mixed-mode oscillations, a scenario observed in many systems (see Sections III–V). Oscillations also occur under galvanostatic conditions, and at p-type electrodes under sufficiently intense illumination, so that one may refer to photocurrent and photopotential oscillations. In the case of potentiostatic current oscillations, the external ohmic potential drop was shown to play a crucial role [162]. If the hydrogen peroxide bulk concentration used in Fig. 65 ($1\,M$) was reduced by a factor of 10, current oscillations disappeared, but they could be induced by a series resistor, the value of which was about 10 times the value of the electrolyte resistance. The importance of having an external series resistance in order to see oscillations is also suggested by the impedance diagrams measured at various potentials under conditions where the entire current–potential curve is stable (see Fig. 66). The corresponding I–V curve is the solid curve in Fig. 68. Note that Fig. 66(e) is galvanostatically unstable with two unstable directions of the steady state, consistent with the observation of galvanostatic potential oscillations.

A mechanism for the cathodic reduction of hydrogen peroxide on GaAs electrodes was suggested by Minks et al. [163] and was recently studied for its ability to explain the oscillatory characteristics [162]. From the observation that, at p-type electrodes, hydrogen peroxide is reduced only under illumination, one can conclude that at least one step in the mechanism involves a (photoexcited) conduction band electron. In addition, the model assumes the existence of an initial chemisorption step, creating a surface intermediate that consists of an electron-deficient GaAs bond. Such a precursor is necessary to explain the experimentally observed (inverse) correlation between the reduction current and the rate of chemical dissolution of the GaAs by hydrogen peroxide. The complete mechanism can be written as

$$H_2O_2 + X_0 \xrightarrow{k_1} X_1^+ \begin{matrix} OH^- \\ OH\cdot \end{matrix} \tag{6.1}$$

$$X_1^+ \begin{matrix} OH^- \\ OH\cdot \end{matrix} + e_{CB}^- \xrightarrow{k_2} X_0\!-\!OH\cdot + OH^- \tag{6.2}$$

$$X_0\!-\!OH\cdot \xrightarrow{fast} X_0 + OH^- + h_{VB}^+ \tag{6.3}$$

$$h_{VB}^+ + e_{CB}^- \rightarrow h\nu, \text{heat} \tag{6.4}$$

In the employed notation, X_0 is a free GaAs surface site, and X_1^+ is an electron-deficient GaAs surface bond. This mechanism is also consistent with the experimentally observed "current doubling" at p-type electrodes, since for every photon absorbed two electrons are transferred across the semiconductor–electrolyte interface, with the electroluminescence (spontaneous light emission) observed at n-type electrodes [161, 163]. The surface state X_1 is also a precursor for chemical dissolution, with an (overall) rate constant k_{ch}.

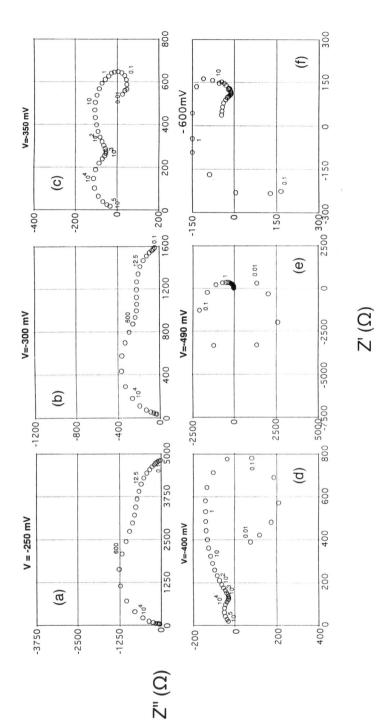

Figure 66. Impedance plots of the complex cell impedance $Z' + iZ''$, $i = \sqrt{-1}$, determined at the indicated potential values (vs. SCE). Indicated frequencies in Hz, $0.1\,M\,H_2O_2$, $1\,M\,H_2SO_4$ at $40°C$ (Reproduced from [162] with permission of the American Chemical Society.)

The two important steps here are Eqs. (6.1) and (6.2), with respective rates v_1 and v_2. The kinetic limitation observed in the experiment must be associated with a small adsorption rate constant k_1. This makes the fractional coverage of the electrode with the X_1 surface complex, denoted θ_1, one of the variables in a phenomenological model of the oscillations. The other variable must be the electrode potential. Here, it is important to take into account the potential distribution appropriate to the semiconductor–electrolyte interface [150, 162]. Since semiconductors have a much lower concentration of free charge carriers than metals, the charge distribution at a semiconductor surface is usually diffuse, extending over a region known as the space-charge layer, or depletion layer if the surface region has a lower concentration of free charges than the bulk region. The typical Debye length of this semiconductor space-charge region is usually larger than the similar space-charge region existing in the electrolyte phase. This results in a specific capacitance of the semiconductor depletion layer (C_{sc}), which is (much) smaller than that of the electrolyte region (the Helmholtz layer, C_H). As a result, a change in the interfacial potential is accommodated almost entirely across the semiconductor depletion layer, provided that the charges in the semiconductor surface states do not change as a result of this potential variation. The product of the electrical potential difference between the surface and the bulk, ϕ_{sc}, and the elementary charge e, is known as the band bending, which is the potential energy difference between electrons at the surface and electrons in the bulk of the semiconductor. This is an important quantity because it regulates through a Boltzmann equilibrium the concentration of free conduction band electrons at the surface n_s (for an n-type electrode), which are available for electrochemical reactions, namely,

$$n_s = n_b \exp(-e\phi_{sc}/kT) \tag{6.5}$$

where n_b is the concentration of conduction band electrons in the bulk. For the time dependence of the interfacial potential ϕ, it follows that

$$C_{sc}\frac{d\phi}{dt} = \frac{V - \phi}{AR_\Omega} + e\Gamma_m v_2 \tag{6.6}$$

and for θ_1, one has

$$\frac{d\theta_1}{dt} = v_1 - v_2 \tag{6.7}$$

where it must be noted that we have followed the usual convention in taking a cathodic current as negative. In these expressions Γ_m is the total number of sites per unit surface area, and $\phi = \phi_{sc} + \phi_H$, where ϕ_H is the potential difference across the Helmholtz layer. Note that the cathodic current is associated with the electron-capture process [Eq. (6.2)], and that the total cathodic current is in fact $2\Gamma_m e v_2$, because the hole injection and recombination processes (Eqs. (6.3) and (6.4)] are assumed to respond instantaneously. In Eq. (6.6), it is assumed that $C_{sc} \ll C_H$ [162].

The negative impedance characteristics of the hydrogen peroxide reduction on GaAs, as observed in both the voltammogram as well as the impedance diagrams, is most likely associated with the concomitant reduction of protons [159, 162], which leads to the formation of adsorbed atomic hydrogen through the so-called Volmer reaction [68],

$$H^+ + e^- + X_0 \leftrightarrow X_0\text{---}H \qquad (6.8)$$

The transition to a hydride-covered electrode with increasing cathodic polarization can in principle lead to a negative impedance in two different ways [162]:

1. Adsorbed hydrogen decreases the number of available surface sites for the adsorption of hydrogen peroxide. If we write for $v_1 = k_1[1 - \theta_1 - \theta_H(\phi)]$, then this competitive adsorption will lead to a positive $\partial v_1 / \partial \phi$, and hence to a negative impedance.

2. Adsorbed hydrogen leads to a decrease of positive surface charge (actually, it is probably more realistic to picture some protons as adsorbed onto the GaAs surface in an acid–base equilibrium [164]), leading to a decrease in the negative countercharge in the semi-conductor phase, and hence to an increase in the band bending. Simple charge considerations show that $\phi_{sc} = \phi + \gamma_H \theta_H + \text{cst}$ [162], with γ_H a parameter that measures the change in the bandbending induced by a monolayer of hydride (assuming that γ_H is independent of potential and coverage and that $C_{sc} \ll C_H$). It can be shown [162] that this may lead to a positive $\partial v_2 / \partial \phi$, that is, a less efficient electron capture with increasing cathodic polarization. Since $v_2 = k_2 n_s \theta_1$, and hence $\partial v_2 / \partial \phi = k_2 \theta_1 (\partial n_s / \partial \phi_{sc})(d\phi_{sc}/d\phi)$, with $\partial n_s / \partial \phi_{sc}$ negative because of Eq. (6.5), it follows that $d\phi_{sc}/d\phi$ must be negative in order to have a negative impedance. This rather uncommon dependence of the band bending on the interfacial potential will be referred to as a band bending anomaly, which in the present model is caused by the transition to a hydride-covered surface with increasingly cathodic potential.

More generally, the negative impedance characteristics can be associated with either a positive $\partial v_1 / \partial \phi$ or a positive $\partial v_2 / \partial \phi$, regardless of whether hydride formation is responsible or not. However, the stability characteristics are not so insensitive to the origin of the negative impedance. This can be seen by calculating the electrical impedance associated with Eqs. (6.6) and (6.7). Assuming that the Volmer reaction is reversible, so that the hydride coverage depends parametrically on ϕ, and taking into account the chemical etching of the GaAs by the X_1 surface state (rate constant k_{ch}), then, by a standard calculation [162], one obtains the equivalent cell circuit of Fig. 67(a), with

$$R_\infty = \left[\frac{e|j|}{kT} \frac{d\phi_{sc}}{d\phi} \right]^{-1} \qquad |j| = 2e\Gamma_m k_2 n_s \theta_1$$

$$C_p = \frac{1}{k_2 n_s R_\infty} \frac{R_H}{R_H + R_\infty} \qquad R_p = \frac{R_\infty(1 + R_\infty/R_H)}{\dfrac{k_1 + k_{ch}}{k_2 n_s} - \dfrac{R_\infty}{R_H}} \qquad (6.9)$$

$$R_H = - \left[2e\Gamma_m k_1 \frac{d\theta_H}{d\phi} \right]^{-1}$$

In the case that both R_∞ and R_p are positive, the impedance plot will result in two semicircles in the first quadrant, as shown in Fig. 67(b). This

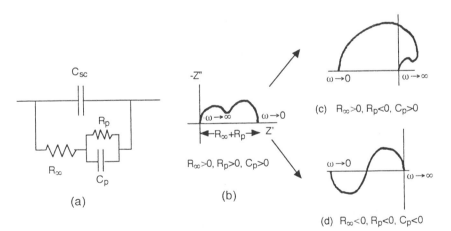

(a)

(b)

(c) $R_\infty > 0$, $R_p < 0$, $C_p > 0$

(d) $R_\infty < 0$, $R_p < 0$, $C_p < 0$

Figure 67. (a) Equivalent cell circuit belonging to Eq. (6.9); symbols are defined in the text; (b) Complex impedance plot for the circuit in the case positive R_∞, R_p, and C_p. (c) and (d) Two alternatives for a transition from a complex impedance plane plot with positive polarization slope $R_\infty + R_p > 0$ to negative polarization slope $R_\infty + R_p < 0$ with $R_p < 0$ and $R_\infty > 0$ (c) or with R_∞, $R_p < 0$ (d). (Reproduced from [162] with permission of the American Chemistry Society.)

is in agreement with the impedance plots observed experimentally at moderate overpotential [Figs. 66(a and b)] (where one may safely assume $R_H \rightarrow \infty$). Now for potentials for which the polarization slope is negative, the impedance must enter into the second quadrant. For the circuit of Fig. 67(a), this can occur in two ways [see Fig. 67(b)]:

(1) If $R_\infty / R_H > (k_1 + k_{ch}) / k_2 n_s$, R_p becomes negative. However, R_∞ and C_p remain positive, so that the impedance will not cross into the third quadrant, which is a necessary condition for having a Hopf bifurcation in the presence of an additional series resistor, as was shown in Section II. The impedance plot will have the form of Fig. 67(c). One concludes that competitive adsorption may explain the negative slope, but not the oscillations.

(2) If R_∞ is negative and smaller in absolute value than R_H, both R_p and C_p will be negative as well and the impedance plot will enter the third quadrant for the lowest frequencies (because of the negative C_p). The impedance plot will have the form of Fig. 67(d). This coupling between a negative R_∞ and other elements (C_p and R_p, in this example) is well known, and also occurs in the Randles circuit as we already noted in Section II. It is clear that only in this case can we anticipate the possibility of a Hopf bifurcation.

The main conclusion of this analysis is that oscillations have to be associated with a band bending anomaly. By measuring the capacitance of the $GaAs/H_2O_2$ interface at high frequencies, the band bending can be obtained experimentally and its potential dependence can be determined [162]. The result is shown in Fig. 68, where the solid line is the voltammogram (which is stable because the ohmic potential drop is small in the present case) and the line with the crosses is band bending. This figure clearly illustrates that the negative polarization slope observed in the return scan is accompanied by a region of $d\phi_{sc}/d\phi < 0$, providing evidence for the band bending anomaly.

Although the above analysis clearly demonstrates the importance of taking into account the semiconductor character of the electrode in order to understand oscillations, this simple extension of the model suggested in Eqs. (6.1–6.4) still suffers from some serious deficiencies. The most obvious drawback is its inability to provide an explanation for the galvanostatic potential oscillations, which is of course fully equivalent to its inability to even qualitatively reproduce the experimental impedance plot of Fig. 66(e) [162]. Also, the experimental evidence for the alleged involvement of the hydride layer in causing the band bending anomaly seems theoretically reasonable but still awaits a sound experimental confirmation. The small positive-impedance loop observed experimentally at high frequencies may be associated with a finite rate of the hydride-

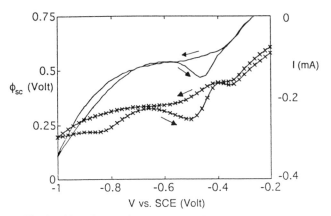

Figure 68. The band bending ϕ_{sc} (line with crosses) as extrapolated from an impedance measurement at 50 kHz and the corresponding reduction current (solid line) as a function of the applied potential V. (Reproduced from [163] with permission of the American Chemical Society.)

layer formation, and can be incorporated into the model at the cost of mathematical transparency.

It is interesting to compare the results on GaAs with the work by Tributsch and co-workers [155–158, 160], who have made extensive studies of the oscillatory hydrogen peroxide reduction on various copper-containing semiconductors (CuS, Cu_2S, Cu_5FeS_4, $CuFeS_2$, and $CuInSe_2$) of both n- and p-type and under both potentiostatic (with external resistor) and galvanostatic conditions. The fact that oscillatory phenomena can also be observed during the hydrogen peroxide reduction on quasimetallic compounds, such as CuS and $CuFeS_2$, indicates that for these materials the semiconductor properties are not of much importance in the generation of the oscillations. In the following, we will restrict our discussion to the authors' most recent work on $CuInSe_2$ [158, 160, 165, 166].

As with GaAs, the cyclic voltammogram for hydrogen peroxide reduction on $CuInSe_2$ displays a region of negative slope, which is most pronounced during the return scan. Figure 69 shows the current–voltage curves for p-$CuInSe_2$ in the dark and under illumination. Figure 70 illustrates the onset of oscillations on inserting a series resistor. From the fact that there is appreciable current flowing in the dark (dashed curve in Fig. 69), one concludes that the electron-transfer mechanism is different from Eqs. (6.1–6.4), and should mainly involve hole injection. Apart from the regular oscillations in Fig. 70, more recently, Neher [166] has also reported chaotic oscillations (and period-doubling bifurcations)

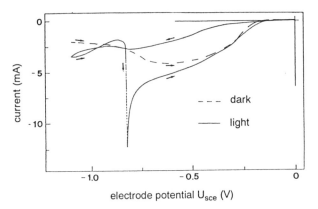

Figure 69. Cyclic voltammograms (20 mV/s) of H_2O_2 reduction at p-CuInSe$_2$ electrode, stationary, under white light (solid line). Fifth cycle after polishing and etching. Dashed line: curve recorded during positive sweep in the dark, after polarization for 20 s under light at negative limit. (Reproduced from [158] with permission of The Electrochemical Society.)

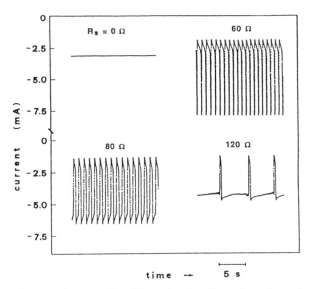

Figure 70. Current–time recordings illustrating the effect of a series resistance on the onset of the oscillating regime. Applied potential −1.0 V (vs. SCE). (Reproduced from [158] with permission of The Electrochemical Society.)

during the hydrogen peroxide reduction on p-CuInSe$_2$, as well as mixed-mode oscillations.

The current plateaus for both n- and p-type electrodes depend on the disk rotation rate, though the Koutecky–Levich plot shows a significant axis intercept, which is evidence of mixed kinetic and diffusion limitation [165]. In order to monitor the flatband potential (= the electrode potential ϕ for which the band bending $\phi_{sc} = 0$) of the CuInSe$_2$—H$_2$O$_2$ interface, Cattarin et al. [165] measured the potential dependence of the electroreflectance spectrum. This technique is suitable for obtaining a quantitative estimate of the band bending. The electroreflectance signal $\Delta R/R_0$ is expected to be linearly proportional to the band bending ϕ_{sc}, where R_0 is the reflectance in the absence of an electric field in the semiconductor, that is, for $\phi_{sc} = 0$. This means that $\Delta R/R_0 = 0$ for the flatband potential ($\Delta R = R - R_0$). It was observed that $\Delta R/R_0$ depends on potential in a fashion that is similar to the way the current depends on potential; that is, in the region of negative slope $\Delta R/R_0$ becomes (anomalously) more positive with increasing cathodic potential. This result may be interpreted as a decreased absorbance in the semiconductor from an increased band bending, and hence as evidence for a band bending anomaly.

Another anomalous effect exhibited by this system can be seen from Fig. 69. In strong contrast to the results obtained on GaAs, the photocurrent is not proportional to the incident photon flux and even changes sign in the potential region from -0.9 to -0.75 V [165]. Since this inverted photoeffect occurs in the same potential region as the negative slope and only with hydrogen peroxide and not with other redox couples on CuInSe$_2$, Cattarin et al. suggest their origin to be the same. A sound theoretical interpretation of this peculiar effect is still lacking, however.

In a more recent study, Pohlmann et al. [160] measured the hydrogen evolution during hydrogen peroxide reduction on p-CuInSe$_2$ by in situ differential electrochemical mass spectrometry. They found hydrogen gas is liberated in an oscillatory fashion during galvanostatic potential oscillations, providing important evidence that the proton reduction is indeed somehow involved in the overall process. Molecular hydrogen may result from the recombination of two adsorbed hydrogen atoms or electrochemically by the Heyrovsky reaction [68]:

$$X_0\text{—H} + H^+ + e^- \rightarrow H_2 + X_0 \qquad (6.10)$$

Based on the above observations, Pohlmann et al. [160] suggested a model for oscillatory hydrogen peroxide reduction on p-CuInSe$_2$. Since they have good reason to assume that the semiconducting nature of the CuInSe$_2$ is not essential for the oscillations, and since mass transport is partially rate determining, they started from a skeletal model given by Eqs. (3.1) and (3.11) [or Eq. (3.12)] from Section III. They derive an expression for the heterogeneous rate constant $k(E)$, which should incorporate the mechanistic origin of the negative slope.

First, the authors assume that the direct reduction rate of hydrogen peroxide is proportional to the number of free active sites

$$X_0 + H_2O_2 + 2H^+ + 2e^- \rightarrow 2H_2O + X_0 \qquad (6.11)$$

(As noted above, it is probably more realistic to regard the two electrons as two holes being injected, but this is not essential.) Second, there is a competition for free sites by two other processes: (1) proton reduction by the Volmer–Heyrovsky mechanism Eqs. (6.8) and (6.10), and (2) reduction of dissolved oxygen (always present in H$_2$O$_2$ solution), following a pathway in which the coverage of adsorbed hydroxyl is assumed to be important. The involvement of the latter reaction is invoked on account of the observation that (photo)current at p-CuInSe$_2$ is no longer proportional to the hydrogen peroxide concentration at very low (<0.1 M) hydrogen peroxide concentrations, which is ascribed to the presence of oxygen in the solution [165]. For unspecified reasons, the authors do not consider the back reaction of Eq. (6.8), so that in order to explain the negative slope the need arises to take into account a recombination reaction between adsorbed hydrogen and hydroxyl radicals. Assuming a quasistationary state for the two coverages, a $k(E)$ function is obtained that includes a region of negative $dk(E)/dE$. Oscillations can be obtained in such a model, as was proved in Section III by employing the Poincaré–Bendixson theorem.

Needless to remark, this model is substantially different from the model proposed for GaAs. First, the semiconductor nature of the electrode is not taken into account in Pohlmann's model. However, as argued above, one may regard the results obtained with the electroreflectance experiments as evidence that the semiconductor band bending displays an anomalous behavior, although this interpretation may not be unique. Second, there is a clear effect of the mass transport with CuInSe$_2$, whereas such an effect is absent for GaAs. Third, the alternative kinetic scheme suggested by Pohlmann is rather involved, and in fact unnecessarily so from a purely modelistic point-of-view if one is willing to accept the reversibility of the Volmer reaction. A deficiency that is shared by the two models is their ability to explain the galvanostatic potential oscillations. As far as this last point is concerned, it is relevant to remark that Cattarin and Tributsch found an impedance plot like that of Fig. 8(f) (i.e., a hidden negative impedance) for the hydrogen peroxide reduction on chalcopyrite $CuFeS_2$ [157].

To close this section, we note that a negative slope in the hydrogen peroxide reduction polarization curve has been observed on several other electrode materials. Gerischer and Mindt [167] attributed a decrease in the hydrogen peroxide reduction current on germanium to a competitive adsorption with hydrogen radicals. On platinum, van Venrooij and Koper [152] observed that the potential region of negative slope shifted cathodically with increasing solution pH, indicating some involvement with the proton reduction; current oscillations were found in the same potential region. Such cathodic passivation effects have been observed for other electrocatalytic reactions on platinum (such as the oscillatory reduction of nitric acid [168]), and have been ascribed to a transition to a hydride-covered electrode surface [169], although a satisfactory quantitative picture is still lacking.

B. Anodic Dissolution of Silicon in Fluoride Media

The electrochemical dissolution of silicon in fluoride-containing solutions is a topic of considerable activity in semiconductor electrochemistry. This finding is chiefly due to the discovery of the technological interesting (photo-)properties of porous silicon, which is formed at low anodic overpotentials, where the formation of oxide layers can be ruled out. Another phenomenon connected with anodic silicon oxidation that has given rise to much excitement recently is the appearance of current oscillations far in the electropolishing potential region ($E > 2.5$ V vs. SCE). These were first observed by Turner in 1958 [170], but have been the subject of numerous studies since their recent "rediscovery" by Gerischer and Lübke [171]. The oscillations have been observed as current oscillations at p-Si or as photocurrent oscillations at illuminated n-Si. Only a few studies have reported the occurrence of potential oscillations under galvanostatic conditions [170].

A typical voltammogram of the anodic dissolution of a rotating p-Si disk ($N_D = 4 \times 10^{15}$ cm^{-3}) for a fluoride concentration of 0.05 M and pH 3 is given in Fig. 71 [172]. Although the currents flowing are strongly dependent on the bulk pH and fluoride concentration, spanning several orders of magnitude, the qualitative shape of the voltammogram is remarkably robust, with essentially two current peaks (a sharp one denoted J_1 and a rather broad one denoted J_3) and two current plateau's (denoted J_2 and J_4) [173]. Oscillations are observed in the J_4 plateau and have also been observed for very different electrolyte compositions. Similar curves are observed for n-Si under sufficiently intense illumination

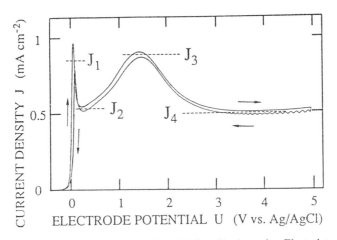

Figure 71. Typical voltammogram of a (111) p-Si electrode. Electrolyte is 0.025 HF + 0.025 NH$_4$F + 0.95 NH$_4$Cl (c_F = 0.05 M, pH 3). Rotation rate 300 rpm; sweep rate 5 mV s^{-1}. Note the small current oscillations in the J_4 plateau. (Reprinted from [172] with permission of the VCH Verlagsgesellschaft mbH.)

[171]. In the following we will concisely outline the electrochemistry that is thought to be associated with the different parts of the voltammogram, and subsequently describe in some detail the most pertinent results from the various studies on the oscillatory behavior.

The steep current rise in Fig. 71 near 0 V (vs. Ag/AgCl/sat.KCl) is associated with porous silicon formation. The silicon oxidation is also accompanied by hydrogen evolution, so that the full oxidation to a tetravalent Si complex at a p-type electrode is considered to follow a mechanism such as [170, 174]

$$Si + 2F^- + 2h^+ \rightarrow SiF_2 \qquad (6.12)$$

$$SiF_2 + 2HF \rightarrow SiF_4 + H_2 \qquad (6.13)$$

where the Si(IV) species in the solution will be further complexed to SiF$_6^{2-}$. At the n-type electrode holes have to be generated by optical excitation:

$$h\nu \rightarrow h_{VB}^+ + e_{CB}^- \qquad (6.14)$$

Quantum efficiences (i.e., the number of electrons transferred per photon absorbed) ranging from four at low-photon intensities to two at high intensities have been reported. This result is explained by assuming a transition from a purely electrochemical oxidation via subsequent electron injection steps at low intensities, to a competition with a chemical oxidation leading to hydrogen evolution [i.e., Reaction (6.12)] at high intensities [174]. It should also be noted that there is an appreciable current flowing in the absence of photogenerated holes, indicating that a reaction pathway exists that involves only electron injection steps [174].

For potentials more anodic than the current peak J_1, a passivating film is formed on the silicon. At relatively low potentials, this film is thought to consist mainly of some silicon hydroxide [175], but at high potentials (in the J_4 region, say) oxidation leads mainly to silica [175], for which one can write a general overall reaction such as [174]

$$Si + 2H_2O + nh^+ \rightarrow SiO_2 + 4H^+ + (4 - n)e^- \qquad (6.15)$$

The oxide film is attacked chemically by HF and HF_2^- in the electrolyte, and by comparing the dissolution rate at various pH values and fluoride concentrations with the equilibria of these complexes in aqueous solution, Chazalviel et al. [173] concluded that the dissolution rate J_4, which is controlled primarily by this chemical attack, approximately follows the rate law

$$J_4 = k_4[HF_2^-][H^+]^{0.5} \qquad (6.16)$$

Apart from these kinetic limitations, Hassan et al. [176] also showed that the dissolution rate J_4 is dependent on the rotation rate of the disk, obeying the Koutecky–Levich law for combined kinetic and mass-transfer control. The authors associate the diffusion limitation with the transport of fluoride ions towards the electrode.

Various authors have attempted to characterize the state of the surface during the oscillations by in situ (spectroscopic) measurements, in an attempt to elucidate the origin of the oscillations. Stumper et al. [177] followed the periodic fluctuations of the interfacial oxide layer with in-situ ellipsometry. By comparing the ellipsometric parameters to those of a thermally grown silicon oxide, they concluded that the anodically grown oxide must be different, and is probably of a mixed-oxyfluoride nature. They also observed that the periodic variations of the ellipsometric parameter Δ, which must be related to the thickness of the surface layer, are so small that they concluded that there is no involvement of a periodic buildup and removal of the oxide layer. The generality of this conclusion has been refuted by at least two other works: (1) Ozanam and Chazalviel [175], using internal IR multiple-reflection spectroscopy, observed about 60% modulation of the oxide-layer thickness during current oscillations by monitoring the νSiOSi absorption peak (they also observed significant variations in the νSiOH peak); (2) Lewerenz and Aggour [178] observed a thickness change of about 10% during oscillations in an already existing oxide film of 35-Å thickness by in situ ellipsometry and ex situ X-ray photoelectron spectroscopy. Ozanam and Chazalviel believe these differences can be explained by the difference in oscillation amplitudes in the various experiments; in their IR measurements, they took care to obtain an oscillation of large amplitude. By carrying out ring-disk voltammetry, Stumper et al. [177] also found significant variations in the rate of hydrogen evolution during oscillations, and associated with this, the ratio Si(II)/Si(IV) in the interfacial layer.

Ozanam, Chazalviel, and co-workers [172, 173, 179–182] emphasized in various papers that the oscillations that are observed in the J_4 plateau are always damped and appear only upon a potential excitation (even very small). Although sustained oscillations can be obtained by inserting an ohmic resistance in series with the working electrode [179], the damped oscillations are still present under conditions where the external ohmic potential drop can be safely neglected. This indicates that the oscillatory, or rather *resonant* behavior is a truly intrinsic interfacial phenomenon. A picture of a heterogeneous surface, consisting of small self-oscillating domains, has been put forward by Chazalviel and Ozanam in order to account for the experimental observations. The oscillating domains, incoherent in the steady state, would become synchronized by a potential excitation. This hypothesis is supported by the observation that a homogeneous oxide layer is grown under oscillatory conditions, whereas a more heterogeneous layer is formed in the steady state [172]. However, the most striking evidence in favor of the conjecture of Chazalviel and Ozanam is found in the impedance data to be discussed below [180, 181].

We will not pay attention to the impedance diagrams obtained at low potentials, in the regime of porous silicon formation [183]. Rather unique impedance characteristics are found in the potential range where the system is prone to current oscillations when a small perturbation is applied [180]. Clearly, if the damped free oscillations have a characteristic frequency ω_f, one expects a resonance peak for the admittance around this frequency [see Eqs. (2.9)–(2.11)]. Surprisingly, the experimental admittance also exhibits resonance peaks for the overtone frequencies $2\omega_f$, $3\omega_f$, and so on, as is shown by the dashed lines in the Bode plot of Fig. 72. Ozanam et al. [180] have taken good care that these overtones are not due to experimental artifacts. They made sure that a possible residual oscillation of the current is damped out before each measurement. They also checked the perseverance of the behavior for very low-excitation amplitudes (down to 0.8 mV peak-to-peak), where the

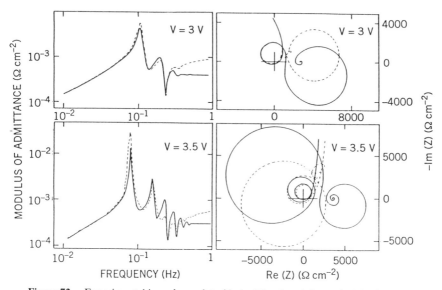

Figure 72. Experimental impedance data (dashed lines) and theoretical fits (solid lines), using Eq. (6.17). (111) p-Si electrode, electrolyte $c_F = 0.05$ M, pH 3. Values of the parameters for the fits: (i) for $V = 3$ V: $C = 4.3$ mF cm^{-2}, $J_0/\sigma = 0.107$ hz, $\Delta T = 1.25$ s, $T = 1.25$ s; for $V = 3.5$ V: $C = 3.7$ mF cm^{-2}, $J_0/\sigma = 0.079$ Hz, $\Delta T = 0.87$ s, $T = 1.4$ s. (Reprinted from [181] with permission of The Electrochemical Society.)

current response is sinusoidal and linearly proportional to the excitation amplitude, proving the linearity of the response. According to the authors, the detection of maxima in the admittance corresponding to overtones of the fundamental resonance is not contradictory with a linear response, but is consistent with the idea of a nonsinusoidal shape of the current oscillations of the individual domains. Therefore, the impedance measurements indicate that the interfacial behavior is intrinsically non-linear.

Chazalviel and Ozanam [181] elaborated a model to reproduce the experimental impedance diagrams. Without going into the mathematical details of this model, the main principles will be outlined and the final expression for the impedance–admittance of the system, called the "synchronization impedance" by the authors, will be given. The model assumes that the contribution of a domain to the current is given by a succession of pulses of arbitrary shape $f(t)$ [Fourier transform $\tilde{f}(\omega)$], two successive pulses being separated by the time necessary to dissolve the amount of oxide created by the first one. The fundamental frequency ω_f of these local current oscillations is then determined by the ratio of the chemical dissolution rate J_0 to the amount of charge σ passed during a single pulse. The potential dependence of this quantity is taken into account by the introduction of a specific capacitance C corresponding to the variation of σ with the electrode potential $(C = d\sigma/dE)$. In order to ensure a linear response regime, it is necessary to invoke a fluctuating component in the time between two successive pulses; this fluctuation is governed by the probability distribution $R(t)$, its Fourier transform denoted by $\tilde{R}(\omega)$. Under these assumptions, one may compute the synchronization admittance as [181]

$$Y(\omega) = \frac{CJ_0}{\sigma} + jC\omega \frac{|\tilde{f}(\omega)|^2}{1 - \exp(j\omega\sigma/J_0)\tilde{R}(\omega)^{-1}} \qquad (6.17)$$

where $j = \sqrt{-1}$ and $\omega_f = 2\pi J_0/\sigma$. This equation shows that one may expect maxima in $Y(\omega)$ if $\omega\sigma/J_0$ is in the vicinity of 2π or multiples thereof. These multiples correspond to the overtones. A Gaussian function is adopted for $R(t)$ and a simple exponential shape for $f(t)$ [i.e., $f(t) = 0$ for $t < 0$ and $f(t) = T^{-1} \exp(-t/T)$ for $t > 0$], so that $|\tilde{f}(\omega)|^2 = 1/(1 + \omega^2 T^2)$ and $\tilde{R}(\omega) = \exp(-\omega^2 \Delta T^2/2)$, where ΔT is the root mean square (rms) width of the Gaussian function. The solid lines in Fig. 72 show the results of the fits for the impedances measured for $E = 3$ and 3.5 V. The agreement is fairly good, considering the complexity of the experimental diagrams and the relatively small number of parameters.

Chazalviel and Ozanam [181] also derived an expression for the noise spectrum generated in the steady state by all the incoherently oscillating domains. By comparing the noise spectrum and the admittance spectrum [182], they can make an estimate of the typical size of a microscope domain. Measurements at various potentials indicate a domain size of approximately $(1000 \text{ Å})^2$. So far, the authors have not been able to obtain direct proof of the existence of the oscillating domains by some kind of spatially resolved microscopy.

Although the agreement between the theoretically predicted and the experimentally measured impedance characteristic is impressive, the model of oscillating microdomains that are completely uncorrelated in the steady state does pose a number of questions. First, one may wonder why domains of such small size should be completely uncoupled under all circumstances. One can think of at least two global coupling effects that should have a tendency to synchronize individual domains: the concentration or diffusion field in the electrolyte solution (recall that the Si dissolution is partially mass-transport controlled in the oscillatory region), and the electric field imposed externally. If locally these fields are distorted by a particularly active or passive domain, this should have an important effect on the surrounding domains. These forms of coupling are similar to the two coupling mechanisms that are usually invoked when the communication between individual catalyst particles in (oscillating) heterogeneously catalyzed reactions is concerned: coupling through concentration changes in the gas phase (gas-phase coupling) and through heat transfer by local temperature gradients (thermal coupling) [184]. Studies on the CO oxidation on different Pt/Al_2O_3 catalyst pellets in a tubular reactor, have shown that it is quite possible that all the pellets oscillate in phase, whereas under different conditions they may be completely desynchronized so that no overall oscillation is observed [185]. Similar observations have been made with the NO/NH_3 reaction on Pt(100), where, upon decreasing the temperature, transitions have been reported from homogeneous rate oscillations, to macroscopic oscillations with microscopic island formation or formation of spirals and target patterns, to a constant overall behavior, while microscopically one observes spirals and target patterns or irregular patterns [186]. In any case, such studies illustrate that one should be careful with disregarding the effect of some kind of coupling between different domains. A possible solution to this problem is that perhaps the microdomains are not incoherent, but instead patterns and structures appear on these microdomain sizes, like target patterns and spiral waves, such as have been observed in heterogeneous catalysis on single-crystal surfaces as noted above [184]. It does not appear so unreasonable to assume that the impedance response in such a case should be similar to that of the uncorrelated oscillating microdomains.

Of course, a major question that remains unanswered in Chazalviel and Ozanam's model concerns the physical origin of the microscopic oscillation. Chazalviel [187], and various other authors [171, 178], speculated that it should be related with processes of ionic transport through the oxide layer. The dissolution rate of the silica SiO_2 is expected to be related to the concentration of charged or neutral defects at the surface, owing to the large substoichiometry of anodically formed SiO_2 (anodic SiO_2 has a much larger dissolution rate than thermal SiO_2). Coupled to the migration of these defects through the oxide layer, Chazalviel [187] claims that this can give rise to a modulation of the dissolution rate, although so far no quantitative proof of this has been given. A similar discussion has been given by Lewerenz and Aggour [178] invoking periodic changes in the number of pores and point defects at the Si/SiO_2 interface where dissolution preferably takes place, and a periodic depletion of the neighboring Si surface region with the charge carrying holes at the n-type material. However, it must be emphasized that as long as no (simple) quantitative explanation has been given, such qualitative explanations do not satisfactorily answer the question of why the system does not settle down in an intermediate stable steady-state situation.

VII.   CONCLUDING REMARKS

This chapter has been an attempt towards a coherent view on the origin and appearance of electrochemical oscillations by stressing the electrical nature of their various manifestations. Since the necessity of some kind of electrical control is a feature common to all electrochemical systems, the qualitative origin of electrochemical instabilities and oscillations is often the same, although they may occur in very different processes. Therefore, in summarizing all the detailed information reviewed in the previous sections, the prime criterion in a categorization of the possible causes of sustained oscillations in electrochemical systems, should lie in the nature of the electrical control.

1. In the truly potentiostatic control, for which the interfacial potential can be considered constant irrespective of the current flowing through the cell, the origin of the oscillations must be purely chemical: that is, autocatalytic surface reactions, (adsorbate-induced) phase transitions of the substrate, deviations from ideal (Langmuirian) adsorption laws, and so on [109]. One may consider such oscillations as intrinsic oscillations, as any coupling to the external circuit must be absent. We have seen in the discussion above that at present there seem to be (only) two candidates to fall within this category: the anodic dissolution of silicon in fluoride media, and the oxidation of formic acid on (single-crystal) platinum surfaces. Such oscillations are in fact comparable to oscillations in heterogeneous catalysis, and the theoretical problems (and their solutions!) should be conceptually similar in many cases. This concerns in particular the question of synchronization of individual domains into a macroscopic oscillation. We have briefly discussed these matters in Section VI.B, when treating the Chazalviel–Ozanam model of incoherently oscillating domains, and indicated what kind of results have been obtained in heterogenous catalysis in this respect.

2. If, in the potentiostatic control, the effect of an external ohmic potential drop cannot be neglected, current and electrode potential oscillations may result from a negative faradaic impedance in conjunction with the ohmic series resistance. It is important to note that not just any physical explanation of such a negative impedance

will suffice; this was illustrated in Section VI.A in our discussion of the oscillations during hydrogen peroxide reduction on GaAs. It was shown that the inhibition leading to the negative slope must work directly on the electron-transfer step (perhaps with some time delay), and not on some preceding nonelectrochemical step such as adsorption. In terms of the complex impedance plane, this amounts to the requirement that the impedance plot should cross the negative real axis for some nonzero frequency. Also, oscillations in the "almost potentiostatic regime" [106] are not restricted to a polarization curve with a negative slope. They can also occur on a branch with a positive slope, provided the real impedance becomes negative for a range of nonzero frequencies. Such systems will also oscillate in the galvanostatic mode (see item 3 below). This possibility has been clearly illustrated recently for the hydrogen oxidation on platinum in the presence of copper and chloride ions [119].

3. In the galvanostatic mode, oscillations typically occur around a branch in the polarization curve with positive slope. Destabilization under constant current load occurs on account of a range of nonzero frequencies for which the real impedance becomes negative (a "hidden" negative impedance). The occurrence of galvanostatic potential oscillations therefore requires the coupling of a negative impedance process that is relatively fast, with a positive impedance process that is slower so that it will dominate the impedance response only for the lowest frequencies, resulting in an overall positive real impedance and polarization slope. This clearly demonstrates that galvanostatic potential oscillations can never be explained by simply referring to a region of negative slope in the polarization curve. The relation between galvanostatic potential oscillations and a hidden negative impedance has been illustrated in Sections IV.C (the Ni electrodissolution in H_2SO_4), V.A (H_2 oxidation on Pt), V.B (HCHO oxidation on Rh) and VI.A (H_2O_2 reduction on semiconductors).

Disregarding the first class, it seems that a classification of electrochemical oscillators should distinguish between potentiostatic and galvanostatic oscillators. However, a galvanostatic oscillator will also oscillate under potentiostatic conditions with an appropriate external resistor, so this classification is not entirely accurate. A better classification distinguishes between electrochemical oscillators with and without a hidden negative impedance. These two classes exhibit different dynamical bifurcations and phase diagrams. A typical electrochemical oscillator without a hidden negative impedance (but with a "nonhidden" negative impedance, of course) is the reduction of In^{3+} on the HMDE from thiocyanate solution; the typical phase diagram for such systems is cross-shaped or fish-shaped [38] (see also [44]). A typical electrochemical oscillator with a hidden negative impedance is the nickel dissolution in sulfuric acid; the typical phase diagram is not cross-shaped but normally involves homoclinic bifurcations of stable (i.e., observable) periodic orbits (Fig. 41). Even in the simplest cases, there are several different phase diagrams possible for these latter systems [188], the diagram shown in Figs. 43 and 44 being one of them; they can be incorporated in the BDK model discussed in Section III.D by simply adding a quadratic term in x to the differential equation for y [188]. This mechanistic classification of electrochemical oscillators fits in with a classification for chemical oscillators based on stoichiometric network analysis [144]. In terms of this classification [144], electrochemical oscillators without a hidden negative impedance are of type 1, and electrochemical oscillators with a hidden negative impedance are of type 2. It is possible to make further subdivisions of the type 1 chemical oscillators on the basis of mechanistic details; it would be interesting to see whether an equivalent subdivision can be made of electrochemical oscillators, based on impedance charac-teristics, for instance.

Despite the important role of the external electrical circuit, the appearance of instabilities and oscillations nonetheless depends critically on the interfacial electrochemical kinetics. This fact has led to a recurrent statement in the literature that studies of oscillations can be used profitably in elucidating the mechanistically important features in a kinetic scheme of the electrochemical system under study. While there may be some truth in such a statement in a general sense, the supple-mental value of the nonlinear dynamics approach in this respect should not be overestimated. The building of two-parameter phase diagrams (like those of Figs, 11, 36, 37, 43, 47, and 48) is important and interesting from the dynamic point-of-view, but adds very little to the already existing methods in the study of electrochemical kinetics. As was shown in Section II, the two most important lines in such a phase diagram, the saddle-node and Hopf bifurcations, also follow directly from the impe-dance method, a method certainly to be preferred to studying oscillations if one is really interested in electrochemical kinetics. What the study of oscillations has to offer, is knowledge about global bifurcations, such as saddle-loop homoclinic bifurcations and transitions to complex or chaotic oscillations. As far as the latter is concerned, chaotic behavior in electrochemical systems is more likely to be associated with effects of (slow) mass transport or the existence of spatial inhomogeneities, than with complicated kinetic mechanisms. The results obtained with the

Figure 73.   Spiral wave patterns during the electrodeposition of a silver–antimony alloy. Wavelength, that is, distance between bands, is approximately 10 μm. Waves move at a velocity of several micrometers per second. (Courtesy of I. Krastev.)

reduction reactions on mercury electrodes (Section III), and the observation of the spatiotemporal period doublings in the iron dissolution in sulfuric acid (Section IV.A), may serve to illustrate this point.

This brings us to the point of the role of spatial effects on electrode surfaces in nonlinear electrochemical systems. This topic has not been discussed in any great detail, except in those cases where it has proved important in the understanding of the overall behavior. Many interesting experimental and theoretical results are emerging at the moment, but a coherent discussion at this point would be premature. It is noteworthy that such spatiotemporal phenomena in electrochemistry, just as oscillations, were already reported and investigated about 100 years ago, notably by Ostwald [65] and his student Heathcote [189], Lillie [190], and Bonhoeffer [191]. These authors studied waves of activity traveling on passive iron wires immersed in nitric acid, and noted the many analogies of this system with impulse propagation along nerves. In the 1960s and 1970s the system has been studied intensively in the Soviet Union, a body of work that is hardly known but that has been reviewed in the first chapter of a recent book by Markin et al. [192]. These authors have also developed a model that is similar to the Hodgkin–Huxley model from neurophysiology and in which the active–passive transition of the iron dissolution and the autocatalytic reduction of the nitric acid play an important role. Several recent studies in metal electrodissolution have shown that oscillations are often accompanied by such wave phenomena. Patterns and structures, but of a very different, that is, smaller, lengthscale, have also been observed in the electrodeposition of some alloys. Krastev and co-workers [193–195] made several studies of the target patterns and spiral waves occurring during the codeposition of silver and antimony (Fig. 73), showing that the patterns are in fact concentration patterns in the silver/antimony ratio. It has been pointed out that these patterns possess characteristics (wavelength and wavespeed) that are very similar to patterns occurring in heterogeneous catalysis [195].

ACKNOWLEDGEMENTS

I am much indebted to my Ph.D. supervisors Professor Jan Sluyters and Dr. Pierre Gaspard for their continuous support and interest in this work, and for their comments on the manuscript. It is also a pleasure to thank Jean-Noël Chazalviel, John Kelly, François Ozanam, Mark Schell, Daniël Vanmaekelbergh, and Willi Wolf for suggesting various modifications and improvements of the manuscript.

REFERENCES

1. S. K. Scott, *Chemical Chaos*, Clarendon Press, Oxford, UK, 1991.
2. J. L. Hudson and T. T. Tsotsis, *Chem. Eng. Sci.*, **49**, 1493 (1994).
3. J. Wojtowicz, in *Modern Aspects of Electrochemistry*, J. O' M. Bockris and B. E. Conway, Eds., Plenum, New York, 1973, Vol. 8, p. 47.
4. For other reviews, see (a) K. S. Indira, S. K. Rangarajan, and K. S. G. Doss, *J. Electroanal. Chem.*, **21**, 57 (1969); (b) P. Poncet, M. Braizaz, B. Pointu, and J. Rousseau, *J. Chim. Phys.*, **74**, 452 (1977); (c) J. L. Hudson and M. R. Bassett, *Rev. Chem. Eng.*, **7**, 109 (1991); (d) V. V. Nechiporuk and I. L. Elgyrt, *Self-organization in Electrochemical Systems*, Nauka, Moscow, 1992 (in Russian); (e) J. L. Hudson, in *Chaos in Chemistry and Biochemistry*, R. J. Field and L. Györgi, Eds., World Scientific, Singapore, 1993, p. 123.
5. M. Sluyters-Rehbach and J. H. Sluyters, in *Electroanalytical Chemistry*, A. J. Bard, Ed., Maracel-Dekker, New York, 1970, Vol. 4, p. 1.
6. I. Epelboin, C. Gabrielli, and M. Keddam, in *Comprehensive Treatise of Electrochemistry*, E. Yeager, Ed., Plenum, New York, 1984, Vol. 9, p. 61.
7. M. Sluyters-Rehbach and J. H. Sluyters, in *Comprehensive Treatise of Electrochemistry*, E. Yeager, Ed., Plenum, New York, 1984, Vol. 9, p. 177.
8. J. R. Macdonald, Ed., *Impedance Spectroscopy*, Wiley, New York, 1987.
9. M. Sluyters-Rehbach, *Pure Appl. Chem.*, **66**, 1831 (1994).

10. E. Kreyszig, *Advanced Engineering Mathematics*, Wiley, New York, 1988.

11. J. Guckenheimer and P. Holmes, *Nonlinear Oscillations, Dynamical Systems, and Bifurcations of Vector Fields*, Springer, Berlin, 1986.

12. S. Wiggins, *Introduction to Applied Nonlinear Dynamical Systems and Chaos*, Springer, Berlin, 1988.

13. R. de Levie, *J. Electroanal. Chem.*, **25**, 257 (1970).

14. C. Gabrielli, Ph.D. Thesis, Université de Paris-VI, Paris, 1973; *Métaux Corr. Ind.*, **573**, 171 (1973); **574**, 223 (1973); **577**, 309 (1973); **578**, 356 (1973).

15. I. Epelboin, C. Gabrielli, and M. Keddam, in *Comprehensive Treatise of Electrochemistry*, J. O' M. Bockris, B. E. Conway, E. Yeager, and R. E. White, Eds., Plenum, New York, 1984, Vol. 4, p. 151.

16. H. Nyquist, *Bell. System Tech. J.*, **11**, 126 (1932).

17. H. W. Bode, *Network Analysis and Feedback Amplifier Design*, D. Van Nostrand Company, Princeton, 1945.

18. R. C. Dorf, *Modern Control Systems*, Addison-Wesley, Reading, MA, 1981.

19. A. Hjelmfelt, I. Schreiber, and J. Ross, *J. Phys. Chem.*, **95**, 6048 (1991).

20. I. Epelboin, C. Gabrielli, M. Keddam, J. C. Lestrade, and H. Takenouti, *J. Electrochem. Soc.*, **119**, 1632 (1972).

21. C. Gabrielli, M. Keddam, E. Stupinek-Lissac, and H. Takenouti, *Electrochim. Acta*, **21**, 757 (1976).

22. M. Sluyters-Rehbach, B. Timmer, and J. H. Sluyters, *Z. Phys. Chem. NF*, **52**, 1 (1967).

23. R. de Levie and L. Pospisil, *J. Electroanal. Chem.*, **22**, 277 (1969).

24. H. Moreira and R. de Levie, *J. Electroanal. Chem.*, **29**, 353 (1971).

25. I. Epelboin, M. Ksouri, E. Lejay, and R. Wiart, *Electrochim. Acta*, **20**, 603 (1975).

26. A. Ya. Gokhstein and A. N. Frumkin, *Dokl. Akad. Nauk. USSR*, **132**, 388 (1960).

27. A. N. Frumkin, O. A. Petrii, and N. V. Nikolaeva-Fedorovich, *Dokl. Akad. Nauk*, *USSR*, **136**, 1158 (1961).

28. Yu. A. Chizmadzev, *Dokl. Akad. Nauk. USSR*, **133**, 745 (1960).

29. A. Ya. Gokhstein, *Dokl. Akad. Nauk USSR*, **140**, 761 (1961); **148**, 136 (1963); **149**, 880 (1963).

30. M. Kolthoff and J. J. Lingane, *Polarography*, Interscience, New York, 1941, p. 275.

31. J. Keizer and D. Scherson, *J. Phys. Chem.*, **84**, 2025 (1980).

32. L. Pospisil and R. de Levie, *J. Electroanal. Chem.*, **25**, 245 (1970).

33. M. T. M. Koper and J. H. Sluyters, *J. Electroanal. Chem.*, **303**, 73 (1991).

34. R. Tamamushi, *J. Electroanal. Chem.*, **11**, 88 (1966).

35. R. Tamamushi and K. Matsuda, *J. Electroanal. Chem.*, **12**, 436 (1966).

36. M. T. M. Koper and J. H. Sluyters, *J. Electroanal. Chem.*, **303**, 65 (1991).

37. M. T. M. Koper and P. Gaspard, *J. Phys. Chem.*, **95**, 4945 (1991).

38. M. T. M. Koper, *Electrochim. Acta*, **37**, 1771 (1992).

39. M. T. M. Koper and P. Gaspard, *J. Chem. Phys.*, **96**, 7797 (1992).

40. M. T. M. Koper, P. Gaspard, and J. H. Sluyters, *J. Phys. Chem.*, **96**, 5674 (1992).

41. M. T. M. Koper, P. Gaspard, and J.H. Sluyters, *J. Chem. Phys.*, **97**, 8250 (1992).

42. J. Heyrovsky and J. Kuta, *Principles of Polarography*, Academic, New York, 1966.

43. A. N. Frumkin, *Z. Elektrochem.*, **59**, 807 (1955).

44. W. Wolf, J. Ye, M. Purgand, M. Eiswirth, and K. Doblhofer, *Ber. Bunsenges. Phys. Chem.*, **96**, 1797 (1992).

45. M. T. M. Koper and J. H. Sluyters, *J. Electroanal. Chem.*, **352**, 51 (1993).

46. S. Minc and J. Adrzejczak, *J. Electroanal. Chem.*, **17**, (1968) 101.

47. F. Argoul, J. Huth, P. Merzeau, A. Arnéodo, and H. L. Swinney, *Physica D*, **62**, 123 (1993).

48. P. Parmananda, H. D. Dewald, and R. W. Rollins, *Electrochim. Acta*, **39**, 917 (1994).

49. J. L. Hudson, M. Hart, and D. Marinko, *J. Chem. Phys.*, **74**, 1601 (1979).

50. M. Hourai, Y. Kotaki, and K. Kuwata, *J. Phys. Chem.*, **89**, 1760 (1985).

51. F. Argoul, A. Arnéodo, P. Richetti, and J. C. Roux, *J. Chem. Phys.*, **86**, 3325 (1987).

52. B. R. Johnson, J. F. Griffiths, and S. K. Scott, *J. Chem. Soc. Faraday Trans.*, **87**, 523 (1991).

53. M. Eiswirth and G. Ertl, *Surf. Sci.*, **177**, 90 (1986).

54. H. Herzel, P. Plath, and P. Svensson, *Physica D*, **48**, 340 (1991).

55. L. P. Shil'nikov, *Sov. Math. Dokl. Akad. USSR*, **6**, 163 (1965); *Math. USSR Sb.*, **10**, 91 (1970).

56. K. Krischer, M. Lübke, M. Eiswirth, W. Wolf, J. L. Hudson, and G. Ertl, *Physica D*, **62**, 123 (1993).

57. D. Barkley, *J. Chem. Phys.*, **89**, 5547 (1988).

58. A. N. Chaudry, P. V. Coveney, and J. Billingham, *J. Chem. Phys.*, **100**, 1921 (1994).

59. V. Petrov, S. K. Scott, and K. Showalter, *J. Chem. Phys.*, **97**, 6506 (1994).

60. M. T. M. Koper, *Physica D*, **80**, 72 (1995).

61. J. Boissonade and P. De Kepper, *J. Phys. Chem.*, **85**, 501 (1980).

62. A. J. Homburg, H. Kokubu, and M. Krupa, *Erg. Theor. Dyn. Syst.*, **14**, 667 (1994).

63. E. S. Hedges and J. E. Myers, *The Problem of Physico-chemical Periodicity*, Arnold, London, 1926.

64. J. P. Joule, *Philos. Mag.*, **24**, 106 (1844).

65. W. Ostwald, *Z. Phys. Chem.*, **35**, 33 (1900); **35**, 204 (1900).

66. J. H. Bartlett and L. Stephenson, *J. Electrochem. Soc.*, **99**, 504 (1952).

67. U. F. Franck, *Z. Elektrochem.*, **62**, 649 (1958).

68. K. J. Vetter, *Elektrochemische Kinetik*, Springer, Berlin, 1961.

69. U. F. Franck and R. FitzHugh, *Z. Elektrochem.*, **65**, 156 (1961).

70. P. P. Russell and J. Newman, *J. Electrochem. Soc.*, **130**, 547 (1983).

71. R. Alkire, D. Ernsberger, and T. R. Beck, *J. Electrochem. Soc.*, **125**, 1383 (1978).

72. M. E. Orazem and M. G. Miller, *J. Electrochem. Soc.*, **134**, 393 (1987).

73. L. Kiss, *Kinetics of Electrochemical Metal Dissolution*, Elsevier, Amsterdam, 1988.

74. I. Epelboin, C. Gabrielli, M. Keddam, and H. Takenouti, *Electrochim. Acta*, **20**, 913 (1975).

75. J.J. Podesta, R. C. V. Piatti, and A. J. Arvia, *J. Electrochem. Soc.*, **126**, 1363 (1979).

76. U. F. Franck, *Werkstoffe Korrosion*, **8/9**, 504 (1958).

77. J. J. Podesta, R. C. V. Piatti, and A. J. Arvia, *Electrochim. Acta*, **24**, 633 (1979).

78. M. C. H. McKubre and D. D. Macdonald, *J. Electrochem. Soc.*, **128**, 524 (1985).

79. D. Sazou and M. Pagitsas, *J. Electroanal. Chem.*, **323**, 247 (1992).

80. M. Pagitsas and D. Sazou, *J. Electroanal. Chem.*, **334**, 81 (1992).

81. R. Otterstedt, Diplomarbeit, Universität Bremen, Bremen (1993).

82. M. T. M. Koper and J. H. Sluyters, *J. Electroanal. Chem.*, **347**, 31 (1993).

83. P. Russell and J. Newman, *J. Electrochem. Soc.*, **133**, 59 (1986).

84. P. Russell and J. Newman, *J. Electrochem. Soc.*, **134**, 1051 (1987).

85. N. Sato, G. Okamoto, in *Comprehensive Treatise of Electrochemistry*, J. O' M. Bockris, B. E. Conway, E. Yeager, and R. E. White, Eds., Plenum, New York, 1984, Vol. 4, p. 193.

86. J. L. Hudson, J. Tabora, K. Krischer, and I. G. Kevrekidis, *Phys. Lett. A*, **179**, 335 (1993).

87. C. Diem and J. L. Hudson, *AIChE J.*, **33**, 218 (1987).

88. D. Sazou, M. Pagitsas, and C. Georgolios, *Electrochim. Acta*, **37**, 2067 (1992).

89. J. Rinzel, *Lect. Notes Biomath.*, **71**, 267 (1987).

90. P. A. Jacquet, *Trans. Electrochem. Soc.*, **69**, (1936) 629.

91. F. N. Albahadily and M. Schell, *J. Chem. Phys.*, **88**, 4312 (1988).

92. F. N. Albahadily, J. Ringland, and M. Schell, *J. Chem. Phs.*, **90**, 813 (1989).

93. M. Schell and F. N. Albahadily, *J. Chem. Phys.*, **90**, 822 (1989).

94. S. H. Glarum and J. H. Marshall, *J. Electrochem. Soc.*, **132**, 2872 (1985).

95. S. H. Glarum and J. H. Marshall, *J. Electrochem. Soc.*, **132**, 2878 (1985).

96. L. T. Tsitsopoulos, T. T. Tsotsis, and I. A. Webster, *Surf. Sci.*, **191** 225 (1987).

97. L. T. Tsitsopoulos, I. A. Webster, and T. T. Tsotsis, *Surf. Sci.*, **220**, 391 (1989).

98. T. P. Hoar and J. A. S. Mowat, *Nature (London)*, **165**, 64 (1950).

99. J. Osterwald and H. G. Feller, *J. Electrochem. Soc.*, **107**, 473 (1960).

100. J. Osterwald, *Electrochim. Acta*, **7**, 523 (1962).

101. O. Lev, A. Wolffberg, M. Sheintuch, and L.M. Pismen, *Chem. Eng. Sci.*, **43**, 1339 (1988).

102. O. Lev, A. Wolffberg, L. M. Pismen, and M. Sheintuch, *J. Phys. Chem.*, **93**, 1661 (1989).

103. D. Haim, O. Lev, L. M. Pismen, and M. Sheintuch, *J. Phys. Chem.*, **96**, 2676 (1992).

104. M. Keddam, H. Takenouti, and N. Yu, *J. Electrochem. Soc.*, **132**, 2561 (1985).

105. I. Epelboin and M. Keddam, *Electrochim. Acta*, **17**, 177 (1972).

106. O. Lev, M. Sheintuch, H. Yarntisky, and L. M. Pismen, *Chem. Eng. Sci.*, **45**, 839 (1990).

107. D. Haim, O. Lev, L. M. Pismen, and M. Sheintuch, *Chem. Eng. Sci.*, **47**, 3907 (1992).

108. A. Pamfilov, A. I. Lopushanskaya, and S. V. Markova, *Elektrokhimiya*, **6**, 849 (1970).

109. M. T. M. Koper and J. H. Sluyters, *J. Electroanal. Chem.*, **371**, 149 (1994).

110. H. A. Liebhavsky and E. J. Cairns, *Fuel Cells and Fuel Batteries*, Wiley, New York, 1968, pp. 428–429.

111. G. Horányi and C. Visy, *J. Electroanal. Chem.*, **103**, 353 (1979).

112. T. Kodera, T. Yamazaki, N. Kubota, *Electrochim. Acta*, **31**, 1477 (1986).

113. T. Kodera, T. Yamazaki, M. Masuda, R. Ohnishi, *Electrochim. Acta*, **33**, 537 (1988).
114. W. Wolf, Ph.D. Thesis, Freie Universität Berlin, Berlin, 1994.
115. K. Krischer, M. Lübke, W. Wolf, M. Eiswirth, and G. Ertl, *Electrochim. Acta*, **40**, 69 (1995).
116. K. Krischer, M. Lübke, W. Wolf, M. Eiswirth, and G. Ertl, *Ber. Bunsenges. Phys. Chem.*, **95**, 820 (1991).
117. M. Eiswirth, M. Lübke, K. Krischer, W. Wolf, J. L. Hudson, and G. Ertl, *Chem. Phys. Lett.*, **192**, 254 (1992).
118. W. Wolf, K. Krischer, M. Lübke, M. Eiswirth, and G. Ertl, *J. Electroanal. Chem.*, **385**, 85 (1995).
119. W. Wolf, M. Lübke, M. T. M. Koper, K. Krischer, M. Eiswirth, and G. Ertl, *J. Electroanal. Chem.*, in press (1996).
120. R. D. Amstrong and M. Henderson, *J. Electroanal. Chem.*, **39**, 81 (1972).
121. D. A. Harrington and B. E. Conway, *Electrochim. Acta*, **32**, 1703 (1987).
122. E. Mueller, *Z. Elektrochem.*, **29**, 264 (1923).
123. F. Raspel and M. Eiswirth, *J. Phys. Chem.*, **98**, 7613 (1994).
124. B. Beden, J.-M. Léger, and C. Lamy, in *Modern Aspects of Electrochemistry*, J. O' M. Bockris, B. E. Conway, and R. E. White, Eds., Plenum, New York, 1992, Vol. 22, p. 97.
125. A. Capon and R. Parsons, *J. Electroanal. Chem.*, **45**, 205 (1973).
126. S. G. Sun, J. Clavilier, and A. Bewick, *J. Electroanal. Chem.*, **240**, 147 (1988).
127. R. Parsons and T. VanderNoot, *J. Electroanal. Chem.*, **257**, 9 (1988).
128. T. G. J. van Venrooij and M. T. M. Koper, unpublished results.
129. H. Angertein-Kozlowska, B. E. Conway, and W. B. A. Sharap, *J. Electroanal. Chem.*, **43**, 9 (1973).
130. C. Lamy, J.-M. Léger, J. Clavilier, and R. Parsons, *J. Electroanal. Chem.*, **150**, 71 (1983).
131. V. S. Bagotsky and Yu. B. Vassiliev, *Electrochim. Acta*, **9**, 869 (1964).
132. D. Gilroy and B. E. Conway, *J. Phys. Chem.*, **69**, 1259 (1965).
133. H. Okamoto, *Electrochim. Acta*, **37**, 37 (1992).
134. F. N. Albahadily and M. Schell, *J. Electroanal. Chem.*, **308**, 151 (1991).
135. J. Wojtowicz, N. Marincic, and B. E. Conway, *J. Chem. Phys.*, **48**, 4333 (1968).
136. M. Schell, F. N. Albahadily, J. Safar, and Y. Xu, *J. Phys. Chem.*, **93**, 4806 (1989).
137. Y. Xu and M. Schell, *J. Phys. Chem.*, **94**, 7137 (1990).
138. M. Orban and I. R. Epstein, *J. Phys. Chem.*, **86**, 3907 (1983).
139. M. Hachkar, B. Beden, and C. Lamy, *J. Electroanal. Chem.*, **287**, 81 (1990).
140. M. Hachkar, M. Choy de Martinez, A. Rakotondrainibe, B. Beden, and C. Lamy, *J. Electroanal. Chem.*, **302**, 173 (1991).
141. M. Hachkar, Ph.D. Thesis, Université de Poitiers, 1988.
142. M. T. M. Koper, M. Hachkar, and B. Beden, submitted to *J. Electroanal. Chem.*
143. G. Inzelt, V. Kertész, and G. Láng, *J. Phys. Chem.*, **97**, 6104 (1993).
144. M. Eiswirth, A. Freund, and J. Ross, *Adv. Chem. Phys.*, **80**, 127 (1991).
145. F. Raspel, R.J. Nichols, and D. M. Kolb, *J. Electroanal. Chem.*, **286**, 279 (1990).
146. A. Tripkovic, K. Popovic, and R. R. Adzic, *J. Chim. Phys.*, **88**, 1635 (1991).
147. N. Markovic and P. N. Ross, *J. Phys. Chem.*, **97**, 9771 (1993).

148. G. Horányi, G. Inzelt, and E. Szetey, *J. Electroanal. Chem.*, **81**, 395 (1977).

149. G. Parida and M. Schell, *J. Phys. Chem.*, **95**, 2356 (1991); M. Schell and X. Cai, *J. Chem. Soc. Faraday Trans.*, **87**, 2255 (1991).

150. H. Gerischer, in *Physical Chemistry, An Advanced Treatise*, H. Eyring, D. Henderson, and W. Jost, Eds., Academic, New York, 1970, Vol. IXA, p. 463.

151. N. Fetner and J. L. Hudson, *J. Phys. Chem.*, **94**, 6506 (1990).

152. T. G. J. van Venrooij and M. T. M. Koper, *Electrochim. Acta*, **40**, 1689 (1995).

153. M. Honda, T. Kodera, and H. Kita, *Electrochim. Acta*, **31**, (1986) 377.

154. S. Strbac and R. R. Adzic, *J. Electroanal. Chem.*, **337**, 355 (1992).

155. H. Tributsch, *Ber. Bunsenges. Phys. Chem.*, **79**, 570 (1975), **79**, 580 (1975).

156. H. Tributsch and J. C. Bennett, *Ber. Bunsenges. Phys. Chem.*, **80**, 321 (1976).

157. S. Cattarin and H. Tributsch, *J. Electrochem. Soc.*, **137**, 3475 (1990).

158. S. Cattarin and H. Tributsch, *J. Electrochem. Soc.*, **139**, 1328 (1992).

159. M. T. M. Koper, E. A. Meulenkamp, and D. Vanmaekelbergh, *J. Phys. Chem.*, **97**, 7337 (1993).

160. L. Pohlmann, G. Neher, and H. Tributsch, *J. Phys. Chem.*, **98**, 11007 (1994).

161. B. P. Minks, G. Oskam, D. Vanmaekelbergh, and J. J. Kelly, *J. Electroanal. Chem.*, **273**, 119 (1989).

162. M. T. M. Koper and D. Vanmaekelbergh, *J. Phys. Chem.*, **99**, 3687 (1995).

163. B. P. Minks, D. Vanmaekelbergh, and J. J. Kelly, *J. Electroanal. Chem.*, **273**, 133 (1989).

164. W. H. Laflère, F. Cardon, and W. P. Gomes, *Surf. Sci.*, **44**, 541 (1974).

165. S. Cattarin, H. Tributsch, and U. Stimming, *J. Electrochem. Soc.*, **139**, 1320 (1992).

166. G. Neher, Ph.D. Thesis, Freie Universität Berlin, Berlin, 1994.

167. H. Gerischer and W. Mindt, *Surf. Sci.*, **4**, 440 (1966).

168. G. Horányi and E. M. Rizmayer, *J. Electroanal. Chem.*, **188**, 347 (1982).

169. O. A. Petrii and T. Ya. Salonova, *J. Electroanal. Chem.*, **331**, 897 (1992).

170. D. R. Turner, *J. Electrochem. Soc.*, **105**, 402 (1958).

171. H. Gerischer and M. Lübke, *Ber. Bunsenges. Phys. Chem.*, **92**, 573 (1988).

172. F. Ozanam, J.-N. Chazalviel, A. Radi, and M. Etman, *Ber. Bunsenges, Phys. Chem.*, **95**, 98 (1991).

173. J.-N. Chazalviel, M. Etman, and F. Ozanam, *J. Electroanal. Chem.*, **297**, 533 (1991).

174. D. J. Blackwood, A. Borazio, R. Greef, L. M. Peter, and J. Stumper, *Electrochim. Acta*, **37**, 889 (1991).

175. F. Ozanam and J.-N. Chazalviel, *J. Electron Spectrosc. Relat. Phenom.*, **64/65**, 395 (1993).

176. H. H. Hassan, J. L. Sculfort, M. Etman, F. Ozanam, and J.-N. Chazalviel, *J. Electroanal. Chem.*, **380**, 55 (1995).

177. J. Stumper, R. Greef, and L. M. Peter, *J. Electroanal. Chem*, **310**, 445 (1991).

178. H. J. Lewerenz and M. Aggour, *J. Electroanal. Chem.*, **351**, 159 (1993).

179. J.-N. Chazalviel, F. Ozanam, M. Etman, F. Paolucci, L. M. Peter, and J. Stumper, *J. Electroanal. Chem.*, **327**, 343 (1992).

180. F. Ozanam, J.-N. Chazalviel, A. Radi, and M. Etman, *J. Electrochem. Soc.*, **139**, 2491 (1992).

181. J.-N. Chazalviel and F. Ozanam, *J. Electrochem. Soc.*, **139**, 2501 (1992).

182. F. Ozanam, N. Blanchard, and J.-N. Chazalviel. *Electrochim. Acta*, **38**, 1627 (1993).

183. P. C. Searson and X. G. Zhang, *J. Electrochem. Soc.*, **137**, 2539 (1990); D. Vanmaekelbergh and P. C. Searson, *ibid.*, **141**, 697 (1994).

184. M. Eiswirth, in *Chaos in Chemistry and Biochemistry*, R. J. Field and L. Györgi, Eds., World Scientific, Singapore, 1993, p. 141.

185. J. Kapicka and M. Marek, *J. Catalysis* **229**, 508 (1989).

186. G. Veser, F. Esch, and R. Imbihl, *Catal. Lett.*, **13**, 371 (1992).

187. J.-N. Chazalviel, *Electrochim. Acta*, **37**, 865 (1992).

188. M. T. M. Koper, *J. Chem. Phys.*, **102**, 5278 (1995).

189. H. L. Heathcote, *Z. Phys. Chem.*, **37**, 368 (1901).

190. R. S. Lillie, *Science*, **50**, 25 (1919); **69**, 473 (1929); *J. Gen. Physiol.*, **32**, 69 (1925).

191. K. F. Bonhoeffer, *J. Gen. Physiol.*, **32**, 69 (1948).

192. V. S. Markin, V. F. Pastushenko, and Yu. A. Chizmadzev, *Theory of Excitable Media*, Wiley, New York, 1987, Chapter 1.

193. I. Krastev and M. Nikolova, *J. Appl. Electrochem.*, **16**, 875 (1986).

194. I. Krastev, M. Nikolova, and I. Nakada, *Electrochim. Acta*, **34**, 1211 (1989).

195. I. Krastev and M. T. M. Koper, *Physica A*, **213**, 199 (1995).

TRANSPORT AND RELAXATION PHENOMENA IN POROUS MEDIA

R. HILFER

Institute of Physics, University of Oslo, 0316 Oslo, Norway

Institut für Physik, Universität Mainz, 55099 Mainz, Germany

CONTENTS

I. Introduction
    A. The Problem
    B. Scope of Review
II. Definition and Examples of Porous Media
    A. Examples of Porous Media
    B. Definition of Porous Media
        1. Deterministic Geometries
        2. Stochastic Geometries
            a. Discrete Space
            b. Continuous Space
III. Geometric Characterization
    A. General Geometric Characterization Theories
        1. Porosity and Other Numbers
            a. Porosity
            b. Specific Internal Surface Area
        2. Correlation Functions
        3. "Pore Size" Distributions
            a. Mercury Porosimetry
            b. Random Point Methods
            c. Erosion Methods
            d. Hydraulic Radius Method
        4. Contact and Chord Length Distributions
        5. Local Geometry Distributions
            a. Local Porosity Distribution
            b. Local Geometry Entropies
            c. Local Specific Internal Surface Distributions

Advances in Chemical Physics, Volume XCII, Edited by I. Prigogine and Stuart A. Rice.
ISBN 0-471-14320-0 © 1996 John Wiley & Sons, Inc.

299

I. INTRODUCTION

A. The Problem

Almost all studies of transport and relaxation in porous media are motivated by one central question. How are the effective macroscopic transport parameters influenced by the microscopic geometric structure of the medium?

My presentation will divide this central question into three subproblems. The first subproblem is to give a quantitative geometric characterization of the complex porous microstructure. It will be discussed in Sections II and III. The second subproblem, treated in Sections IV and V, is to calculate effective macroscopic transport properties from the geometric characterization and the equations of motion for the phenomenon of interest. The third subproblem, known as the problem of "upscaling," runs as a common thread through all sections, and consists in defining and controlling the macroscopic limit. The macroscopic limit is a limit in which the ratio of a typical macroscopic length scale (e.g., the sample size) to a typical microscopic length scale (e.g., grain or pore size) diverges.

Distinguishing between the first and the second problem is conceptually convenient and important. Geometric properties of porous media are determined exclusively by the complex system of internal boundaries that defines the microstructure. Geometric properties can be calculated from a complete specification of the microstructure alone. Physical transport and relaxation properties on the other hand also require exact or approximate equations of motion describing the physical phenomenon of interest. Often, this involves the relaxation of small perturbations or the steady-state transport of physical quantities such as mass, energy, charge, or momentum. The distinction between geometric and physical properties of porous media has become blurred in the literature [1, 2], and physical transport properties such as fluid flow permeability or formation factors are sometimes referred to as geometric characteristics. This can be understood because practically employed experimental methods for observing the pore space geometry often involve observations of relaxation and transport phenomena from which the geometry is inferred by inversion techniques.

Geometric properties of porous media are observed in practice either directly using light microscopy, electron microscopy, scanning tunneling microscopes (STM), or indirectly from interpreting experimental measurements of transport and relaxation processes such as fluid flow,

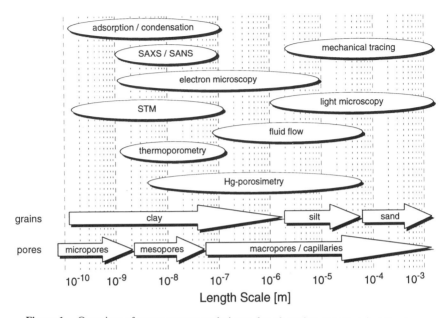

Figure 1. Overview of measurement techniques for observing porous microstructures and their range of sensitivity. The two lines of arrows at the bottom represent the DIN definition of grain sizes and the IUPAC definition of pore sizes.

electrical conduction, mercury intrusion, or small angle X-ray/small angle neutron scattering (SAXS/SANS). An overview over commonly used methods is given in Fig. 1. The different methods are represented on a length scale grid to indicate their ranges of applicability. The arrows in the lower part of Fig. 1 represent the IUPAC recommendation for classifying porous media into microporous, mesoporous, or macroporous media, and the DIN classification of porous media into clay, silt, and sand according to the grain size rather than pore size. Granular media with grains larger than 1 mm are called gravels or boulders.

Loosely speaking, a porous medium may be characterized as a medium containing a complex system of internal surfaces and phase boundaries. These internal interfaces define pores with a finite pore volume [3] and frequently a large surface area. Such a rough characterization applies also to other *heterogeneous media* and *composites*. Therefore some authors restrict the definition of porous media by requiring permeability [2], connectedness [4, 5], or randomness [5] as defining properties for porous media. Others [6–8], however, generalize the definition by including liquids into the purview. The precise definition of porous media adopted for the present review will be given in Section II.B.

B. Scope of Review

The study of transport and relaxation in porous media is scattered throughout many fields of science and technology ranging from mathematics [9, 10], through solid state physics [11–13] and materials science [14–16], to applications in geology [17–19], hydrology [20, 21], geophysics [22, 23], environmental technology [24–26], petroleum engineering [27–29], or separation technology [30]. In recent years a large number of books [2, 5, 31–38, 38a] and comprehensive reviews [8, 17, 22, 39–45] have discussed transport and relaxation in porous media. Therefore, this chapter will try to emphasize those aspects of the central questions that are complementary to the existing discussions.

As pointed out by Landauer [46] more emphasis is often placed on calculational schemes for effective transport properties than on finding general geometric characterizations of the medium that can be used as input for such calculations. Consequently, this chapter puts more emphasis on the first subproblem of geometric characterization than on the second subproblem of solving equations of motion for media with correlated disorder. Geometrical characterizations of porous media are treated in Section III, and they fall into two categories discussed in Sections III.A and III.B. The first category contains general theories which attempt to identify general and well-defined geometric quantities that can be used to distinguish between different classes of porous media. The second category consists of specific models which attempt to idealize one particular class of porous media by abstracting its most essential geometrical features and incorporating them into a detailed model. The main difference between the two categories is the degree to which they specify the geometric microstructure. In general theories the microstructure remains largely unspecified while it is specified completely in modeling approaches.

Dielectric relaxation and fluid transport are discussed in Section V as representative examples for more general physical processes in porous media. Transport and relaxation processes in porous media invariably involve the disordered Laplacian operator $\nabla^T \cdot (\mathbf{C}(\mathbf{r})\nabla)$, where ∇ is the Nabla operator, the superscript T denotes transposition, and the second rank tensor field $\mathbf{C}(\mathbf{r})$ gives the fluctuating local transport coefficients. Dielectric relaxation and single-phase fluid transport in porous media are problems of practical and scientific interest, which show unexpected experimental behavior, such as permeability–porosity correlations, Archie's law, or dielectric enhancement. The discussion in Section V will specifically address these issues. Methodically, the discussion in Section V will emphasize homogenization and local porosity theory because they allow one to control the macroscopic limit.

The upscaling problem as the controlled transition from microscopic to macroscopic length scales will then become the focus of attention in Section VI, which discusses two-phase fluid transport. The upscaling problem for two-phase flow is largely unresolved. Recent work [47, 48] has revisited the fundamental dimensional analysis [49] dating back more than 50 years, and uncovered a tacit assumption in the analysis that could help to resolve the upscaling difficulties. Thus the upscaling problem is treated in Section VI merely by comparing the microscopic and macroscopic dimensional analysis, and not by calculating effective relative permeabilities. Despite its simplicity the revisited dimensional analysis allows quantitative estimates of fluid transport rates and gravitational relaxation times based on the balance of viscous, capillary, and gravitational forces in the macroscopic limit.

II.   DEFINITION AND EXAMPLES OF POROUS MEDIA

A.   Examples of Porous Media

Most materials are porous when viewed at an appropriate length scale. Examples range from porous silicon, which is porous on the subnanometer scale, to limestone caves and underground river systems on the kilometer scale. An anthology of examples illustrates the variability of length scales and microstructures found in porous media.

Highly porous electrochemically etched silicon has recently found much interest because of its visible luminescence, which is attributed to its porous microstructure [12, 13, 50–54]. Figure 2 shows a transmission electron micrograph of the microstructure of porous silicon [13]. Other examples for microporous solids are zeolites.

Figure 3 shows a scanning electron micrograph (SEM) of a critical point dried gel consisting of ultrahigh molecular weight polyethylene [55]. The gel was obtained from 2% solutions in decalin under agitated conditions prior to gelation, and the fibrillar structure of the polyethylene crystals reflects flow prior to gelation. Thermoreversible gels consist of a macroscopic mechanically coherent network of macromolecules that is formed and stabilized through interconnected crystals. These materials are of considerable importance in the processing of high-performance polymers with very high tensile strength [55].

Figure 4 displays the microstructure of silicon nitride ceramics consisting of elongated Si_3N_4 grains embedded in a matrix of finer grains and a

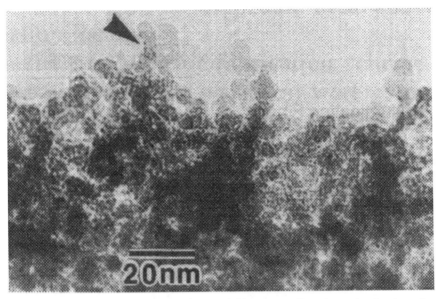

Figure 2. Transmission electron image of thin porous silicon layer showing irregular columnar microstructure of solid silicon. Some columns, such as the one indicated by the arrow, have cross-sectional diameter significantly less than 5 nm. [Reprinted with permission from *Nature*, A. Cullis and L. Canham, **353**, 335 (1991). Copyright © Macmillan Magazine Limited.]

grain boundary phase. Silicon nitride ceramics with high strength have a fine grained elongated microstructure, while materials with a high fracture toughness are more coarse grained [15].

Wood is a strongly anisotropic natural porous medium exhibiting cylindrical pores. Figure 5 shows an SEM of a partially cut and fractured surface from Malaysian Nemesu wood, which gives an impression of the irregularities within the material. The image shows tracheid cells of about 25 μm in diameter and vessel elements that are roughly an order of magnitude larger.

The large surface/volume ratio, which is characteristic for porous media (see Section III.A), is essential for the function of the lung consisting of some 300 million small air chambers. Figure 6 shows the foamlike structure formed by the respiratory air chambers in the lung. As emphasized by Weibel, biological porous media should *not be viewed as random* [56]. The general difficulty of modeling porous media as random or disordered media is further discussed in Section II.B.2.

Building and construction materials such as cements and concrete are porous media. Figure 7 shows an SEM of cement and lime mortar. Each

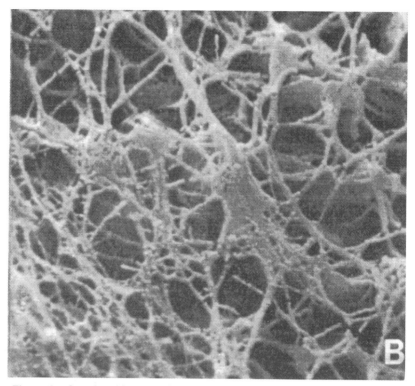

Figure 3. Scanning electron micrograph of critical point dried gel of ultrahigh molecular weight polyethylene. (Reproduced with permission from [55].)

dash in the dashed line visible in the image corresponds to 10 μm. The micrograph shows an extended network of fissures whose properties govern the moisture transfer and sorption properties of such materials [57].

An understanding of transport and relaxation in rocks, soils, and other geologically important heterogeneous media is of crucial importance in hydrology, exploration geology, petroleum engineering, and environmental research. Figure 8 shows a thin section of a clastic sandstone formed by fluvial deposits. Sandstones and other sedimentary rocks have attracted much research interest, and are among the best investigated examples of porous media.

The variety of two component porous media is enormous. Most of the following discussion will focus on irregular, random media. Regular and

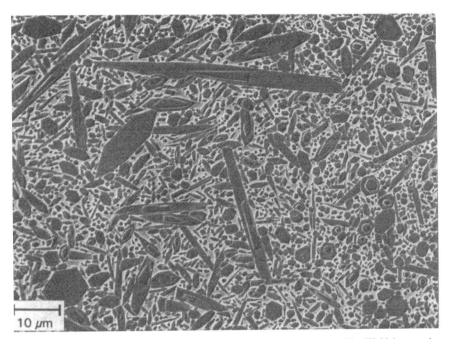

Figure 4. Scanning electron micrograph of a sintered silicon nitride (Si_3N_4) ceramic [15].

ordered microstructures (such as in zeolites) may be considered as a special case of irregular porous media.

B. Definition of Porous Media

1. Deterministic Geometries

An n-component *porous medium* in d dimensions is defined as a compact and singly connected subset $\mathbb{S}$ of $\mathbb{R}^d$, which contains n closed subsets $\mathbb{P}_i \subset \mathbb{S}$ such that

$$\mathbb{S} = \mathbb{P}_1 \cup \cdots \cup \mathbb{P}_n \tag{2.1}$$

$$0 = V_d(\partial \mathbb{P}_i) \tag{2.2}$$

for all $1 \leq i \leq n$. The set $\mathbb{S}$ is called the sample space and may represent, for example, a piece of porous rock. The subsets $\mathbb{P}_i$ $(i = 1, \ldots, n)$ represent n different phases or components, such as different minerals or

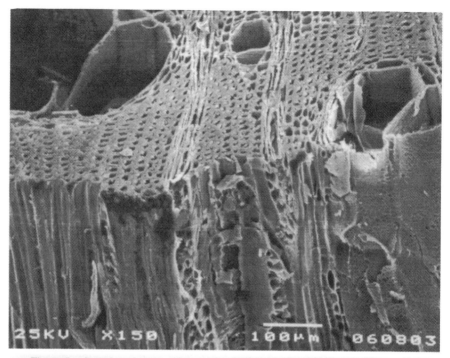

Figure 5. Scanning electron micrograph of Malaysian Nemesu wood (Shorea Pauci-flora).

fluid phases contained in a rock. The symbol $V_d(\mathbb{G})$ denotes the d-dimensional *volume* of a set $\mathbb{G} \subset \mathbb{R}^d$. It is defined as

$$V_d(\mathbb{G}) = \int \chi_\mathbb{G}(\mathbf{r}) d^d\mathbf{r} \qquad (2.3)$$

where $\mathbf{r}$ is a d-dimensional vector, and $d^d\mathbf{r}$ is the d-dimensional Lebesgue measure. Thus V_2 denotes an area, and V_1 is a length. When there is no danger of confusion $V_3(\mathbb{G}) = V(\mathbb{G})$ will be used below. The *characteristic (or indicator) function* of a set $\mathbb{G}$ is defined as

$$\chi_\mathbb{G}(\mathbf{r}) = \begin{cases} 1 & \text{for} \quad \mathbf{r} \in \mathbb{G} \\ 0 & \text{for} \quad \mathbf{r} \notin \mathbb{G} \end{cases} \qquad (2.4)$$

and it indicates when a point is inside or outside of $\mathbb{G}$. The sets $\partial \mathbb{P}_i$ are the phase boundaries separating the different components. The boundary

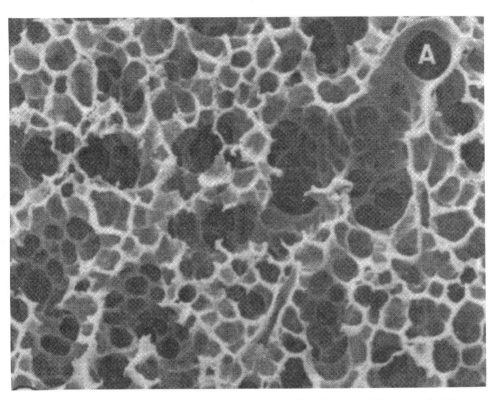

Figure 6. Scanning electron micrograph of air chambers in a lung. (Reproduced with permission from E. Weibel, "The non-statistical nature of biological structure and its implications on sampling for stereology," in *Geometrical Probability and Biological Structures: Buffon's 200th Anniversary*, E. Miles and J. Serra, Eds., Springer, Berlin, 1978, p. 171. Copyright © Springer-Verlag, 1978).

operator ∂ is defined on a general set $\mathbb{G}$ as the difference[1] $\partial\mathbb{G} = \overset{\bullet}{\mathbb{G}} \backslash \overset{\circ}{\mathbb{G}}$. The set $\overset{\circ}{\mathbb{G}}$, called the *interior* of $\mathbb{G}$, is defined as the union of all open sets contained in $\mathbb{G}$. The set $\overset{\bullet}{\mathbb{G}}$ is called the *closure* of $\mathbb{G}$ and is defined as the intersection of all closed sets containing $\mathbb{G}$. The condition (2.2) excludes fractal boundaries or cases in which a boundary set is dense. By replacing the Lebesgue measure in Eq. (2.3) with Hausdorff measures some of these restrictions may be relaxed [58–61].

---

[1] The settheoretic difference operation explains the use of the differential symbol ∂, which may also be motivated by the fact that the derivative operator $\nabla\chi_{\mathbb{P}}(\mathbf{r})$ applied to the characteristic function $\chi_{\mathbb{P}}(\mathbf{r})$ of a closed set $\mathbb{P}$ yields the Dirac distribution concentrated on the set $\partial\mathbb{P}$.

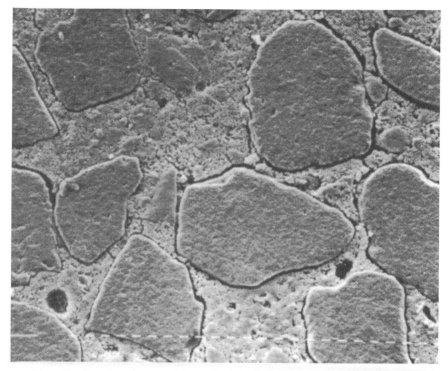

Figure 7. Scanning electron micrograph of a polished surface of mortar (magnification 200×). Reproduced with permisstion from [57].

Frequently, the different phases or components may be classified into solid and fluid phases. An example is a porous rock. In this case, it is convenient to consider the two-component medium in which all the solid phases are collectively denoted as *matrix space* $\mathbb{M}$, and the fluid phases are denoted collectively as *pore space* $\mathbb{P}$. The union of the pore and matrix space $\mathbb{S} = \mathbb{P} \cup \mathbb{M}$ gives the full porous *sample space* $\mathbb{S}$ and the intersection of $\mathbb{P}$ and $\mathbb{M}$ defines the boundary set $\mathbb{P} \cap \mathbb{M} = \partial\mathbb{P} = \partial\mathbb{M}$. It will usually be assumed that the boundary $\partial\mathbb{P}$ of the pore or matrix space is a surface in $\mathbb{R}^3$. This implies that $V_3(\partial\mathbb{P}) = 0$ in agreement with Eq. (2.2).

As an example consider a clean quartz sandstone filled with water. The sets $\mathbb{P}$ and $\mathbb{M}$ can be defined using the density contrast between the density of water $\rho_\mathrm{w} \approx 1\,\mathrm{g\,cm}^{-3}$, and that of quartz $\rho_\mathrm{Q} \approx 2.65\,\mathrm{g\,cm}^{-3}$ [62].

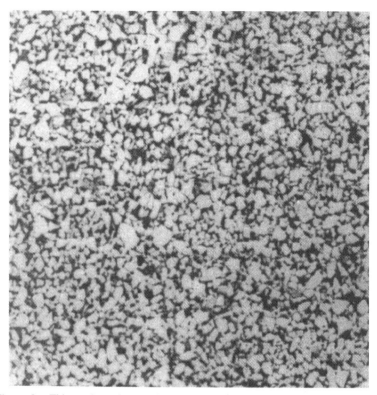

Figure 8. Thin-section micrograph of clastic sandstone from fluvial deposits in the Statfjord formation 3130 m below the bottom of the North Sea. The displayed section measures roughly 2 cm across.

Let $\rho(\mathbf{r}, \varepsilon)$ denote the total density in a small closed ball

$$\mathbb{B}(\mathbf{r}, \varepsilon) = \{\mathbf{q} \in \mathbb{R}^3 : |\mathbf{q} - \mathbf{r}| \le \varepsilon\} \tag{2.5}$$

of radius ε around $\mathbf{r}$. If $\lim_{\varepsilon \to 0} \rho(\mathbf{r}, \varepsilon)$ exists then the matrix and pore space are defined as $\mathbb{M} = \overset{\circ}{\mathbb{Q}}$ and $\mathbb{P} = \overset{\circ}{\mathbb{W}}$. Here

$$\mathbb{Q} = \{\mathbf{r} \in \mathbb{S} : |\lim_{\varepsilon \to 0} \rho(\mathbf{r}, \varepsilon) - \rho_Q| < \Delta\rho_Q\} \tag{2.6}$$

$$\mathbb{W} = \{\mathbf{r} \in \mathbb{S} : |\lim_{\varepsilon \to 0} \rho(\mathbf{r}, \varepsilon) - \rho_W| < \Delta\rho_W\} \tag{2.7}$$

are the regions occupied by quartz and water, and $\Delta\rho_Q$, $\Delta\rho_W$ are the uncertainties in the values of their densities ρ_Q and ρ_W. In practice, it may happen that $\mathbb{P} \cup \mathbb{M} \neq \mathbb{S}$ and even the case $\mathbb{P} \cap \mathbb{M} = \emptyset$ is conceivable. An example would be a muscovite overgrowth with a density of $2.82\,\mathrm{g\,cm}^{-3}$ surrounding the quartz grains of the sandstone.

A generalization of the definition above is necessary if the pore space is filled with two immiscible fluids (see Section VI). In this case the fluid–fluid interface is mobile, and thus all the sets above may in general become time dependent. The same applies when the matrix $\mathbb{M}$ is not rigid and represents a deformable medium such as a gel.

Finally, it is of interest to estimate the amount of information contained in a complete specification of a porous geometry according to the definitions above. This will depend on the spatial resolution a and on the size L of the system. Assuming that the resolution is limited by $a \approx 10^{-10}$ m and that $L \approx 10^{-2}$ m, the configuration of an n-component medium in d dimensions is completely specified by roughly $(L/a)^d \approx 10^{8d}$ numbers. For $d = 3$ these are $O(10^{24})$ numbers.

2. Stochastic Geometries

a. Discrete Space. The irregular geometry of porous media frequently appears to be random or to have random features. This observation suggests the use of probabilistic methods. The idealization underlying the use of statistical methods is that the irregular geometry is a realization drawn at random from an ensemble of possible geometries. It must be emphasized that an idealization is involved in discussing an emsemble rather than individual geometries. It assumes that there exists some form of recognizable statistical regularity in the irregular fluctuations and heterogeneities of the microstructure. This idealization is modeled after statistical mechanics where the microstructure corresponds to a full specification of the positions and momenta of all particles in a fluid while the recognizable regularities are contained in its macroscopic equation of state or thermodynamic potentials. The statistical idealization assumes that the recognizable regularities of porous media can be described by a suitable probability distribution on the space of all possible geometries. Such a description may not always be the most obvious or most advantageous [56], and in fact the merit of the stochastic description does not lie in its improved practicability. The merit of the stochastic description lies in the fact that it provides the necessary framework to define typical or average properties of porous media. The typical or average properties, it is hoped, will provide a more practical geometric characterization of porous media.

Before embarking on the definition of stochastic porous media I wish to emphasize a recent development in the foundations of statistical mechanics [63–69], which concerns the concept of *stationarity* or *homogeneity*. Stationarity is often invoked in the statistical characterization of porous media, although many media are known to be *heterogenous on all scales*. Heterogeneity on all scales means that the geometrical or physical properties of the medium never approach a large-scale limit but continue to fluctuate as the length scale is increased from the microscopic resolution to some macroscopic length scale. Homogeneity or stationarity assumes the absence of macroscopic fluctuations, and postulates the existence of some intermediate length scale beyond which fluctuations decrease [5]. Recent developments in statistical mechanics [63–69] indicate that the traditional concept of stationarity is too narrow, and that there exists a generalization that describes *stationary but heterogeneous macroscopic* behavior. Although these new concepts are still under development they have already been applied in the context of local porosity theory discussed in Section III.A.5.

Consider a porous sample (e.g., cubically shaped) of extension or side length L, and let a be the microscopic resolution. (For concreteness let $a = 10^{-10}$ m as before.) Then there are $N = (L/a)^d$ volume elements inside the sample space that are conveniently addressed by their position vectors

$$\mathbf{r}_i = \mathbf{r}_{i_1 \cdots i_d} = (ai_1, \ldots, ai_d) \tag{2.8}$$

with integers $1 \leq i_1, \ldots, i_d \leq L/a$. Here $\mathbf{r}_i$ is a shorthand notation for $\mathbf{r}_{i_1 \cdots i_d}$. A random configuration or *random geometry* G of an n-component medium is then given as an N-tuple $G = (X_1, \ldots, X_N) = (X(\mathbf{r}_1), \ldots, X(\mathbf{r}_N))$, where the random variables $X_i \in \mathbb{I}_n = \{\rho_{\mathbb{P}_1}, \ldots, \rho_{\mathbb{P}_n}\}$ defined as

$$X_i = X(\mathbf{r}_i) = \sum_{j=1}^{n} \rho_{\mathbb{P}_j} \chi_{\mathbb{P}_j}(\mathbf{r}_i) \tag{2.9}$$

indicate the presence of phase $\mathbb{P}_i$ for the volume element $\mathbf{r}_i$ as identified from its density value $\rho_{\mathbb{P}_i}$. The set $\mathbb{I}_n = \{\rho_{\mathbb{P}_1}, \ldots, \rho_{\mathbb{P}_n}\}$ is a set of indicators, here the densities, which are used to label the phases. Of course the density could be replaced by other quantities characterizing or labeling the components. The discretization is always chosen such that $\mathbf{r}_i \notin \partial \mathbb{P}_j$ for all $1 \leq i \leq N$ and $1 \leq j \leq n$.

An n-component *stochastic porous medium* is defined as a discrete probability density on the set of geometries through

$$\mu(x_1, \ldots, x_N) = \text{Prob}\{G = (x_1, \ldots, x_N)\}$$

$$= \text{Prob}\{(X_1 = x_1) \wedge \cdots \wedge (X_N = x_N)\} \qquad (2.10)$$

where $x_i \in \mathbb{I}_n = \{\rho_{\mathbb{P}_1}, \ldots, \rho_{\mathbb{P}_n}\}$. *Expectation values* of functions $f(G) = f(x_1, \ldots, x_N)$ of the random geometry are defined as

$$\langle f(G) \rangle = \langle f(x_1, \ldots, x_N) \rangle = \sum_{x_1 \in \mathbb{I}_n} \cdots \sum_{x_N \in \mathbb{I}_n} f(x_1, \ldots, x_N) \mu(x_1, \ldots, x_N)$$

$$(2.11)$$

where the sum is over all configurations of the geometry. Note the analogy between Eq. (2.11) and expectation values in statistical mechanics. The analogy becomes an equivalence if μ is a finite dimensional normalized Boltzmann–Gibbs measure.

A stochastic porous medium is called *stationary* and *homogeneous* (in the traditional sense) if its distribution $\mu(x_1, \ldots, x_N) = \mu(x(\mathbf{r}_1), \ldots, x(\mathbf{r}_N))$ is translation invariant, that is,

$$\mu(x(\mathbf{r}_1), \ldots, x(\mathbf{r}_N)) = \mu(x(\mathbf{r}_1 + \mathbf{q}), \ldots, x(\mathbf{r}_N + \mathbf{q})) \qquad (2.12)$$

for all $N \in \mathbb{N}$, $\mathbf{q} \in \mathbb{R}^d$. This traditional definition of stationarity is a special case of the more general concept of fractional stationarity [63–69], which is currently being developed to describe macroscopic heterogeneity.

A stochastic porous medium is called *isotropic* if its distribution is invariant under all rigid euclidean motions, that is,

$$\mu(x(\mathbf{r}_1), \ldots, x(\mathbf{r}_N)) = \mu(x(\mathbf{R}\mathbf{r}_1), \ldots, x(\mathbf{R}\mathbf{r}_N)) \qquad (2.13)$$

for all $N \in \mathbb{N}$ where $\mathbf{R}$ denotes a combination of rotation and translation.

The set of possible geometries contains n^N elements. For a two-component porous cube of side length $L = 1$ cm there are $2^{10^{24}}$ possible configurations at the chosen resolution $a = 10^{-10}$ m. Thus the complete specification of a stochastic porous medium through $\mu(x_1, \ldots, x_N)$ is even less practical than specifying all the volume elements of a particular sample. This does not diminish the theoretical importance of the microscopic geometry distribution $\mu(x_1, \ldots, x_N)$. In fact, it is even useful to generalize it to continuous space where the required amount of data to specify the distribution becomes infinite.

b. Continuous Space. Instead of discretizing the space it is possible to work directly with the notion of random sets in continuous space. The mathematical literature about random sets [10, 70, 71] is based on pioneering work by Choquet [72].

To define random sets, recall first the concepts of a *probability space* and a *random variable* [73–75]. An event $\mathbb{E}$ is a subset of a set $\mathbb{O}$ representing all possible outcomes of some experiment. The probability $\mathrm{Pr}(\mathbb{E})$ of an event is a set function obeying the fundamental rules of probability $\mathrm{Pr}(\mathbb{O}) = 1$, $\mathrm{Pr}(\mathbb{E}) \geq 0$ and $\mathrm{Pr}(\cup_{i=1}^{\infty} \mathbb{E}_i) = \Sigma_{i=1}^{\infty} \mathrm{Pr}(\mathbb{E}_i)$ if $\mathbb{E}_i \cap \mathbb{E}_j = \emptyset$ for $i \neq j$. Formally, the probability Pr is a function on a class $\mathfrak{D}$ of subsets of a set $\mathbb{O}$, called the sample space. If the collection of sets $\mathfrak{D}$ for which the probability is defined is closed under countable unions, complements, and intersections, then the triple $(\mathbb{O}, \mathfrak{D}, \mathrm{Pr})$ is called a probability space. The family of sets $\mathfrak{D}$ is called a σ algebra. The *conditional probability* of an event $\mathbb{E}$ given the event $\mathbb{G}$ is defined as

$$\mathrm{Pr}(\mathbb{E}|\mathbb{G}) = \frac{\mathrm{Pr}\{\mathbb{E} \cap \mathbb{G}\}}{\mathrm{Pr}\{\mathbb{G}\}}, \qquad \mathrm{Pr}\{\mathbb{G}\} \neq 0 \qquad (2.14)$$

A *random variable* is a real valued function on a probability space.

Random sets are generalizations of random variables. The mathematical theory of random sets is based on the "hit-or-miss" idea that a complete characterization of a set can be obtained by intersecting it sufficiently often with an arbitrary compact test set and recording whether the intersection is empty or not [10, 70]. Suppose that $\mathcal{F}$ denotes the family of all closed sets in $\mathbb{R}^d$ including the empty set $\emptyset$. Let $\mathcal{K}$ denote the set of all compact sets. $\mathfrak{F}$ is the smallest σ algebra of subsets of $\mathcal{F}$ that contains all the hitting sets $\mathcal{F}_{\mathbb{K}} = \{\mathbb{F} \in \mathcal{F} : \mathbb{F} \cap \mathbb{K} \neq \emptyset\}$, where $\mathbb{K}$ is a compact test set. An event in this context is the statement whether or not a random set hits a particular countable family of compact subsets.

A *random set* $\mathbb{X}$ (more precisely a random closed set) is defined as a measurable map from a probability space $(\mathbb{O}, \mathfrak{D}, \mu)$ to $(\mathcal{F}, \mathfrak{F})$ [10]. This allows us to assign probabilities to countable unions and intersections of the sets $\mathcal{F}_{\mathbb{K}}$ which are the elements of $\mathfrak{F}$. For example,

$$\mathrm{Pr}(\mathcal{F}_{\mathbb{K}}) = \mu(\mathbb{X}^{-1}(\mathcal{F}_{\mathbb{K}})) \qquad (2.15)$$

is the probability that the intersection $\mathbb{X} \cap \mathbb{K}$ is not empty. This probability plays an important role in the geometric characterizations of porous media based on capacity functionals [10, 72] discussed below. Note that there exists no simple mathematical analogue of the expectation value defined for the discrete case in Eq. (2.11). Its definition, which will not be needed here, requires the introduction of functional integrals on an infinitely dimensional space, or the study of random measures associated with the random set $\mathbb{X}$ [10].

While the expectation value is not readily carried over to the continuous case, the concepts of stationarity and isotropy are straightforwardly generalized. A random set $\mathbb{X}$ is called *stationary* if

$$\Pr\{\mathbb{X} \cap \mathbb{K} \neq \emptyset\} = \Pr\{(\mathbb{X} + \mathbf{r}) \cap \mathbb{K} \neq \emptyset\} \tag{2.16}$$

for all vectors $\mathbf{r} \in \mathbb{R}^d$ and all compact sets $\mathbb{K}$. The notation $\mathbb{G} + \mathbf{r}$ denotes the *translated set* defined as

$$\mathbb{G} + \mathbf{r} = \{\mathbf{q} + \mathbf{r} : \mathbf{q} \in \mathbb{G}\} \tag{2.17}$$

for $\mathbf{r} \in \mathbb{R}^d$ and $\mathbb{G} \subset \mathbb{R}^d$. Using the analogous notation

$$\mathbf{R}\mathbb{G} = \{\mathbf{R}\mathbf{q} : \mathbf{q} \in \mathbb{G}\} \tag{2.18}$$

for $\mathbf{R}$ a rigid euclidean motion allows to define a random set to be *isotropic* if

$$\Pr\{\mathbb{X} \cap \mathbb{K} \neq \emptyset\} = \Pr\{(\mathbf{R}\mathbb{X}) \cap \mathbb{K} \neq \emptyset\} \tag{2.19}$$

for all rigid motions $\mathbf{R}$ and compact sets $\mathbb{K}$. For later reference the notation

$$c\mathbb{G} = \{c\mathbf{q} : \mathbf{q} \in \mathbb{G}\} \tag{2.20}$$

is introduced to denote the multiplication of sets by real numbers. The traditional definition of stationarity presented in Eq. (2.16) is restricted to macroscopically homogeneous porous media. It is a special case of the more general concept of fractional stationarity that describes macroscopic heterogeneity and is currently under development [63–69].

The mathematical definition of random sets in continuous space is even less manageable from a practical perspective than its definition for a discretized space. A complete specification of a random set would require the specification "all" compact or "all" closed subsets of $\mathbb{R}^d$, which is in practice impossible. Nevertheless, the definition is important to clarify the concept of a random set.

III.   GEOMETRIC CHARACTERIZATION

A complete specification of the microstructure of a realistic porous medium is impractical. This section discusses possibilities for characterizing porous media without specifying a porous geometry in all its detail. As emphasized in the introduction, this is the main theoretical task. There are two general approaches: One approach constructs simplified geometric models for each specific porous medium of interest. The other approach attempts to find general characterizations for large classes of porous media that are less specific but more widely applicable. This latter approach will be discussed first.

A. General Geometric Characterization Theories

A general geometric characterization of porous media should satisfy the following requirements:

- It should be well defined in terms of geometric quantities.
- It should involve only parameters that are directly observable or measurable in an experiment independent of the phenomenon of interest.
- It should not require the specification of too many parameters. The required independent experiments should be simple and economical to carry out. What is economical depends on the available data processing technology. With current data processing technology a characterization requiring more than 10^{12} numbers must be considered uneconomical.
- The characterization should be usable in exact or approximate solutions of the equations of motion governing the phenomenon of interest.

The following sections discuss methods based on porosities $\bar{\phi}$, correlation functions $S_n(\mathbf{r})$, $C_n(\mathbf{r})$, local porosity distributions $\mu(\phi)$, pore size distributions $\Pi(r)$, and capacities $T(\mathbb{K})$. Table I collects the advantages and disadvantages of these methods according to the specified criteria.

1. Porosity and Other Numbers

a. Porosity. The porosity of a porous medium is its most important geometrical property. Most physical properties are influenced by the porosity.

The *porosity* $\phi(\mathbb{S})$ of a two-component porous medium $\mathbb{S} = \mathbb{P} \cup \mathbb{M}$ consisting of a pore space $\mathbb{P}$ (component one) and a matrix space $\mathbb{M}$ (component two) is defined as the ratio

$$\phi(\mathbb{S}) = \frac{V_3(\mathbb{P})}{V_3(\mathbb{S})} \tag{3.1}$$

TABLE I

Advantages and Disadvantages of Different Geometric Characterization Methods for Porous Media[a]

Characterization	Well Defined	Predictive	Economical	Easily Usable
$\bar{\phi}, \ldots$	Yes	Yes	Yes	Yes
$S_2(\mathbf{r}), \ldots$	Yes	Yes	Yes	Yes
$S_n(\mathbf{r}), \ldots, (n \geq 3)$	Yes	Yes	No	Yes
$\Pi(r)$	No	No	Yes	Yes
$\mu(\phi)$	Yes	Yes	Yes	Yes
$T(\mathbb{K})$	Yes	No	No	No

[a] A geometrical characterization is called economical if it requires the specification of less than 10^{12} numbers. The parameter $\bar{\phi}, \ldots,$ stands for porosity and other numbers defined in Section III.A.1. The correlation functions $S_n(\mathbf{r})$ and $C_n(\mathbf{r})$ are discussed in Section III.A.2. The pore size distribution of mercury porosimetry $\Pi(r)$ is defined in Section III.A.3. The local porosity distributions $\mu(\phi)$ are discussed in Section III.A.5. The capacities $T(\mathbb{K})$ are defined in Section III.A.6.

which gives the volume fraction of pore space. Here $V_3(\mathbb{P})$ denotes the volume of the pore space defined in Eq. (2.3) and $V_3(\mathbb{S})$ is the total sample volume. In the following, the shorthand notation $V(\mathbb{G}) = V_3(\mathbb{G})$ will often be employed.

The definition (3.1) is readily extend to stochastic porous media. In that case $V(\mathbb{P})$ and ϕ are random variables. If the medium is stationary, then one finds using Eqs. (2.3) and (2.11)

$$
\begin{aligned}
\langle \phi \rangle &= \frac{\langle V(\mathbb{P}) \rangle}{V(\mathbb{S})} \\
&= \frac{1}{V(\mathbb{S})} \left\langle \int_{\mathbb{S}} \chi_{\mathbb{P}}(\mathbf{r}) \, d^3\mathbf{r} \right\rangle \\
&= \frac{1}{V(\mathbb{S})} \int_{\mathbb{S}} \langle \chi_{\mathbb{P}}(\mathbf{r}) \rangle \, d^3\mathbf{r} \\
&= \frac{1}{V(\mathbb{S})} \int_{\mathbb{S}} \Pr\{\mathbf{r} \in \mathbb{P}\} \, d^3\mathbf{r} \\
&= \Pr\{\mathbf{r}_0 \in \mathbb{P}\} = \langle \chi_{\mathbb{P}}(\mathbf{r}_0) \rangle
\end{aligned}
\tag{3.2}
$$

where the last line holds only if the medium is stationary. The vector $\mathbf{r}_0$ in the last line is an arbitrary point. Although the use of the expectation value $\langle \cdots \rangle$ from Eq. (2.11) requires an underlying discretization, a continuous notation was used to indicate that the result holds also in the continuous case. If the stochastic porous medium is not only stationary but also *mixing* or *ergodic*, and if it can be thought of as being infinitely extended, then the limit

$$\bar{\phi} = \lim_{R(\mathbb{S})\to\infty} \phi(\mathbb{S}) = \langle\phi\rangle \qquad (3.3)$$

exists and equals $\langle\phi\rangle$. Here the *diameter* $R(\mathbb{G})$ of a set $\mathbb{G}$ is defined as $R(\mathbb{G}) = \sup\{|\mathbf{r}_1 - \mathbf{r}_2| : \mathbf{r}_1, \mathbf{r}_2 \in \mathbb{G}\}$, the supremum of the distance between pairs of points. The notation $\bar{\phi} = \overline{\chi_{\mathbb{P}}(\mathbf{r})}$ indicates a spatial average while $\langle\phi\rangle = \langle\chi_{\mathbb{P}}(\mathbf{r})\rangle$ is a configurational average.

Equation (3.3) always represents an idealization. Geological porous media, for example, are often *heterogeneous on all scales* [5]. This means that their composition or volume fraction $\phi(R(\mathbb{S}))$ does not approach a limit for $R(\mathbb{S})\to\infty$. Equation (3.3) assumes the existence of a length scale beyond which fluctuations of the porosity decrease. This scale is used traditionally to define so-called "representative elementary volumes" [5, 76]. The problem of macroscopic heterogeneity is related to the remarks in the discussion of stationarity in Section II.B.2. It will be taken up again in Section III.A.5.

The definition of porosity in Eq. (3.1) gives the so-called *total porosity*, which has to be distinguished from the *open porosity* or *effective porosity*. Open porosity is the ratio of accessible pore volume to total volume. Accessible means connected to the surface of the sample.

The porosity of a simple porous medium $\mathbb{S} = \mathbb{P} \cup \mathbb{M}$ is related to the bulk density $\rho_{\mathbb{S}}$, the density of the matrix material $\rho_{\mathbb{M}}$, and the density of the pore space material $\rho_{\mathbb{P}}$ through

$$\phi = \frac{\rho_{\mathbb{M}} - \rho_{\mathbb{S}}}{\rho_{\mathbb{M}} - \rho_{\mathbb{P}}} \qquad (3.4)$$

Therefore, porosity is conveniently determined from measuring densities using liquid buoyancy or gas expansion porosimetry [1–3, 77]. Other methods of measuring porosity include small angle neutron, small angle X-ray scattering, and quantitative image analysis for total porosity [2, 43, 44, 77, 78]. Open porosity may be obtained from xylene and water impregnation, liquid-metal impregnation, nitrogen adsorption, and air or helium penetration [44, 77].

Porosity in rocks originates as *primary porosity* during sedimentation or organogenesis and as *secondary porosity* at later stages of the geological development [1]. In sedimentary rocks the porosity is further classified as intergranular porosity between grains, intragranular or

intercrystalline porosity within grains, fracture porosity caused by me-
chanical or chemical processes, and cavernous porosity caused by organ-
isms or chemical processes.

b. *Specific Internal Surface Area.* Similar to the porosity the specific
internal surface area is an important geometric characteristic of porous
media. In fact, a porous medium may be loosely defined as a medium
with a large "surface to volume" ratio. The specific internal surface area
is a quantitative measure for the surface/volume ratio. Often this ratio is
so large that it has been idealized as infinite [43, 78–85] and the
application of fractal concepts has found much recent attention [42, 58,
84, 86–91].

The *specific internal surface S* of a two-component porous medium is
defined as

$$S = \frac{V_2(\partial \mathbb{P})}{V_3(\mathbb{S})} \tag{3.5}$$

where $V_2(\partial \mathbb{P})$ is the surface area, defined in Eq. (2.3), of the boundary
set $\partial \mathbb{P}$. The surface area $V_2(\partial \mathbb{P})$ exists only if the internal surface or
interface $\partial \mathbb{P}$ fulfills suitable smoothness requirements. Fractal surfaces
would have $V_2(\partial \mathbb{P}) = \infty$ and in such cases it is necessary to replace the
Lebesgue measure in Eq. (2.3) with the Hausdorff measure or another
suitable measure of the "size" of $\partial \mathbb{P}$ [58–61].

The specific internal surface is a characteristic inverse length giving the
surface/volume ratio of a porous medium. Typical values for unconsoli-
dated sand are $2 \times 10^4 \, \mathrm{m}^{-1}$, and range from 10^5 to $10^7 \, \mathrm{m}^{-1}$ for sandstones
[3, 92]. A piece of sandstone measuring 10 cm on each side and having a
specific internal surface of $10^7 \, \mathrm{m}^{-1}$ contains the same area as a sports
arena of dimensions $100 \times 100 \, \mathrm{m}$. This illustrates the importance of
surface effects for all physical properties of porous media.

Specific internal surface area can be measured by similar techniques as
porosity. Some commonly employed methods are given in Fig. 1 together
with their ranges of applicability. Particularly important methods are
based on physisorption isotherms [93, 94]. The interpretation of the
BET-method [93] is restricted to certain types of isotherms, and its
interpretation requires considerable care. In particular, if micropores are
present these will be filled spontaneously and application of the BET-
analysis will lead to wrong results [44]. Other methods to determine S
measure the two-point correlation function. As discussed further in
Section II.A.2 the specific internal surface area can for statistically

homogeneous media be deduced from the slope of the correlation function at the origin [95].

2. Correlation Functions

Porosity and specific internal surface area are merely two numbers characterizing the geometric properties of a porous medium. Obviously, these two numbers are not sufficient for a full statistical characterization of the system. A full characterization can be given in terms of multipoint correlation functions [6, 8, 96–112].

The average porosity $\langle \phi \rangle$ of a stationary two-component porous medium is given by Eq. (3.2) as

$$\langle \phi \rangle = \langle \chi_{\mathbb{P}}(\mathbf{r}_0) \rangle = \Pr\{\mathbf{r}_0 \in \mathbb{P}\} \tag{3.6}$$

in terms of the expectation value of the random variable $\chi_{\mathbb{P}}(\mathbf{r}_0)$ taking the value 1 if the point $\mathbf{r}_0$ lies in the pore space and 0 if not. This is an example of a so-called one-point function. An example of a two-point function is the *covariance function* $C_2(\mathbf{r}_0, \mathbf{r})$, defined as the covariance of two random variables $\chi_{\mathbb{P}}(\mathbf{r})$ and $\chi_{\mathbb{P}}(\mathbf{r}_0)$ at two points $\mathbf{r}_0$ and $\mathbf{r}$,

$$C_2(\mathbf{r}_0, \mathbf{r}) = \langle [\chi_{\mathbb{P}}(\mathbf{r}_0) - \langle \chi_{\mathbb{P}}(\mathbf{r}_0) \rangle][\chi_{\mathbb{P}}(\mathbf{r}) - \langle \chi_{\mathbb{P}}(\mathbf{r}) \rangle] \rangle . \tag{3.7}$$

For a stationary medium the covariance function depends only on the difference $\mathbf{r} - \mathbf{r}_0$, which allows to set $\mathbf{r}_0 = \mathbf{0}$ without loss of generality. This gives

$$C_2(\mathbf{r}) = \langle \chi_{\mathbb{P}}(\mathbf{0})\chi_{\mathbb{P}}(\mathbf{r}) \rangle - \langle \phi \rangle^2 \tag{3.8}$$

Because $\chi_{\mathbb{P}}^2(\mathbf{r}) = \chi_{\mathbb{P}}(\mathbf{r})$ it follows that $C_2(\mathbf{0}) = \langle \phi \rangle(1 - \langle \phi \rangle)$. The *correlation coefficient* of two random variables X and Y is in general defined as the ratio of the covariance $\text{cov}(X, Y)$ to the two standard deviations of X and Y [73, 74]. It varies between 1 and -1 corresponding to complete correlation or anticorrelation. The covariance function is often normalized analogous to the correlation coefficient by division with $C_2(\mathbf{0})$ to obtain the *two-point correlation function*

$$G_2(\mathbf{r}) = \frac{C_2(\mathbf{r})}{C_2(\mathbf{0})} = \frac{C_2(\mathbf{r})}{\langle \phi \rangle(1 - \langle \phi \rangle)} \tag{3.9}$$

An illustration of a two-point correlation function can be seen in Fig. 14. The porosity in Eq. (3.6) is an example of a moment function. The general nth *moment function* is defined as

$$S_n(\mathbf{r}_1, \ldots, \mathbf{r}_n) = \left\langle \prod_{i=1}^{n} [\chi_\mathbb{P}(\mathbf{r}_i)] \right\rangle \tag{3.10}$$

where the average is defined in Eq. (2.11) with respect to the probability density of microstructures given in Eq. (2.10). The covariance function in Eqs. (3.7) or (3.8) is an example of a *cumulant function* (also known as Ursell or cluster functions in statistical mechanics). The nth cumulant function is defined as

$$C_n(\mathbf{r}_1, \ldots, \mathbf{r}_n) = \left\langle \prod_{i=1}^{n} [\chi_\mathbb{P}(\mathbf{r}_i) - \langle \chi_\mathbb{P}(\mathbf{r}_i) \rangle] \right\rangle = \left\langle \prod_{i=1}^{n} [\chi_\mathbb{P}(\mathbf{r}_i) - \langle \phi \rangle] \right\rangle$$

$$\tag{3.11}$$

where the second quality assumes stationarity. The cumulant functions are related to the moment functions. For $n = 1$, 2, or 3 one has the relations

$$C_1(\mathbf{r}_1) = 0 \tag{3.12}$$

$$C_2(\mathbf{r}_1, \mathbf{r}_2) = S_2(\mathbf{r}_1, \mathbf{r}_2) - S_1(\mathbf{r}_1)S_1(\mathbf{r}_2) \tag{3.13}$$

$$C_3(\mathbf{r}_1, \mathbf{r}_2, \mathbf{r}_3) = S_3(\mathbf{r}_1, \mathbf{r}_2, \mathbf{r}_3) - S_2(\mathbf{r}_1, \mathbf{r}_2)S_1(\mathbf{r}_3) - S_2(\mathbf{r}_1, \mathbf{r}_3)S_1(\mathbf{r}_2)$$

$$- S_2(\mathbf{r}_2, \mathbf{r}_3)S_1(\mathbf{r}_1) + 2S_1(\mathbf{r}_1)S_1(\mathbf{r}_2)S_1(\mathbf{r}_3) \tag{3.14}$$

The analogous moment functions may be defined for the matrix space $\mathbb{M}$ by replacing $\chi_\mathbb{P}$ in all formulas with $\chi_\mathbb{M}$. These functions have been called n-point matrix probability functions [105, 107] or simply correlation functions [111]. From Eqs. (2.10), (2.11) and (3.10) the probabilistic meaning of the moment functions is found as

$$S_n(\mathbf{r}_1, \ldots, \mathbf{r}_n) = \langle \chi_\mathbb{P}(\mathbf{r}_1) \cdot \ldots \cdot \chi_\mathbb{P}(\mathbf{r}_n) \rangle = \Pr\{\mathbf{r}_1 \in \mathbb{P}, \ldots, \mathbf{r}_n \in \mathbb{P}\}$$

$$\tag{3.15}$$

Therefore $S_n(\mathbf{r}_1, \ldots, \mathbf{r}_n)$ is the probability that all the points $\mathbf{r}_1, \ldots, \mathbf{r}_n$ fall into the pore space.

The case $n = 2$ of the second moment is of particular interest. If the pore space is stationary and isotropic then $S_2(\mathbf{r}) = S_2(|\mathbf{r}|)$ and one has $S_2(0) = \langle \phi \rangle$. If the porous medium is also mixing then $S_2(\infty) = \langle \phi \rangle^2$. If the pore space $\mathbb{P}$ is three dimensional, and does not contain flat two-dimensional surfaces of zero thickness, then its derivative at the origin is related to the specific internal surface area S through

$$S_2'(0) = \left.\frac{dS_2(r)}{dr}\right|_{r=0} = -\frac{S}{4} \tag{3.16}$$

In two dimensions an analogous formula holds in which S is replaced with a "specific internal length" and the denominator 4 is replaced with π.

The practical measurement of two-point correlation functions is based on Minkowski addition and subtraction of sets [10, 37]. The *Minkowski addition* of two sets $\mathbb{A}$ and $\mathbb{B}$ in $\mathbb{R}^d$ is defined as the set

$$\mathbb{A} \oplus \mathbb{B} = \{\mathbf{q} + \mathbf{r} : \mathbf{q} \in \mathbb{A}, \mathbf{r} \in \mathbb{B}\} \tag{3.17}$$

Note that $\mathbb{A} \oplus \{\mathbf{r}\} = \mathbb{A} + \mathbf{r}$ is the translation defined in Eq. (2.17). Therefore $\mathbb{A} \oplus \mathbb{B} = \cup_{\mathbf{q} \in \mathbb{B}} (\mathbb{A} + \mathbf{q}) = \cup_{\mathbf{r} \in \mathbb{A}} (\mathbb{B} + \mathbf{r})$ is the union of the translates $(\mathbb{B} + \mathbf{r})$ as $\mathbf{r}$ runs through $\mathbb{A}$. The dual operation to Minkowski addition is *Minkowski subtraction* defined as

$$\mathbb{A} \ominus \mathbb{B} = \bigcap_{\mathbf{q} \in \mathbb{B}} (\mathbb{A} + \mathbf{q}) = C(C\mathbb{A} \oplus \mathbb{B}) \tag{3.18}$$

where $C\mathbb{A}$ denotes the complement of $\mathbb{A}$. With these definitions the two-point function is given as

$$S_2(\mathbf{r}) = \Pr\{\mathbf{0} \in \mathbb{P}, \mathbf{r} \in \mathbb{P}\} = \Pr\{\mathbf{0} \in \mathbb{P} \ominus (-\mathbb{B})\} \tag{3.19}$$

where $\mathbb{B} = \{\mathbf{0}, \mathbf{r}\}$ is the set consisting of the origin and the point $\mathbf{r}$. This formula is the basis for the statistical estimation of $S_2(\mathbf{r})$ and $G_2(\mathbf{r})$ in image analyzers from the area of the "eroded" set $\mathbb{P} \ominus (-\mathbb{B})$. The operation $\mathbb{P} \rightarrow \mathbb{P} \ominus (-\mathbb{B})$ is called *erosion* of the set $\mathbb{P}$ with the set $\mathbb{B}$ and it has been used in methods to define pore size distributions, which will be discussed in Section III.A.3. An example for the erosion of a pore space image by a set $\{\mathbf{0}, \mathbf{r}\}$ is shown in Fig. 9. The original image (shown Fig. 12) is obtained from a cross-section micrograph of a Savonnier oolithic sandstone. Two copies of the image are displaced relative to each other by a vector $\mathbf{r}$ as indicated in Fig. 3.1. The two images are rendered in gray, and their intersection is colored black. The area of the intersection is an estimate for $S_2(\mathbf{r})$.

The main advantage of the correlation function method for characterizing porous media is that it provides a set of well-defined functions of increasing complexity for the geometric description. In practice, one truncates the hierarchy of correlation functions at the two-point functions. While this provides much more information about the geometry than the porosity and specific surface area alone, many important

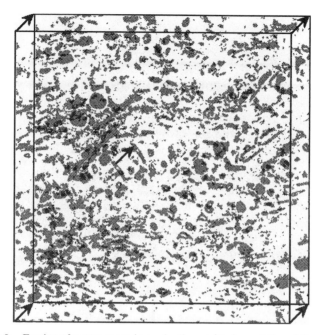

Figure 9. Erosion of a pore space image by the set $\{\mathbf{0}, \mathbf{r}\}$ consisting of the origin $\mathbf{0}$ and a vector $\mathbf{r}$. The vector $-\mathbf{r}$ is shown as an arrow in the image. The overlap of the two images is rendered darker than the rest.

properties of the medium (such as its connectivity) are buried in higher order functions.[2] Depending on the required accuracy a simple two-point function for a three-dimensional stationary but anisotropic two-component medium could be specified by 10^6–10^9 data points, which would be economical according to the criterion adopted previously. An n point function with the same accuracy would require $10^{6(n-1)}$–$10^{9(n-1)}$ data points. Specifying five or higher point functions quickly becomes just as impractical as specifying a given geometry completely.

3. *"Pore Size" Distributions*

In certain porous materials, such as wood (see Fig. 5), it is natural to identify cylindrically shaped pores and to represent their disorder through a distribution of pore diameters. In other media, such as systems with

---

[2] Two points are called connected if there exists a path between them that lies completely inside the pore space. Therefore the probabilistic description of connectedness properties requires multipoint correlation functions involving all the points that make up the path.

cavernous or oomoldic porosity, it is possible to identify roughly convex pore bodies analogous to convex sand grains dispersed in a uniform background. If the radius R of the cylindrical capillaries or spherical pore bodies in such media is randomly distributed, then the *pore size distribution* $\Pi(r)$ can be defined as

$$\Pi(r) = \Pr\{R \le r\} \qquad (3.20)$$

giving the probability that the random radius R of the cylinders or spheres is smaller than r. For general porous microstructures, however, it is difficult to define "pores" or "pore bodies," and the concept of pore size distribution remains ill defined.

Nevertheless many authors have introduced a variety of well-defined probability distributions of length for arbitrary media which intended to overcome the stated difficulty [2, 3, 113–119]. The concept of pore size distributions enjoys continued popularity in most fields dealing with porous materials. Recent examples can be found in chromatography [120, 121], membranes [122–124], polymers [125], ceramics [126–129], silica gels [130–132], porous carbon [133, 134], cements [135–137], rocks and soil science [138–143], fuel research [144], separation and adhesion technology [30, 145], or food engineering [146]. The main reasons for this popularity are adsorption measurements [30, 147, 148] and mercury porosimetry [149–152].

a. Mercury Porosimetry. The "pore size distribution" of mercury porosimetry is not a geometric but a physical characteristic of a porous medium. Mercury porosimetry is a transport and relaxation phenomenon [43, 153], and its discussion would find a more appropriate place in Section V. On the other hand "pore size distributions" are routinely measured in practice using mercury porosimetry, and many readers will expect its discussion in a section on pore size distributions. Therefore, pore size distributions from mercury porosimetry are discussed already here together with other definitions of this important concept.

Mercury porosimetry is based on the fact that mercury is a strongly nonwetting liquid on most substrates, and that it has a high surface tension. To measure the "pore size distribution" a porous sample $\mathbb{S}$ with pore space $\mathbb{P}$ is evacuated inside a pycnometer pressure chamber at elevated temperatures and low pressures [43]. Subsequently, the sample is immersed into mercury and an external pressure is applied. As the pressure is increased, mercury is injected into the pore space occupying a subset $\mathbb{P}_{Hg}(P)$ of the pore space which depends on the applied external pressure P. The experimenter records the injected volume of mercury

$V(\mathbb{P}_{Hg}(P))$ as a function of the applied external pressure. If the volume of the pore space $V(\mathbb{P})$ is known independently then this gives the saturation $S_{Hg}(P) = V(\mathbb{P}_{Hg}(P))/V(\mathbb{P})$ as a function of pressure. The cumulative "pore size" distribution function $\Pi_{Hg}(r) = \Pr\{R \le r\}$ of mercury porosimetry is now defined by

$$\Pi_{Hg}(r) = 1 - S_{Hg}\left(\frac{2\sigma_{Hg}\cos\theta}{r}\right) \tag{3.21}$$

For rocks a contact angle $\theta_{Hg} \approx 135°$ and surface tension with a vacuum of $\sigma_{Hg} \approx 0.48\,\mathrm{Nm}^{-1}$ are commonly used [1, 43, 153]. The definition of $\Pi_{Hg}(r)$ is based on the equation

$$P_{cap} = \frac{2\sigma_{Hg}\cos\theta}{r} \tag{3.22}$$

for the capillary pressure P_{cap}, which expresses the force balance in a single cylindrical capillary tube. Equation (3.21) follows from Eq. (3.22) if it is assumed that the saturation history $S_{Hg}(P)$ is identical to that obtained from the so-called capillary tube model discussed in Section III.B.1. The capillary tube model is a hypothetical porous medium consisting of parallel nonintersecting cylindrical capillaries of random diameter.

The fact that $\Pi_{Hg}(r)$ is not a geometrical quantity but a capillary pressure function for drainage is obvious from its definition. It depends on physical properties such as the nature of the injected fluid or wetting properties of the walls. For a suitable choice of tube diameter the function $S_{Hg}(P)$ of pressure could equally well be translated into a distribution of the wetting angles θ. The parameter $S_{Hg}(P)$ shows hysteresis implying that the pore size distribution $\Pi_{Hg}(r)$ is process dependent.

Although $\Pi_{Hg}(r)$ is not a geometrical quantity, it contains much useful information about the microstructure of the porous sample. An example for the information obtained from mercury porosimetry is shown in Fig. 10 together with an image of the rock for which it was measured [1].[3] The rock is an example of a medium with hollow pores. The correct interpretation of the saturation history $S_{Hg}(P)$ obtained from mercury porosimetry continues to be an active research topic [154–160].

b. *Random Point Methods.* Several authors [3, 113, 116] suggest defining the "pore size" by first choosing a point $\mathbf{r} \in \mathbb{P}$ at random in the pore

---

[3] 1 MPa = 10 bar = 10^7 dyn cm^{-2} = 9.869 atm = 145.04 psi.

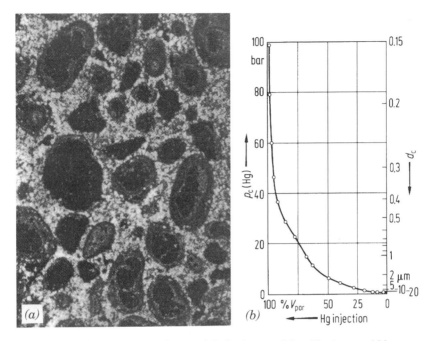

Figure 10. (*a*) Crossed nicols image of Dolomite oncolithe. The image width corresponds to 2.3 mm. The porosity is 11%, the permeability 11 md. (*b*) Capillary pressure curve for the rock shown in the image. The right axis shows the "pore size" scale. (Reproduced with permission from J. Schopper, "Porosität und Permeabilitat," in *Landolt–Börnstein: Physikalische Eigenschaften der Gesteine*, K.-H. Hellwege, Ed., Vol. V/1a, Springer, Berlin, 1982, p. 184. Copyright © Springer-Verlag, 1982.)

space, then choosing a compact set $\mathbb{K}$ containing $\mathbf{r}$, and finally enlarging $\mathbb{K}$ until it first intersects the matrix space $\mathbb{M}$. In its simplest version [3, 116] the set $\mathbb{K}$ is chosen as a small sphere $\mathbb{B}(\mathbf{r}, \varepsilon)$ of radius ε [see Eq. (2.5)]. Then the pore size distribution $\Pi_{rp}(x)$ of the random point method is defined as the distribution function $\Pi_{rp}(x) = \Pr\{R(\mathbf{r}) \leq x\}$ of the random variable $R(\mathbf{r})$ defined as

$$R(\mathbf{r}) = \inf\{\varepsilon : \mathbb{B}(\mathbf{r}, \varepsilon) \cap \mathbb{M} \neq \emptyset\} \tag{3.23}$$

In a deterministic porous medium, R is a random variable because $\mathbf{r} \in \mathbb{P}$ is chosen at random.

In a more sophisticated version of the same idea the set $\mathbb{K}$ is chosen as a small coordinate cross whose axes are then increased independently until they first touch the matrix space [2, 113]. This gives direction dependent pore size distributions.

The main weakness of such a definition is that it is imprecise. This becomes apparent from the fact that the randomness of the pore sizes $R(\mathbf{r})$ does not arise from the irregularities of the pore space, but from the random placement of $\mathbf{r}$. Consider a regular pore space consisting of nonintersecting spheres of equal radii R_0 centered at the vertices of a simple (hyper)cubic lattice. Assuming that the points $\mathbf{r}$ are chosen at random with a uniform distribution it follows that the pore size distribution Π_{rp} is not given by a δ function at R_0 but instead as a uniform distribution on the interval $[0, R_0]$. More dramatically, exactly the same pore size distribution is obtained for every pore space made from nonintersecting spheres of equal radii, regardless of whether they are placed randomly or not. In addition, $\Pi_{rp}(x)$ can be changed arbitrarily by changing the distribution function governing the random placement of $\mathbf{r}$.

A more precise formulation of the same idea is possible in terms of conditional probabilities. In Section III.A.4 it will be seen that the vague ideas underlying random point methods are given a precise definition in the form of contact and chord length distributions [10, 37].

c. Erosion Methods. Another approach to the definition of a geometrical pore size distribution [117–119] is borrowed from the erosion operation in image processing [161, 162]. Erosion is defined in terms of Minkowski addition and subtraction of sets introduced above in Eqs. (3.17) and (3.18). The *erosion* of a set $\mathbb{A}$ by a set $\mathbb{B}$ is defined as the map $\mathbb{A} \rightarrow \mathbb{A} \ominus (-\mathbb{B}) = \cap_{\mathbf{q} \in \mathbb{B}} (\mathbb{A} - \mathbf{q})$. The erosion was illustrated in Fig. 9 for a pore space image and a set $\mathbb{B} = \{\mathbf{0}, \mathbf{r}\}$.

The method for locating "pore chambers," "pore channels," and "pore throats" suggested in [117] is based on eroding the matrix space $\mathbb{M}$ of a two component porous medium. A ball $\mathbb{B}\,(\mathbf{0}, \varepsilon)$ is chosen as the structuring element. The erosion $\mathbb{M} \ominus (-\mathbb{B}(\mathbf{0}, \varepsilon))$ shrinks the matrix space $\mathbb{M}$. The erosion operation is repeated until the matrix space decomposes into disconnected fragments. By continuing the erosion the pieces may either fragment again or become convex. If a piece becomes convex it is called a "grain." The centroid of the convex grain is called a "grain center." Reversing the erosion process allows one to locate the point of respectively first last contact of two fragments. Connecting neighboring grain centers by a path through their last contact point produces a network model of the grain space. After having defined a network of grain centers and last contact points, the authors of [117] and their followers [118, 119] suggest to erect contact "surfaces" in each contact point. A contact plane is defined as a "minimum area cross section" of $\mathbb{M}$. Subsequently, a ball is placed at each grain center and continually enlarged. When the enlargement encounters a surface plane the ball is truncated at the surface plane and only its nontruncated pieces continue to grow until the sample space is completely filled with the inflated grains. The intersection points of three or more planes in the resulting tesselation of space are defined to be *pore chambers*. The intersection lines of two planes are called *pore channels*, and "minimal-area cross sections" of the pore space $\mathbb{P}$ along the pore channels are called *pore throats*. The pore throats are not unique and are sensitive to details of the local geometry. The pore chambers and channels will in general not lie in the pore space.

A drawback of this procedure is that it is less unique than it seems at first sight. The network constructed from eroding the pore space is not unique because the erosion operation involves the set $\mathbb{B}$ as a structuring element, and hence there are infinitely many erosions possible. The resulting grain network depends on the choice of the set $\mathbb{B}$, a fact that is not discussed in [117, 119]. Figure 11 shows an example where erosion with a sphere produces three grains while erosion with an ellipsoid produces only two grains. The original set consists of the gray and black region, the eroded part is coloured gray, and the residual set is colored black. The theoretically described procedure for determining the network was not carried out in practice [117]. Instead "subjective human pre-processing" ([117], p. 4158) was used to determine the network.

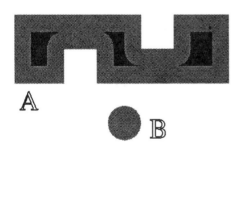

Figure 11. Erosion $\mathbb{A} \oplus (-\mathbb{B})$ of a set $\mathbb{A}$ by a set $\mathbb{B}$ for the cases where $\mathbb{B}$ is a sphere or an ellipsoid. For a sphere the eroded set (shown in black) has three components while for an ellipsoid it consists of only two connected components.

d.   Hydraulic Radius Method. The hydraulic radius method [2, 114, 115] for determining pore size distributions is based on the idea of "symbolically closing pore throats." The definition of pore throats is given in terms of "cross sections" of the pore space. A cross-section $\mathbb{C}$ could be defined as the intersection of a plane $\mathbb{E}(\mathbf{q}, \mathbf{r}_0)$, characterized by its unit normal $\mathbf{q} \in \partial\mathbb{B}(\mathbf{0}, 1)$ and a point $\mathbf{r}_0$ in the plane, with the pore space $\mathbb{P}$ and some suitable set $\mathbb{G}(\mathbf{q}, \mathbf{r}_0)$ that represents the region of interest and could depend on the choice of plane. In symbols $\mathbb{C}(\mathbf{q}, \mathbf{r}_0) = \mathbb{P} \cap \mathbb{E}(\mathbf{q}, \mathbf{r}_0) \cap \mathbb{G}(\mathbf{q}, \mathbf{r}_0)$. A *pore throat containing the point* $\mathbf{r}_0$ is then defined as

$$\mathbb{C}^*(\mathbf{r}_0) = \min_{\mathbf{q} \in \partial\mathbb{B}(\mathbf{0},1)} \frac{V_2(\mathbb{C}(\mathbf{q}, \mathbf{r}_0))}{V_1(\partial\mathbb{C}(\mathbf{q}, \mathbf{r}_0))} \tag{3.24}$$

where the minimum is taken over the unit sphere of orientations of the planes. A *pore throat* is now defined as a local minimum of the function $\mathbb{C}^*(\mathbf{r}_0)$ as $\mathbf{r}_0$ is varied over the pore space. This ideal definition has in practice been replaced with a subjective choice of orientations based on the assumption of isotropy [2, 114, 115]. After constructing all pore throats of a medium the pore space becomes divided into separate compartments called "pore bodies" whose "size" can then be measured by a suitable measure such as the volume to the power one-third.

The definition of pore throats in hydraulic radius methods is very sensitive to surface roughness. This is readily seen from an idealized spherical pore with a few spikes. Another problem as remarked in [115], page 586, is that "the size of a pore body is not readily related in a unique manner to any measurable physical quantity."

4.   Contact and Chord Length Distributions

Chord length distributions [163–166] are special cases of so-called contact distributions [10, 37]. Consider the random matrix space $\mathbb{M}$ of a two component stochastic porous medium and choose a compact set $\mathbb{K}$ containing the origin $\mathbf{0}$. Then the *contact distribution* is defined as the conditional probability

$$\Pi_{\mathbb{K}}(x) = 1 - \mathrm{Pr}\{\mathbf{0} \notin \mathbb{M} \oplus (-x\mathbb{K}))|\mathbf{0} \notin \mathbb{M}\}$$
$$= 1 - \frac{\mathrm{Pr}\{\mathbf{0} \notin \mathbb{M} \oplus (-x\mathbb{K})\}}{1 - \phi} \tag{3.25}$$

for $x \geq 0$. Here ϕ denotes the bulk porosity as usual. Two special choices of the compact set $\mathbb{K}$ are of particular importance. These are the unit sphere $\mathbb{K} = \mathbb{B}(\mathbf{0}, 1) \subset \mathbb{R}^3$ and the unit interval $\mathbb{K} = [0, 1] \subset \mathbb{R}$.

For a unit sphere $\mathbb{K} = \mathbb{B}(\mathbf{0}, 1)$ the quantity $1 - \Pi_{\mathbb{B}(\mathbf{0},1)}(x)$ is the conditional probability that a fixed point in the pore space is the center of a sphere of radius x contained completely in the pore space, under the condition that the chosen point does not belong to $\mathbb{M}$. If $\mathbb{M}$ is isotropic and its boundary is sufficiently smooth, then the specific internal surface S can be obtained from the derivative at the origin as [37]

$$S = \phi \frac{d}{dx} \Pi_{\mathbb{B}(0,1)}(x) \bigg|_{x=0} \tag{3.26}$$

The spherical contact distribution $\Pi_{\mathbb{B}(0,1)}(x)$ provides a more precise formulation of the random point generation methods for pore size distributions [3, 116] discussed in Section III.A.3.b.

For the unit interval $\mathbb{K} = [0, 1]$ the contact distribution $\Pi_{[0,1]}(x)$ is related to the *chord length distribution* $\Pi_{cl}(x)$ giving the probability that an interval in the intersection of $\mathbb{P}$ with a straight line containing the unit interval has a length smaller than x. This provides a more precise formulation of the random point generation ideas in [2, 113]. The relation between the contact distribution and the chord length distribution is given by the equation

$$\Pi_{[0,1]}(x) = 1 - \frac{\int_x^\infty (y - x)\, d\Pi_{cl}(y)}{\int_0^\infty y\, d\Pi_{cl}(y)} \tag{3.27}$$

where the denominator on the right-hand side gives the mean chord length $\bar{\ell} = \int_0^\infty y\, d\Pi_{cl}(y)$. The mean chord length is related to the specific internal surface area through $\bar{\ell} = 4\phi/S$. In Section III.A.2 it was mentioned that the specific internal surface area can be obtained from the two-point function [see Eq. (3.16)]. Therefore the mean chord length can also be related to the correlation function through

$$\bar{\ell} = -\frac{\phi}{S_2'(0)} \tag{3.28}$$

Along these lines it has been suggested in [167] that the full chord length distribution can be obtained directly from small-angle scattering experiments.

Contact and chord length distributions provide much more geometrical information about the porous medium than the porosity and specific surface area, and are at the same time not as unnecessarily detailed as the complete specification of the deterministic or stochastic geometry. Depending on the demands on accuracy a contact or chord length distribution may be specified by 10–1000 numbers irrespective of the microscopic resolution. This should be compared with 10^{24} or $2^{10^{24}}$ numbers for a fully deterministic or stochastic characterization.

5. Local Geometry Distributions

a. Local Porosity Distribution. Local porosity distributions, or more generally local geometry distributions, provide a well-defined general geometric characterization of stochastic porous media [168–175]. Local porosity distributions were mainly developed as an alternative to pore size distributions (see Section III.A.3). They are intimately related with the theory of finite size scaling in statistical physics [64, 176–178]. Although fluctuations in the porosities have been frequently discussed [2, 5, 10, 37, 76, 179–182], the concept of local porosity distributions and its relation with correlation functions was developed only recently [168–175]. More applications are being developed [183, 184].

Local porosity distributions can be defined for deterministic as well as for stochastic porous media. For a single deterministic porous medium consider a partitioning $\mathscr{K} = \{\mathbb{K}_1, \ldots, \mathbb{K}_M\}$ of the sample space $\mathbb{S}$ into M mutually disjoint subsets, called *measurement cells* $\mathbb{K}_j$. Thus $\cup_{j=1}^{M} \mathbb{K}_j = \mathbb{S}$ and $\mathbb{K}_i \cap \mathbb{K}_j = \emptyset$ if $i \neq j$. A particular partitioning was used in the original publications [169–171, 178], where the $\mathbb{K}_j$ are unit cells centered at the vertices of a Bravais lattice superimposed on $\mathbb{S}$. This has the convenient feature that the $\mathbb{K}_j$ are translated copies of one and the same set, and they all have the same shape. An example is illustrated in Fig. 12 showing a quadratic lattice as the measurement grid in two dimensions superposed on a thin section of an oolithic sandstone. The *local porosity* inside a measurement cell $\mathbb{K}_j$ is defined as

$$\phi(\mathbb{K}_j) = \frac{V(\mathbb{P} \cap \mathbb{K}_j)}{V(\mathbb{K}_j)} = \frac{1}{M_j} \sum_{\mathbf{r}_i \in \mathbb{K}_j} \chi_{\mathbb{P}}(\mathbf{r}_i) \qquad (3.29)$$

where the second equality applies in case of discretized space and M_j denotes the number of volume elements or voxels in $\mathbb{K}_j$. Thus the *empirical one-cell local porosity density function* is defined as

$$\tilde{\mu}(\phi; \mathscr{K}) = \frac{1}{M} \sum_{j=1}^{M} \delta(\phi - \phi(\mathbb{K}_j)) \qquad (3.30)$$

where $\delta(x)$ is the Dirac δ distribution. Obviously, the distribution depends on the choice of partitioning the sample space. Two extreme partitions are of immediate interest. The first arises from setting $M = N$, and thus each $\mathbb{K}_j$ contains only one individual volume element $\mathbb{K}_j = \{\mathbf{r}_j\}$ with $1 \leq j \leq M = N$. In this case $\phi(\mathbb{K}_j) = 0$ or $\phi(\mathbb{K}_j) = 1$, depending on

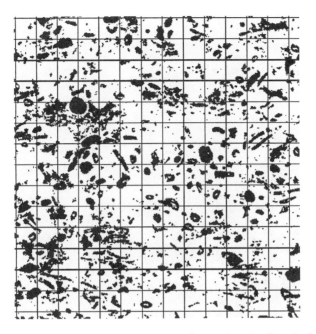

Figure 12. Measurement lattice of squares superimposed on the discretized thin section image of a sandstone. The pore space $\mathbb{P}$ is colored black, the matrix space is rendered white.

whether the volume element falls into matrix space (0), or pore space (1). This immediately gives

$$\tilde{\mu}(\phi; \{\{\mathbf{r}_1\}, \ldots, \{\mathbf{r}_N\}\}) = \phi(\mathbb{S})\delta(\phi - 1) + (1 - \phi(\mathbb{S}))\delta(\phi) \quad (3.31)$$

where $\phi(\mathbb{S})$ is the total porosity. The other extreme arises for $M = 1$. In this case $\mathbb{K}_1 = \mathbb{S}$ and the measurement cell coincides with the sample space. Hence

$$\tilde{\mu}(\phi; \{\mathbb{S}\}) = \delta(\phi - \phi(\mathbb{S})) \quad (3.32)$$

Note that in both extreme cases the local porosity density is completely determined by the total porosity $\phi(\mathbb{S})$, which equals $\phi(\mathbb{S}) = \bar{\phi}$ if the sample is sufficiently large and mixing or ergodicity Eq. (3.3) holds.

For a stochastic porous medium the *one-cell local porosity density function* is defined for each measurement cell as

$$\mu(\phi; \mathbb{K}_j) = \langle \delta(\phi - \phi(\mathbb{K}_j)) \rangle \quad (3.33)$$

where $\mathbb{K}_j \in \mathcal{K}$ is an element of the partitioning of the sample space. For the finest partition with $M = N$ and $\mathbb{K}_j = \{\mathbf{r}_j\}$ one finds now using Eq. (3.2)

$$\mu(\phi; \{\mathbf{r}_j\}) = \Pr\{X(\mathbf{r}_j) = 1\}\delta(\phi - 1) + \Pr\{X(\mathbf{r}_j) = 0\}\delta(\phi)$$
$$= \langle \phi \rangle \delta(\phi - 1) + (1 - \langle \phi \rangle)\delta(\phi) \qquad (3.34)$$

independent of j. If mixing Eq. (3.3) holds, then $\phi(\mathbb{S}) = \langle \phi \rangle = \bar{\phi}$ if the sample becomes sufficiently large, and the result becomes identical to Eq. (3.31) for deterministic media. In the other extreme of the coarsest partition one finds

$$\mu(\phi; \mathbb{S}) = \langle \delta(\phi - \phi(\mathbb{S})) \rangle \qquad (3.35)$$

which may in general differ from Eq. (3.32) even if the sample becomes sufficiently large, and mixing holds. This is an important observation because it emphasizes the necessity to consider more carefully the infinite volume limit $\mathbb{S} \to \mathbb{R}^d$.

If a large deterministic porous medium is just a realization of a stochastic medium obeying the mixing property, and if the sets $\mathbb{K}_j$ are chosen such that the random variables $\phi(\mathbb{K}_j)$ are independent, then Gliwenkos theorem [185] of mathematical statistics guarantees that the empirical one-cell distribution approaches $\mu(\phi; \mathbb{K})$ in the limit $M \to \infty$. In symbols

$$\lim_{M \to \infty} \tilde{\mu}(\phi; \mathcal{K}) = \mu(\phi; \mathbb{K}_j) \qquad (3.36)$$

where the right-hand side is independent of the choice of $\mathbb{K}_j$. Therefore μ and $\tilde{\mu}$ are identified in the following. This identification emerges also from considering average and variance as shown next.

Define the *average local porosity* $\bar{\phi} = \int_0^1 \phi \mu(\phi) \, d\phi$ as the first moment of the local porosity distribution. For a stationary (=homogeneous) porous medium the definitions (3.33) and (3.29) immediately yield

$$\overline{\phi(\mathbb{K}_j)} = \int_0^1 \phi \mu(\phi; \mathbb{K}_j) \, d\phi$$
$$= \langle \phi(\mathbb{K}_j) \rangle$$
$$= \frac{1}{M_j} \sum_{\mathbf{r}_i \in \mathbb{K}_j} \langle \chi_{\mathbb{P}}(\mathbf{r}_i) \rangle$$
$$= \langle \phi \rangle \qquad (3.37)$$

where $\mathbb{K}_j$ is a measurement cell. Similarly, the variance of local porosities reads

$$\overline{(\phi(\mathbb{K}_j) - \overline{\phi(\mathbb{K}_j)})^2} = \int_0^1 [\phi - \overline{\phi(\mathbb{K}_j)}]^2 \mu(\phi; \mathbb{K}_j)\, d\phi$$

$$= \int_0^1 \phi^2 \langle \delta(\phi - \phi(\mathbb{K}_j)) \rangle\, d\phi - \bar{\phi}^2$$

$$= \langle \phi(\mathbb{K}_j)^2 \rangle - \langle \phi(\mathbb{K}_j) \rangle^2$$

$$= \frac{1}{M_j^2} \left\langle \left[M_j \phi - \sum_{i=1}^{M_j} \chi_{\mathbb{P}}(\mathbf{r}_i) \right]^2 \right\rangle$$

$$= \frac{\langle \phi \rangle}{M_j} (1 - \langle \phi \rangle) + \frac{2}{M_j^2} \sum_{\substack{i \neq k \\ r_i, r_k \in \mathbb{K}_j}} C_2(\mathbf{r}_i - \mathbf{r}_k).$$

$$(3.38)$$

This result is important for two reasons. First, it relates local porosity distributions to correlation functions discussed in Section III.A.2. Second, it shows that the variance depends inversely on the volume M_j of the measurement cell. This observation reconciles Eq. (3.35) for stochastic media with Eq. (3.32) for deterministic media because it shows that $\mu(\phi; \mathbb{K}_j)$ approaches a degenerate δ distribution in the limit $M_j \to \infty$. Together with Eq. (3.37) and $M_j = N$ and $M = 1$ for $\mathbb{K}_j = \mathbb{S}$ this shows that Eqs. (3.35) and (3.32) become equivalent.

The *n-cell local porosity density function* $\mu_n(\phi_1, \ldots, \phi_n; \mathbb{K}_1, \ldots, \mathbb{K}_n)$ is the probability density to find local porosity ϕ_1 in measurement cell $\mathbb{K}_1$, ϕ_2 in $\mathbb{K}_2$, and so on, until n. Formally, it is defined by generalizing Eq. (3.33) to read

$$\mu_n(\phi_1, \ldots, \phi_n; \mathbb{K}_1, \ldots, \mathbb{K}_n) = \langle \delta(\phi_1 - \phi(\mathbb{K}_1)) \cdots \delta(\phi_n - \phi(\mathbb{K}_n)) \rangle$$

$$(3.39)$$

where the sets $\mathbb{K}_1, \ldots, \mathbb{K}_n \in \mathcal{K}$ are a subset of measurement cells in the partition $\mathcal{K}$. Note that the *n*-cell functions μ_n are only defined for $M > n$, that is, if there are a sufficient number of cells. In particular, for the extreme case $M = 1$ only the one-cell function is defined. In the extreme case $M = N$ of highest resolution ($\mathbb{K}_j = \{\mathbf{r}_j\}$) the moments of the *n*-cell local porosity distribution reproduce the moment functions (3.10) of the correlation function approach (Section III.A.2) as

$$\overline{\phi_{i_1} \cdots \phi_{i_m}} = \int_0^1 \cdots \int_0^1 \phi_{i_1} \cdots \phi_{i_m} \mu_n(\phi_1, \ldots, \phi_n; \{\mathbf{r}_1\}, \ldots, \{\mathbf{r}_n\})$$

$$\times \, d\phi_1 \cdots d\phi_n$$

$$= \langle \chi_\mathbb{P}(\mathbf{r}_{i_1}) \cdots \chi_\mathbb{P}(\mathbf{r}_{i_m}) \rangle$$

$$= S_m(\mathbf{r}_{i_1}, \ldots, \mathbf{r}_{i_m}) \qquad (3.40)$$

where $m \leq n$ must be fulfilled. This provides a connection between local porosity approaches and correlation function approaches. Simultaneously, it shows that the local porosity distributions are significantly more general than correlation functions. Already the one-cell function μ contains more information than the two-point correlation functions S_2 or C_2 as demonstrated on test images in [171].

The most important practical aspect of local porosity distributions is that they are easily measurable in an independent experiment. Experimental determinations of local porosity distributions have been reported in [173–175, 186]. They were obtained from two-dimensional sections through a sample of sintered glass beads. An example from [175] is shown in Fig. 13. The porous medium was made from sintering glass beads of roughly $250 \, \mu$m diameter. Scanning electron migrographs obtained from the specimen were then digitized with a spatial resolution of 4.1 μm per pixel. The pore space is represented by black in Fig. 13 and the total porosity is $\bar{\phi} = 10.7\%$. The pixel–pixel correlation function of the pore space image is shown in Fig. 14. The local porosity distribution was then measured using a square lattice with lattice constant L. The results are shown in Fig. 15 for measurement cell sizes of $L = 40$, 50, 60 and $L = 80$ pixels.

Measurements of the local porosity distribution from three-dimensional pore space images are currently carried out [183]. The pore space is reconstructed from serial sections that are a standard, but costly, technique to obtain three-dimensional pore space representations [2, 113, 114, 117, 162, 183, 188]. The advent of synchrotron microtomography [4, 119, 188a] promises to reduce the cost and effort.

b. Local Geometry Entropies. The linear extension L of the measurement cells is the length scale at which the pore space geometry is described. The length L can be taken as the side length of a hypercubic measurement cell, or more generally as the diameter $R(\mathbb{K}) = \sup\{|\mathbf{r}_1 - \mathbf{r}_2| : \mathbf{r}_1, \mathbf{r}_2 \in \mathbb{K}\}$ of a cell $\mathbb{K}$ defined as the supremum of the distance

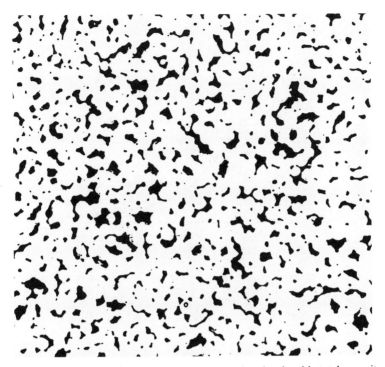

Figure 13. Pore space image of sintered 250-μm glass beads with total porosity of 10.7%. The pore space is rendered in black, the matrix space in white.

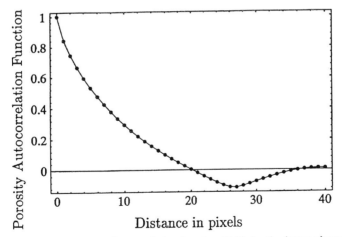

Figure 14. Pixel–pixel porosity correlation function $G_2(r)$ for the image shown in Fig. 13. The distance r is measured in pixels corresponding to 4.1 μm per pixel [175].

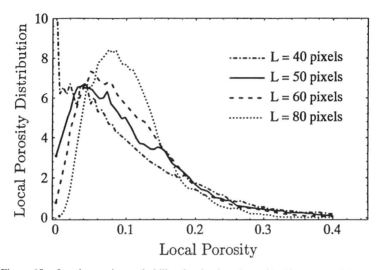

Figure 15. Local porosity probability density function $\mu(\phi; L)$ measured from square shaped measurement cells from the image shown in Fig. 13. The dash–dotted line corresponds to $L = 40$ pixels, the solid line to $L = 50$, the dashed line to $L = 60$ and the dotted line to $L = 80$ pixels. One pixels corresponds to 4.1 μm [175]

between pairs of points. As $L \to \infty$ the local porosity distribution approaches a δ distribution concentrated at $\bar{\phi}$ according to Eqs. (3.32) and (3.35). On the other hand, for $\ell \to 0$ it approaches two δ distributions concentrated at 0 and 1 according to Eqs. (3.31) and (3.34). In both limits the local porosity distribution contains only the bulk porosity $\bar{\phi} = \langle \phi \rangle = \phi(\mathbb{S})$ as a geometric parameter. At intermediate scales the distribution contains additional information, such as the variance of the porosity fluctuations. This suggests a search for an intermediate scale L^*, which provides an optimal description. Several criteria for determining L^* were discussed in [171]. One interesting possibility is to optimize an information measure or entropy associated with $\mu(\phi; L)$ and to define the entropy function

$$I(L) = \int_0^1 \mu(\phi; L) \log \mu(\phi; L) \, d\phi \qquad (3.41)$$

relative to the conventional a priori uniform distribution. The so-called entropy length is then obtained from the extremality condition

$$\left. \frac{d\mu(\phi; L)}{dL} \right|_{L=L^*} = 0 \qquad (3.42)$$

That the entropy length L^* exists and is well defined was first demon-

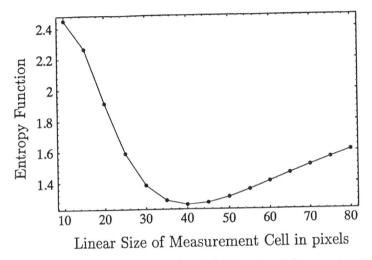

Figure 16. Entropy function as a function of the linear size of the measurement cell in pixels for the local porosity probability density functions obtained from the image shown in Fig. 13. One pixel corresponds to 4.1 μm [175].

strated in [171] using synthetic computer generated images. Figure 16 shows the function $I(L)$ calculated for the image displayed in Figure 13. A clear minimum appears at $L \approx 40$ pixels corresponding to 164 μm.

A similar entropy analysis was recently discussed in [189] for thin-film morphologies of composites. The entropy function $H^*(L)$ defined in [189] is not equivalent to $I(L)$ above because it is not additive. Nevertheless, it was recently found [184] that $H^*(L)$ and $I(L)$ give the same value of L^*.

c. Local Specific Internal Surface Distributions. Local specific internal surface area distributions are a natural generalization of local porosity distributions. They were first introduced in [171] in the study of fluid transport in porous media. Define the local specific internal surface area in a cell $\mathbb{K}_j$ as

$$S(\mathbb{K}_j) = \frac{V_2(\partial \mathbb{P} \cup \mathbb{K}_j)}{V_3(\mathbb{K}_j)} = \frac{1}{M_j} \sum_{\mathbf{r}_i \in \mathbb{K}_j} \chi_{\partial \mathbb{P}}(\mathbf{r}_i) \qquad (3.43)$$

which is analogous to Eq. (3.29). Generalizing Eqs. (3.30) and (3.33) the *local specific internal surface area probability density* is defined as

$$\mu(S; \mathbb{K}_j) = \langle \delta(S - S(\mathbb{K}_j)) \rangle \qquad (3.44)$$

in analogy with Eq. (3.33). The joint probability density $\mu(\phi, S; \mathbb{K}_j)$ to find a local porosity ϕ and local specific internal surface areas in the range ϕ to $\phi + d\phi$ and S to $S + dS$ will be called *local geometry distribution*. It is defined as the probability density

$$\mu(\phi, S; \mathbb{K}_j) = \langle \delta(\phi - \phi(\mathbb{K}_j)) \, \delta(S - S(\mathbb{K}_j)) \rangle \qquad (3.45)$$

The average specific internal surface area in a measurement region $\mathbb{K}$ is then obtained from the local geometry distribution as

$$\overline{S(\mathbb{K})} = \int_0^\infty \int_0^\infty S\mu(\phi, S; \mathbb{K}) \, d\phi \, dS \qquad (3.46)$$

and it represents an important local length scale.

Of course, local geometry distributions can be extended to include other well-defined geometric characteristics such as mean curvature or topological invariants. The definition of the generalized local geometry distribution is then obtained by generalizing Eq. (3.45).

d. Local Percolation Probability. In addition to the local porosity distributions and local specific internal surface area distributions it is necessary to characterize the geometrical connectivity properties of a porous medium. This is important for discussing transport properties that depend critically on the connectedness of the pore space, but are less sensitive to its overall porosity or specific internal surface.

Two points inside the pore space $\mathbb{P}$ of a two-component porous medium are called *connected* if there exists a path contained entirely within the pore space which connects the two points. By using this connectivity criterion a cubic measurement cell $\mathbb{K}$ is called percolating if there exist two points on opposite surfaces of the cell which are connected to each other. The *local percolation probability* $\lambda(\phi, S; \mathbb{K})$ is defined as the probability to find a percolating geometry in a measurement cell $\mathbb{K}$ whose local porosity is ϕ and whose local specific internal surface area is S. In practice, the estimator for $\lambda(\phi, S; \mathbb{K})$ is the fraction of percolating measurement cells which have the prescribed values of ϕ and S.

The average local percolation probability defines an important global geometric characteristic

$$p(\mathbb{K}) = \int_0^\infty \int_0^1 \lambda(\phi, S; \mathbb{K})\mu(\phi, S; \mathbb{K}) \, d\phi \, dS \qquad (3.47)$$

which gives the *total fraction of percolating local geometries*. It will be seen in Sections V.B.4 and V.C.6 that $p(\mathbb{K})$ is an important control parameter for the connectivity of the porous medium.

e. Large-Scale Local Porosity Distributions. This section reviews the application of recent results in statistical physics [63–69, 178, 190] to the problem of describing the *macroscopic heterogeneity on all scales*. The original definition (3.33) of the local porosity distributions depends on the size and shape of the measurement cells, that is, on the partitioning $\mathcal{H}$ of the sample space. This dependence on the choice of a test set or "structuring element" is characteristic for many methods of mathematical morphology [10, 37, 71], and many of those discussed in Sections III.A.3 and III.A.4. On the other hand, Section III.A.5.a has shown that in the limit of large measurement cells $\mathscr{R}(\mathbb{K}) \to \infty$ the form of the local porosity distribution $\mu(\phi; \mathbb{K})$ becomes independent of $\mathbb{K}$ and approaches one and the same universal limit given by $\delta(\phi - \bar{\phi})$. This behavior is an expression of the central limit theorem. Local porosity distributions have support in the unit interval, hence their second moment is always finite, and the average local porosities must become sharp in the limit. It will be seen now that this behavior is indeed characteristic for *macroscopically homogeneous* porous media, while other limiting distributions may arise for *macroscopically heterogeneous* media.

Consider a convex measurement cell $\mathbb{K}$ of volume V_0. Let b be the random scale factor at which the pore space volume $V(\mathbb{P} \cap b\mathbb{K})$ of the inflated measurement cell $b\mathbb{K}$ first exceeds V_0, that is, define b as

$$b = \inf\{c \geq 1 : V(\mathbb{P} \cap c\mathbb{K}) \geq V_0\} \qquad (3.48)$$

Consider N mutually disjoint measurement cells $\mathbb{K}_i$ $(1 \leq i \leq N)$ all having the same volume $V(\mathbb{K}_i) = V_0$. Let b_i denote the scale factors associated with the cells, and let $V_i = V(\mathbb{P} \cap b_i \mathbb{K}_i)$ be the N values of the pore space volumes. If the medium is homogeneous there exists a finite correlation length beyond which fluctuations decrease. Then the nonoverlapping cells $\mathbb{K}_i$ can be chosen such that the inflated cells remain nonoverlapping, and such that they are separated more than the correlation length. Then, the N local porosities $\phi(\mathbb{K}_i) = V_0/V_i$ are uncorrelated random variables. For macroscopically heterogeneous media the correlation length may be infinite, and thus it is necessary to consider the limit $V_0 \to \infty$ of infinitely large cells to obtain uncorrelated porosities. In [63–65, 178, 190] the resulting *ensemble limit* $N, V_0 \to \infty$ has been defined and studied in detail. In the present context the ensemble limit can be used to study the limiting distribution of the N-cell porosity

$$\phi(\mathbb{K}_1, \ldots, \mathbb{K}_N) = \frac{NV_0}{V_1 + \cdots + V_N} = N\left[\sum_{i=1}^{N} (\phi(\mathbb{K}_i))^{-1}\right]^{-1} \quad (3.49)$$

obtained from the N measurements. In the ensemble limit the N porosities $\phi(\mathbb{K}_i)$ become independent but ill defined, because $V(\mathbb{K}_i) \to \infty$. This suggests that one instead considers the limiting behavior of the renormalized sums of positive random variables

$$\varphi_N = \frac{1}{D_N} \left(\sum_{i=1}^{N} (\phi(\mathbb{K}_i))^{-1} - C_N\right) \quad (3.50)$$

where $D_N > 0$ and C_N are renormalization constants. Note that $\phi(\mathbb{K}_i)^{-1} \geq 1$ for all i.

If the sequence of distribution functions of the random variables φ_N converges in the ensemble limit $N \to \infty$ and $V_0 \to \infty$, then the limiting distribution is given by a stable law [74]. The existence of this limit is an indication of fractional stationarity [63–65, 68, 69, 178]. The limiting probability density function for the variables φ_N is obtained along the lines of [64] as

$$h(\varphi; \varpi, C, D) = \frac{1}{\varpi(\varphi - C)} H_{11}^{10}\left(\frac{D^{1/\varpi}}{\varphi - C} \middle| \begin{array}{c} (0, 1) \\ (0, 1/\varpi) \end{array}\right) \quad (3.51)$$

where $\varphi \geq C$ and the parameters obey the restrictions $0 < \varpi < 1$, $D \geq 0$ and $C \geq 1$. The function H_{11}^{10} appearing on the left-hand side is a generalized hypergeometric function that can be defined through a Mellin–Barnes contour integral [191]. The limiting local porosity density is obtained as the distribution of the random variable $\phi = 1/\varphi$, and reads

$$\mu(\phi; \varpi, C, D) = \frac{1}{\varpi\phi(1 - C\phi)} H_{11}^{10}\left(\frac{D^{1/\varpi}\phi}{1 - C\phi} \middle| \begin{array}{c} (0, 1) \\ (0, 1/\varpi) \end{array}\right) \quad (3.52)$$

for $0 \leq \phi \leq 1/C$ and $\mu(\phi) = 0$ for $\phi > 1/C$. The universal limiting local porosity distributions $\mu(\phi; \varpi, C, D)$ depend on only three parameters, and are independent of the diameter or size of the measurement cells $\mathbb{K}_i$. It is plausible that the limiting distributions will also be independent of the shape of the measurement cells, at least for the classes and sequences of convex measurement sets usually employed in studying the thermodynamic limit [192].

Because the limiting distributions have their support in the interval $0 \leq \phi \leq 1/C$ all its moment exist, and one has $\overline{\phi^n} = \phi^n(\varpi, C, D)$. If the moments for $n = 1, 2, 3$ can be inverted, then the parameters ϖ, C, D can be written as

$$\varpi = \varpi(\overline{\phi}, \overline{\phi^2}, \overline{\phi^3}) \qquad C = C(\overline{\phi}, \overline{\phi^2}, \overline{\phi^3}) \qquad D = D(\overline{\phi}, \overline{\phi^2}, \overline{\phi^3})$$

$$(3.53)$$

in terms of the first three integer moments. Figure 17 displays the form of $\mu(\phi; \frac{1}{2}, 2, D)$ for various values of D. In the limit of small porosities $\phi \to 0$ the result (3.52) behaves as a power law

$$\mu(\phi) \propto \phi^{\varpi - 1} \tag{3.54}$$

Within local porosity theory this behavior can give rise to scaling laws in transport and relaxation properties of porous media [170]. The importance of universal limiting local geometry distributions arises from the fact that there exists a class of limit laws which remains broad even after taking the macroscopic limit $N \to \infty$, $V_0 \to \infty$. This is the signature of *macroscopic heterogeneity*, and it occurs for $\varpi < 1$. Macroscopically homogeneous systems, corresponding to $\varpi = 1$, converge instead towards a δ distribution concentrated at the bulk porosity $\overline{\phi}$.

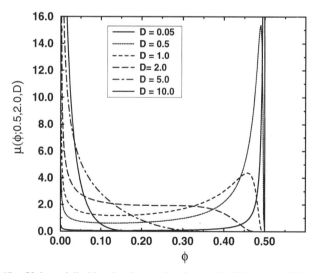

Figure 17. Universal limiting local porosity density (3.52) for $\varpi = 0.5$, $C = 2$, and $D = 0.05, 0.5, 1.0, 2.0, 5.0, 10.0$. For $D \to 0$ the density function is concentrated at $\phi = 1/C$, and it vanishes generally for $\phi > 1/C$.

6. Capacities

While local porosity distributions (in their one-cell form) give a useful practical characterization of stochastic porous media they do not characterize the medium completely. A complete characterization of a stochastic medium is given by the so-called Choquet capacities [10, 72]. Although this characterization is very important for theoretical and conceptual purposes it is not practical because it requires to specify the set of "all" compact subsets (see discussion in Section II.B.2).

Consider the pore space $\mathbb{P}$ of a stochastic two-component porous medium as a random set. Let $\mathscr{F}$ denote the family of all closed sets and $\mathscr{H}$ the set of all compact sets as in Section II.B.2. For any $\mathbb{K} \in \mathscr{H}$ the "hitting function" or *capacity functional* is defined as

$$T(\mathbb{K}) = \Pr\{\mathscr{F}_{\mathbb{K}}\} = \Pr\{\mathbb{P} \cap \mathbb{K} \neq \emptyset\} \tag{3.55}$$

where Pr is the probability law governing the random set $\mathbb{P}$. Then T has the following properties: (1) $T(\emptyset) = 0$ and $0 \le T(\mathbb{K}) \le 1$. (2) If $\mathbb{K}_1, \ldots, \mathbb{K}_n, \ldots$ is a sequence of compact subset, then $\mathbb{K}_1 \downarrow \mathbb{K}$ implies $T(\mathbb{K}_n) \downarrow \mathbb{K}$. (3) For all n the numbers $S_n(\mathbb{K}; \mathbb{K}_1, \ldots, \mathbb{K}_n) \ge 0$ are nonnegative where the S_n are determined by the recursion relation

$$S_n(\mathbb{K}; \mathbb{K}_1, \ldots, \mathbb{K}_n) = S_{n-1}(\mathbb{K}; \mathbb{K}_1, \ldots, \mathbb{K}_{n-1})$$
$$- S_{n-1}(\mathbb{K} \cup \mathbb{K}_n; \mathbb{K}_1, \ldots, \mathbb{K}_{n-1}) \tag{3.56}$$

and $S_0(\mathbb{K}) = 1 - T(\mathbb{K})$. The number $S_n(\mathbb{K}; \mathbb{K}_1, \ldots, \mathbb{K}_n)$ give the probability that $\mathbb{P} \cap \mathbb{K}$ is empty, but $\mathbb{P} \cap \mathbb{K}_i$ is not empty for all i. The functional $T(\mathbb{K})$ is called an *alternating Choquet capacity of infinite order* [10, 71].

Choquet's theorem says that the converse is also true. Explicitly, if T is a functional on $\mathscr{H}$, then there exists a necessarily unique distribution Pr in $\mathfrak{F}$ with

$$\Pr(\mathscr{F}_{\mathbb{K}}) = T(\mathbb{K}) \tag{3.57}$$

for all $\mathbb{K} \in \mathscr{H}$, if and only if T is an alternating Choquet capacity of infinite order. This theorem shows that capacity functionals play the same defining role for random sets in continuous space as do the numbers $\mu(x_1, \ldots, x_N)$ in Eq. (2.10) in the discrete case.

The main problem with this theoretically important result is that the family of hitting sets $\mathscr{F}_{\mathbb{K}}$ is much too large for both practical and theoretical purposes.

B.   Specific Geometric Models

1.   Capillary Tubes and Slits

A simple model for porous media is the capillary tube model in which the pore space is represented as an array of cylindrical tubes. The crucial assumption of the model is that the tubes do not intersect each other. Often it is also assumed that the tubes are straight or parallel to each other. Consider a cubic sample $\mathbb{S}$ with side length L and volume $V(\mathbb{S}) = L^3$. If there are N tubes of length L_i that have circular cross sections of radii a_i ($i = 1, \ldots, N$) the porosity becomes

$$\phi = \frac{\pi}{L^3} \sum_{i=1}^{N} L_i a_i^2 \qquad (3.58)$$

The specific internal surface on the other hand is given as

$$S = \frac{2\pi}{L^3} \sum_{i=1}^{N} L_i a_i \qquad (3.59)$$

In a stochastic model the radii a_i and tube lengths L_i are chosen at random according to a joint probability density $\Pi(a, L)$.

Several special cases are of particular interest. If the random radii a_i and tube lengths L_i are chosen statistically independently, then the joint density $\Pi(a, L)$ factorizes as $\Pi(a, L) = \Pi_a(a)\Pi_L(L)$ into a "pore width distribution" $\Pi_a(a)$ and a "pore length distribution" $\Pi_L(L)$. The average porosity and average specific internal surface area become in this case

$$\langle S \rangle = 2\pi \langle \mathcal{T} \rangle \langle a \rangle \frac{N}{L^2} \qquad (3.60)$$

$$\langle \phi \rangle = \pi \langle \mathcal{T} \rangle \langle a^2 \rangle \frac{N}{L^2} \qquad (3.61)$$

where N/L^2 is the number of capillaries per unit area, and $\langle \mathcal{T} \rangle = \langle L \rangle / L$ is the *average tortuosity* factor obtained by averaging the dimensionless *tortuosity* $\mathcal{T}_i$ defined for each tube as

$$\mathcal{T}_i = \frac{L_i}{L} \qquad (3.62)$$

Moreover, in this case the ratio

$$\frac{\langle \phi \rangle}{\langle S \rangle} = \frac{\langle a^2 \rangle}{2 \langle a \rangle} \tag{3.63}$$

is a characteristic length independent of the tortuosity and sample size. Two further special cases arise from setting all radii equal to each other, $a_i = a$ or all lengths to the system size, $L_i = L$. If $L_i = L$ the tortuosity factor in Eqs. (3.60) and (3.61) is unity, and Eq. (3.63) holds unchanged. If the a_i are chosen statistically independently, then for large N the average porosity is related to the variance of the specific internal surface according to

$$\langle \phi \rangle = \frac{L^2}{4\pi} (\langle S^2 \rangle - \langle S \rangle^2) \tag{3.64}$$

In the case where all radii are equal, $a_i = a$, the relation (3.63) simplifies to

$$\frac{\langle \phi \rangle}{\langle S \rangle} = \frac{a}{2} \tag{3.65}$$

Although this relation holds only in a special case it has become the basis for defining the so-called *hydraulic radius*

$$R_H = \frac{\phi}{S} = \frac{\text{area of cross section}}{\text{perimeter of cross section}} \tag{3.66}$$

as a characteristic length scale of porous media [1, 2]. The hydraulic radius concept was used in Section III.A.3.d for defining pore size distributions. The capillary tube model and the hydraulic radius concept play an important role for fluid flow through porous media, and will be discussed further in Section V.C.2.

A model that is closely related to the capillary tube model is obtained by considering N slits, that is, N parallel planes of width b_i, instead of tubes. All slits are assumed to be parallel, and hence nonintersecting. The resulting capillary slit model has a porosity

$$\phi = \frac{1}{L} \sum_{i=1}^{N} b_i \tag{3.67}$$

and specific internal surface area

$$S = \frac{2N}{L} \tag{3.68}$$

independent of the widths of the slits. Here the specific internal surface is independent of the distribution of widths. As in the capillary tube model one also finds a relation

$$\frac{\langle \phi \rangle}{S} = \frac{\langle b \rangle}{2} \tag{3.69}$$

similar to Eq. (3.65) for the model of capillary slits. The model may be generalized by allowing small undulations and smooth fluctuations of the slits.

2. Grain Models

Grain models of various sorts have long been studied in optics [179, 193], colloids [194, 195], phase transitions [196], and disordered systems [197]. An important class of grain models are random bead packs [198–204], which provide a reasonable starting point for modeling unconsolidated sediments. In grain models, either the pore or the matrix space are represented as an array of convex grains [107–109, 205–209]. The grains could be regularly shaped such as spheres, cubes or ellipoids, or more irregularly shaped convex sets. They may be positioned randomly or regularly in space, and they may have equal or varying diameters. If the grains are placed randomly their centroids are assumed to form a stochastic point process. For a *Poisson point process* the centers or centroids of the grains are placed randomly and independently in space such that the number $N(\mathbb{B})$ of points inside a set $\mathbb{B}$ is Poisson distributed with point density ϱ. For a Poisson point process the whole distribution is determined by its density ϱ. If $\mathbb{B}_1, \ldots, \mathbb{B}_n$ are n disjoint bounded sets, then the numbers $N(\mathbb{B}_1), \ldots, N(\mathbb{B}_n)$ are independent Poisson random variables with joint probability distribution

$$\Pr\{N(\mathbb{B}_1) = N_1, \ldots, N(\mathbb{B}_n) = N_n\}$$
$$= \frac{\varrho^{N_1 + \cdots + N_n} V_d(\mathbb{B}_1)^{N_1} \cdots V_d(\mathbb{B}_n)^{N_n}}{N_1! \cdots N_n!} \exp(-\varrho(V_d(\mathbb{B}_1) + \cdots + V_d(\mathbb{B}_n)))$$

$$\tag{3.70}$$

The Poisson point process with constant density is stationary and isotropic. The contact distribution $\Pi_{\mathbb{K}}(x)$, defined in Section III.A.4, for the Poisson point process can be obtained from the so-called void probability that there is no point inside $\mathbb{K}$ as

$$\Pi_{\mathbb{K}}(x) = 1 - \Pr\{N(x\mathbb{K}) = 0\} = 1 - \exp(-\varrho V_d(x\mathbb{K})) \tag{3.71}$$

A simple class of grain models is obtained by attaching compact sets to the points of a Poisson point process. The compact sets are called *primary grains*. Spheres of constant radius are examples. Important generalizations are obtained by randomizing the primary grains. An example would be spherical grains with random radii. More generally, it is possible to use as grains independent realizations of a random compact set as defined in Section II.B.2.b. If $\mathbb{G}_i$ denote the independent realizations of the grains and $\mathbf{r}_i$ the points of a Poisson point process, then a grain model is obtained as the set

$$\mathbb{P} = \bigcup_{i=1}^{\infty} (\mathbb{G}_i + \mathbf{r}_i) \tag{3.72}$$

if the grains are interpreted as pores. If the grains are matrix then $\mathbb{P}$ has to be replaced by $\mathbb{M}$. The grain model is uniquely characterized by its capacity functional $T(\mathbb{K}) = \Pr\{\mathbb{P} \cap \mathbb{K} \neq \emptyset\}$ defined in Section III.A.6. If $\mathbb{G}_0$ denotes an independent realization of the primary grains, and $\mathbb{K}$ a compact set, then the capacity functional can be shown to have the form [10, 37]

$$T(\mathbb{K}) = 1 - \exp[-\varrho \langle V_d((-\mathbb{G}_0) \oplus \mathbb{K}) \rangle] \tag{3.73}$$

where $\langle \cdots \rangle$ denotes the expectation value with respect to the distribution of primary grains, and the Minkowski addition $\oplus$ of sets was defined in Eq. (3.17). The porosity of a grain model is obtained as

$$\langle \phi \rangle = 1 - \exp[-\varrho \langle V_d(\mathbb{G}_0) \rangle] \tag{3.74}$$

where $\mathbb{G}_0$ is again a typical primary grain. The covariance function defined in Eqs. (3.7) and (3.8) reads

$$C_2(\mathbf{r}) = 2\langle \phi \rangle - 1 + (1 - \langle \phi \rangle)^2 \exp[\varrho \langle V_d(\mathbb{G}_0 \cap (\mathbb{G}_0 - \mathbf{r})) \rangle] \tag{3.75}$$

and it determines the specific internal surface according to Eqs. (3.16) and (3.13). The contact distribution $\Pi_{\mathbb{K}}(x)$ for a compact set $\mathbb{K}$ is given as

$$\Pi_{\mathbb{K}}(x) = 1 - \exp[-\varrho(\langle V_d(\mathring{\mathbb{G}}_0 \oplus x\mathbb{K}) \rangle - \langle V_d(\mathbb{G}_0) \rangle)] \tag{3.76}$$

where the interior $\mathring{\mathbb{G}}$ of a set was defined in Section II.B.1 and $x \geq 0$.

Two simple classes of grain models are obtained by randomly placing penetrable or impenetrable spheres of radius R and number density ρ. Specializing Eq. (3.74) to spherical grains of equal diameter one obtains

$$\langle \phi \rangle = \exp(-4\pi R^3 \varrho / 3) \tag{3.77}$$

for the porosity of fully penetrating spheres. The relation

$$\langle \phi \rangle = 1 - 4\pi R^3 \varrho/3 \tag{3.78}$$

applies to hard spheres of radius R [108]. The specific internal surface area of overlapping spheres reads

$$S = 4\pi R^2 \varrho \exp(-4\pi R^3 \varrho/3) \tag{3.79}$$

and for hard spheres

$$S = 4\pi R^2 \varrho \tag{3.80}$$

is obtained [108].

A basic question of stereology concerns the "unfolding" of three-dimensional information from planar sections [210]. For a stationary Poisson distributed grain model with spherical grains of random diameter the problem of calculating the probability density $p_3(r)$ of the diameters of spheres from the probability density $p_2(r)$ of section circles was solved long ago [211, 212]. The spatial and planar distributions are related through an Abel integral equation

$$p_2(r) = \frac{r}{a} \int_r^\infty \frac{p_3(x)}{\sqrt{(x^2 - r^2)}} \, dx \tag{3.81}$$

where the mean-sphere diameter a is given by

$$a = \frac{\pi}{2} \left[\int_0^\infty \frac{1}{r} p_2(r) \, dr \right]^{-1} \tag{3.82}$$

The solution to this equation is

$$p_3(r) = -\frac{2ar}{\pi} \int_r^\infty \frac{1}{\sqrt{(x^2 - r^2)}} \frac{d}{dx} \left[\frac{p_2(x)}{x} \right] dx \tag{3.83}$$

for $r \geq 0$. In the special case where all spheres have the same constant diameter a the probability density of section circle diameters is given as

$$p_2(r) = \frac{r}{a\sqrt{(a^2 - r^2)}} \tag{3.84}$$

with $0 \leq r \leq a$. The average diameter of the section circles is $a_2 = \pi a/4$, and its variance is $\sigma^2 = (32 - 3\pi^2)a^2/48$. Another interesting special case is when the diameters of the spheres are distributed according to

$$p_3(r) = \frac{r}{b^2} \exp\left[-\frac{r^2}{(2b^2)} \right] \tag{3.85}$$

for $r \geq 0$, which gives a mean-sphere diameter $a = b\sqrt{\pi/2}$. In this case, the distribution reproduces itself, that is, $p_3(r) = p_2(r)$.

Even the simplest grain model with penetrable spheres of equal diameter and Poisson distributed centers still poses unsolved problems. At low-dimensionless densities $\rho = (4\pi/3)R^3\varrho$ the grains form isolated bounded sets. Here R is the radius of the spheres and ϱ their number density. As the dimensionless density is increased the grains begin to overlap and ultimately "percolate," which means that an unbounded connected component appears. This *continuum percolation transition* between a state without unbounded connected component and a state where the strains percolate to infinity is a phase transition of order larger than 2 in the sense of statistical mechanics [64, 213]. It continues to be the subject of much research in recent years [214–219].

A different parallel with statistical mechanics emerges if the grains are identified with the particles in statistical mechanics [6–8, 208]. This identification suggests generalizations of the underlying uncorrelated Poisson point process, which corresponds to an ideal gas of noninteracting particles, by adding interactions between the points. A large variety of new models such as hard-sphere models or Gibbs point fields [34, 37] emerges from this generalization.

3. Network Models

Network models represent the most important and widely used class of geometric models for porous media [155, 157, 187, 220–233]. They are not only used in theoretical calculations but also in the form of micromodels in experimental observations [157, 234–238]. For random bead packs a random network model has recently been derived starting from the microstructure [200, 204]. Network models arise generally and naturally from discretizing the equations of motion using finite difference schemes. As such they will be discussed in more detail in Section V.

A network is a graph consisting of a set of vertices or sites connected by a set of bonds. The vertices or sites of the network could, for example, represent the grain centers of a grain model. If the grains represent pore bodies the bonds represent connections between them. The vertices can be chosen deterministically as for the sites of a regular lattice or randomly as in the realization of a Poisson or other stochastic point process. Similarly, the bonds connecting different vertices may be chosen according to some deterministic or random procedure. Finally, the vertices are "dressed" with convex sets such as spheres representing pore bodies, and the bonds are dressed with tubes providing a connecting path between the pore bodies. A simple ordered network model consists of a regular lattice with spheres of equal radius centered at its vertices that are connected through cylindrical tubes of equal diameter. Very often the diameters of spheres and tubes in a regular network model are chosen at random. If a finite fraction of the bond diameters is zero, one obtains the *percolation model*.

4. Percolation Models

The purpose of this section is not to review percolation theory but to introduce the concept of a percolation transition, and to collect for later reference the values of percolation thresholds. The name "percolation" derives from fluid flow through a coffee percolator, and it has been used extensively to model various aspects of flow through porous media [41, 113, 153, 154, 156, 226, 239–243]. Invasion percolation has become a frequently studied model for displacement processes in porous media [42, 244–246, 246a]. Percolation theory itself is a well-developed branch in the theory of disordered systems and critical phenomena, and the reader is referred to [197, 213, 247, 248] for thorough information on the subject.

Percolation as a geometrical model for porous media is closely related both to grain models and to network models. The model of spherical grains attached to the points of a Poisson process is also known as continuum percolation or the "swiss cheese" model [213]. Site percolation is an abstract version of a grain model, while bond percolation may be seen as an abstraction from network models. The distinguishing feature of percolation theory from other models is its focus on a sudden phase transition associated with the connectivity of random media.

The simplest model of percolation is bond percolation on a lattice. In *bond percolation* the bonds of a regular (e.g., simple cubic) lattice are occupied randomly with connecting (=conducting) elements (e.g., tubes) with a certain occupation probability p. Alternatively, one removes a fraction $1 - p$ of bonds at random from a fully occupied infinite lattice. Two lattice sites are called *connected* if there exists a path between them traversing only bonds occupied by connecting elements. A set of connected bonds is called a *cluster*. In an infinitely large system there exists a critical occupation probability p_{bc} above which there exists an unbounded connected cluster, while below p_{bc} all clusters are finite and bounded. If the connecting elements are cylindrical tubes of fixed or variable diameter, then the resulting network model of a porous medium would be permeable above p_{bc} and impermeable below p_{bc}.

Similarly, *site percolation* may be viewed as a lattice version of a grain model. In site percolation the sites of the underlying regular lattice are occupied randomly with spherical pore bodies of radius at least one-half the lattice constant. Two nearest-neighbor sites are called connected if they are both occupied. As in bond percolation there exists a critical occupation probability p_{sc} separating a permeable connected regime from an impermeable regime.

These basic bond and site percolation models may be modified in many ways. The underlying lattice may be replaced with an arbitrary regular or random graph. The radii of tubes and pores may be randomized, and the connectivity criterion may be changed. Table II shows the values of the critical occupation probabilities (thresholds) for bond and site percolation for some common two- and three-dimensional lattices [213]. The table also lists the coordination number of each lattice defined as the number of bonds meeting at an interior lattice site.

The transition between the permeable and impermeable regime becomes a phase transition in the limit of an infinitely large lattice. The role of the order parameter is played by the *percolation probability* $P(p)$ defined as the probability that a given point belongs to an infinite cluster. The correlation length $\xi(p)$ is defined from the correlation function giving the probability that a site at distance r from a given site is occupied and connected to the given site. The correlation length measures the typical size of a cluster, and it diverges as p approaches p_c.

Similar to network models percolation arises naturally in the study of transport and relaxation phenomena. Effective medium theories dis-

TABLE II

Values of the Bond p_{bc} and Site p_{sc} Percolation Thresholds for Various Two- and Three-Dimensional Lattices[a]

Lattice Type	Dimension	Coordination	p_{bc}	p_{sc}
Honeycomb	2	3	$1 - 2\sin(\pi/18) \approx 0.6527$	0.6962
Square	2	4	$\frac{1}{2}$	0.592746
Triangular	2	6	$2\sin(\pi/18) \approx 0.3473$	$\frac{1}{2}$
Diamond	3	4	0.3886	0.4299
Simple cubic	3	6	0.2488	0.3116
Body centered cubic	3	8	0.180	0.246
Face centered cubic	3	12	0.119	0.198

[a] Also given is the coordination number defined as the number of bonds meeting at an interior lattice site.

cussed in Sections V.B.3 or V.C.6 contain an underlying percolation transition.

5. Filtering and Reconstruction

Recently, a new type of geometric models has appeared [118, 171, 249–252], which is based on image processing techniques. These models attempt to reconstuct a porous medium with prespecified statistical characteristics such as its porosity $\bar{\phi}$ and two-point correlation function $S_2(\mathbf{r})$ [249, 250]. This is achieved through real space or Fourier space filtering of random fields. To this end a two- or three-dimensional random function assigning an independent random number to each point in space is convoluted with a smoothing kernel. Alternatively, the Fourier transforms of the two functions are multiplied and backtransformed to obtain the smoothed image. In [249, 250, 252] it has been argued that such reconstructed images resemble the structures observed in sedimentary rocks or Vycor glass. In [171] such models have been used to test the microstructural sensitivity of local porosity distributions (discussed in Section III.A.5) on models with identical two-point correlation functions.

The motivation for studying reconstructed porous media is to generate precisely known microgeometries whose transport properties can then be calculated numerically [249, 252]. As shown in [171] two media with the same porosity and correlation function may still show significant differences in their geometric characteristics. More importantly, the porosity and two-point correlation function are not sufficient to determine the connectivity of a medium. The connectivity controls the transport, and therefore it is unclear to what extent reconstructed porous media are useful for predicting transport or relaxation properties.

6. Process Models

The geometric models discussed in the previous sections do not account for the fact that the pore space configuration is often the result of a physical process. This suggests the use of dynamic *process models* that describe the formation of the porous medium. While such models are employed routinely for unconsolidated bead packs [199, 201, 202, 253] they are relatively rare for consolidated porous media. The so-called *bond shrinkage model* [254] was developed for sedimentary rocks. In this model one starts from a random resistor network on a simple cubic lattice in d dimensions. Each resistor in the model represents a cylindrical tube with radius r_i in a corresponding network model on the lattice. Next, a tube element is randomly chosen and its radius is reduced by a fixed factor x with $0 < x < 1$. The radius of the shrunk element is then xr_i. The shrinkage process may be repeated as often as necessary to reach a specified porosity or until some other criterion is satisfied depending on the modeling purpose. The condition $x > 0$ guarantees that the model remains connected at all times if the shrinkage is repeated a finite number of times. Because $x > 0$ the bond shrinkage model does not give rise to a percolation threshold.

The so-called grain consolidation model [255, 256] does give rise to a percolation threshold. In the grain consolidation model one starts from a grain model with nonoverlapping grains $\mathbb{G}_i$. The grains are then allowed to grow. In the simplest version [255] the grains form a regular cubic lattice of spheres. As the grains grow the porosity is reduced and more and more narrow constrictions between grains are closed. Thus the system becomes impenetrable at a finite porosity. For a regular simple cubic lattice of spheres the percolation threshold appears at a critical porosity of $\phi_c = 0.0349$ [255]. This value is much smaller than the values $p_{sc} \approx 0.2488$ or $p_{bc} \approx 0.3116$ in Table II for bond and site percolation. If the grains are grown from a random bead pack the critical porosity is found to be $\phi_c \approx 0.03$. This value is again much smaller than the threshold $\phi_c \approx 0.17$ in continuum percolation [257, 258].

Another process model, called *local porosity reduction model*, was introduced in [170]. Consider a simple regular lattice (e.g., simple cubic) with lattice constant L superimposed on a porous medium as shown in Fig. 12. The lattice cells are the measurement cells of local porosity theory defined in Section III.A.5. In the simplest example, one may assume that all measurement cells have the same initial porosity ϕ_0 and specific internal surface area S_0. The consolidation process is modeled by picking at random a particular cell and reducing its porosity by a factor r, and its specific surface by a factor s. This operation is repeated until a desired average porosity $\bar{\phi}$ has been reached. It is assumed that the local percolation probability function $\lambda(\phi, S)$ is unaffected by the consolidation operations. The final state of the porous medium can be described by assigning to each cell the random integer n giving the number of times a local consolidation operation was performed on that cell. The random variable n is Poisson distributed, and this observation allows one to find the local porosity distribution as [170]

$$\mu(\phi, L) = \frac{1}{\Gamma(\ln(r\phi/\phi_0)/\ln r)} \left(\frac{\phi}{\phi_0} \right)^{\ln \bar{n}/\ln r} \exp(-\bar{n}) \qquad (3.86)$$

where

$$\bar{n} = \frac{\ln \bar{\phi} - \ln \phi_0}{r - 1} \qquad (3.87)$$

and $\Gamma(x)$ denotes Eulers gamma function. A similar consideration could be performed for the specific internal surface distribution. In many consolidation processes the reduction factors are not independent. For grain consolidation models this is illustrated by Eqs. (3.77)–(3.80), relating the porosity and specific surface through the sphere radius. Similarly, if a crack is closed by an applied external pressure this will reduce the porosity but not the surface area, as long as the possibility of crushing the crack profile is neglected. Therefore, in this case $s = r^0 = 1$. Similarly, if a void space is uniformly cemented by precipitation of minerals from the pore fluids this implies the relation $s = r^{2/3}$ between the reduction factors. The simple relation

$$s = r^\alpha \tag{3.88}$$

summarizes several idealized processes such as crack compaction, $\alpha = 0$, shrinkage of capillaries, $\alpha = \frac{1}{2}$, shrinkage of voids, $\alpha = \frac{2}{3}$, and void filling, $\alpha = 1$.

IV. DEFINITION AND EXAMPLES OF TRANSPORT

A. Examples

This section begins the discussion of physical processes in porous media involving the transport or relaxation of physical quantities, such as energy, momentum, mass, or charge. As discussed in the introduction, physical properties require equations of motion describing the underlying physical processes. Recurrent examples of experimental, theoretical, and practical importance include:

- The disordered *diffusion equation*

$$\frac{\partial T(\mathbf{r}, t)}{\partial t} = \nabla^T \cdot (\mathbf{D}(\mathbf{r})\nabla T(\mathbf{r}, t)) \tag{4.1}$$

where $\mathbf{r} \in \mathbb{S}$, $\mathbf{D}(\mathbf{r}) = \kappa(\mathbf{r})/(c_p(\mathbf{r})/\rho(\mathbf{r})) > 0$ is the thermal diffusivity tensor, $T(\mathbf{r}, t)$ is the space–time dependent temperature field, $\rho(\mathbf{r})$ is the density, $\kappa(\mathbf{r})$ the thermal conductivity, and $c_p(\mathbf{r})$ the specific heat at constant pressure. The superscript T denotes transposition. If the tensor field $\mathbf{D}(\mathbf{r})$ is sufficiently often differentiable the equations are completed with boundary conditions at the sample boundary $\partial \mathbb{S}$. For the microscopic description of diffusion in a two-component porous medium $\mathbb{S} = \mathbb{P} \cup \mathbb{M}$ whose components have diffusivities $\mathbf{D}_\mathbb{P}$ and $\mathbf{D}_\mathbb{M}$ the diffusivity field $\mathbf{D}(\mathbf{r})$ has the form $D(r) = D_\mathbb{P}\chi_\mathbb{P}(r) + D_\mathbb{M}\chi_\mathbb{M}(r)$, which is not differentiable at $\partial \mathbb{P}$. In such cases, additional boundary conditions are required at the internal interface $\partial \mathbb{P}$, and the equation is interpreted in the sense of distributions [259]. Typical values for sedimentary rocks are $\kappa_M \approx 1 \ldots 6 \ \text{Wm}^{-1} \ \text{K}^{-1}$, $\rho_M \approx 1 \ldots 3 \ \text{g cm}^{-3}$, and $c_{\rho M} \approx 0.8 \ldots 1.2 \ \text{kJ kg}^{-1} \ \text{K}^{-1}$.

- The Laplace equation with variable coefficients

$$\nabla^T \cdot (\mathbf{C}(\mathbf{r}) \nabla u(\mathbf{r})) = 0 \tag{4.2}$$

where $\mathbf{C}(\mathbf{r})$ is again a second rank tensor field of local transport coefficients, $u(\mathbf{r})$ is a scalar field, $\mathbf{r} \in \mathbb{S}$, and the same remarks apply as for the diffusion equation with respect to differentiability of $\mathbf{C}(\mathbf{r})$ and boundary conditions. For constant $\mathbf{C}(\mathbf{r})$ the equation reduces to the Laplace equation $\Delta u = 0$. If the medium is random the coefficient matrix $\mathbf{C}$ is a random function of $\mathbf{r}$. Equation (4.2) is the basic equation for Section IV.B. It is frequently obtained as the steady-state limit of the time-dependent equations such as the diffusion equation (4.1). Other examples of Eq. (4.2) occur in fluid flow, dielectric relaxation, or dispersion in porous media. In dielectric relaxation u is the electric potential and $\mathbf{C}$ is the matrix of local (spatially varying) dielectric permittivity. In diffusion problems or heat flow u is the concentration field or temperature, and $\mathbf{C}$ the local diffusivity. In Darcy flow through porous media u is the pressure and $\mathbf{C}$ is the tensor of locally varying absolute permeabilities.

- The elastic wave equation is a system of equations for the three components $u_i(\mathbf{r}, t)$ ($i = 1, 2, 3$) of a vector displacement field

$$\frac{\partial^2 u_i(\mathbf{r}, t)}{\partial t^2} = v_s^2 \, \Delta u_i(\mathbf{r}, t) + (v_p^2 - v_s^2) \frac{\partial}{\partial r_i} \left(\sum_{j=1}^{3} \frac{\partial u_j(\mathbf{r}, t)}{\partial r_j} \right) \tag{4.3}$$

where v_p is the compressional and v_s the shear wave velocity of the material.

- Maxwells equations in SI units for a medium with real dielectric constant ε', magnetic permeability μ', real conductivity σ' and charge density ρ

$$\nabla \cdot (\varepsilon' \varepsilon_0 \mathbf{E}(\mathbf{r}, t)) = \rho(\mathbf{r}, t) \tag{4.4}$$

$$\nabla \cdot (\mu' \mu_0 \mathbf{H}(\mathbf{r}, t)) = 0 \tag{4.5}$$

$$\nabla \times \mathbf{E}(\mathbf{r}, t) = -\frac{\partial}{\partial t} (\mu' \mu_0 \mathbf{H}(\mathbf{r}, t)) \tag{4.6}$$

$$\nabla \times \mathbf{H}(\mathbf{r}, t) = \sigma' \mathbf{E}(\mathbf{r}, t) + \frac{\partial}{\partial t} (\varepsilon' \varepsilon_0 \mathbf{E}(\mathbf{r}, t)) \tag{4.7}$$

for the electric field $\mathbf{E}(\mathbf{r}, t)$, magnetic field $\mathbf{H}(\mathbf{r}, t)$ supplemented by boundary conditions and the continuity equation

$$\frac{\partial \rho(\mathbf{r}, t)}{\partial t} + \nabla \cdot \sigma' \mathbf{E}(\mathbf{r}, t) = 0 \tag{4.8}$$

Here $\varepsilon_0 = 8.8542 \times 10^{-12} \, \text{F m}^{-1}$ is the permittivity and $\mu_0 = 4\pi \times 10^{-7} \, \text{H m}^{-1}$ is the magnetic permeability of empty space.

- The Navier–Stokes equations for the velocity field $\mathbf{v}(\mathbf{r}, t)$ and the pressure field $P(\mathbf{r}, t)$ of an incompressible liquid flowing through the pore space

$$\rho \frac{\partial \mathbf{v}}{\partial t} + \rho(\mathbf{v}^T \cdot \nabla)\mathbf{v} = \eta \, \Delta \mathbf{v} + \rho g \nabla z - \nabla P \tag{4.9}$$

$$\nabla^T \cdot \mathbf{v} = 0 \tag{4.10}$$

where ρ is the density and μ the viscosity of the liquid. The coordinate system was chosen such that the acceleration of gravity g points in the z direction. These equations have to be supplemented with the no-slip boundary condition $\mathbf{v} = 0$ on the pore-space boundary.

In the following, mainly the equations for fluid transport and Maxwells equation for dielectric relaxation will be discussed in more detail. Combining fluid flow and diffusion into convection–diffusion equations yields the standard description for solute and contaminant transport [21, 24, 26, 260–262].

B. General Formulation

Transport and relaxation processes in a three-dimensional two-component porous medium $\mathbb{S} = \mathbb{P} \cup \mathbb{M}$ (defined above in Section II) may be formulated very broadly as a system of partial differential equations

$$F_{\mathbb{P}}\left(r_1, r_2, \ldots, u_1(\mathbf{r}), u_2(\mathbf{r}), \ldots, \frac{\partial u_i(\mathbf{r})}{\partial r_j}, \ldots, \frac{\partial^2 u_i(\mathbf{r})}{\partial r_j \partial r_k}, \ldots\right) = 0 \quad \mathbf{r} \in \mathbb{P}$$

$$F_{\mathbb{M}}\left(r_1, r_2, \ldots, u_1(\mathbf{r}), u_2(\mathbf{r}), \ldots, \frac{\partial u_i(\mathbf{r})}{\partial r_j}, \ldots, \frac{\partial^2 u_i(\mathbf{r})}{\partial r_j \partial r_k}, \ldots\right) = 0 \quad \mathbf{r} \in \mathbb{M}$$

$$F_{\partial \mathbb{P}}\left(r_1, r_2, \ldots, u_1(\mathbf{r}), u_2(\mathbf{r}), \ldots, \frac{\partial u_i(\mathbf{r})}{\partial r_j}, \ldots, \frac{\partial^2 u_i(\mathbf{r})}{\partial r_j \partial r_k}, \ldots\right) = 0 \quad \mathbf{r} \in \partial \mathbb{P}$$

$$(4.11)$$

for n unknown functions $u_i(\mathbf{r})$ with $i = 1, \ldots, n$ and $\mathbf{r} = (r_1, \ldots, r_d) \in \mathbb{R}^d$. Here the unknown functions u_i describe properties of the physical process (such as displacements, velocities, temperatures, pressures, electric fields, etc.), and the given functions $F_{\mathbb{P}}$, $F_{\mathbb{M}}$, and $F_{\partial \mathbb{P}}$ depend on a finite number of its derivatives. The function $F_{\partial \mathbb{P}}$ provides a coupling between the processes in the pore and matrix space. The main difficulty arises from the irregular structure of the boundary. The formulation may be generalized to porous media with more than two components. For a stochastic porous medium the solutions $u_i(\mathbf{r})$ of Eq. (4.11) depend on the random realization, and one is usually interested in the averages $\langle u_i(\mathbf{r}) \rangle$. Some authors [2, 41] have recently emphasized the difference between continuum descriptions such as Eq. (4.11) and discrete descriptions such as network models. Section V will show that discrete formulations arise as approximations and reduce to the continuum description of the same phenomenon in an appropriate limit.

V. TRANSPORT AND RELAXATION IN TWO COMPONENT MEDIA

The calculation of effective macroscopic physical properties from a geometrical characterization of the microstructure is the second subproblem of the central question discussed in the introduction. This section will focus on single-phase fluid transport and dielectric relaxation in porous media as representative examples for this general problem. The third subproblem of passage between microscopic and macroscopic length scales that has played an important role in Section III.A.5 will become more prominent in this and Section VI. The upscaling problem appears in the present section as the need to find effective macroscopic equations of motion from averaging the underlying microscopic equations of motion.

Successive spatial averaging allows passage to larger and larger length scales, and it can be carried out using systematic expansions in the ratio of length scales or self-consistent effective medium theories. The idea of a self-consistently determined homogeneous reference medium is cental to the definition of an effective macroscopic physical property. Asymptotic expansions in the ratio of a microscopic length scale to a macroscopic scale are known as homogenization theory, and will be discussed in Sections V.C.3 and V.C.4. Their purpose is to provide a systematic method of identifying useful macroscopic reference properties. Once a useful macroscopic description is identified, a generalized form of effective medium theory can be employed to calculate the effective macroscopic properties.

A. Effective Transport Coefficients

1. Definition

A large number of transport and relaxation processes in porous media are governed by the disordered Laplace equation (4.2) with variable coefficients $C(\mathbf{r})$ for a scalar field $P(\mathbf{r})$

$$\nabla^T \cdot (C(\mathbf{r})\nabla P(\mathbf{r})) = 0 \qquad (5.1)$$

within the sample region $\mathbb{S} = \mathbb{P} \cup \mathbb{M}$. This "equation of motion" for P must be supplemented with suitable boundary conditions on the sample boundary $\partial \mathbb{S}$, and, if $C(\mathbf{r})$ is discontinuous across $\partial \mathbb{P}$, also on the internal boundary $\partial \mathbb{P}$. By introducing the vector field $\mathbf{v}(\mathbf{r})$, Eq. (5.1) may be rewritten as

$$\mathbf{v}(\mathbf{r}) = -C(\mathbf{r})\nabla P(\mathbf{r})$$
$$\nabla^T \cdot \mathbf{v}(\mathbf{r}) = 0 \qquad (5.2)$$

These equations can be used as the microscopic starting point although, as shown in Section V.C.3 for the case of fluid flow, they may hold only in a macroscopic limit starting from a different underlying microscopic description. Equations (5.1) or (5.2) appear in many transport and relaxation problems for porous and heterogeneous media. For Darcy flow in porous media P is the pressure, $C = K/\eta$ is the quotient of absolute hydraulic permeability and fluid viscosity, and $\mathbf{v}$ is the fluid velocity field. For dielectric relaxation P becomes the electrostatic potential, $\mathbf{v}$ becomes the dielectric displacement, and C becomes the dielectric permittivity tensor. In diffusion or dispersion problems P is the concentration field, $\mathbf{v}$

TABLE III

Quantities Corresponding to P, $\mathbf{v}$, and $\mathbf{C}$ in Eq. (5.2) for Different Transport and Relaxation Problems in Porous Media

Problem Type	P	$\mathbf{v}$	$\mathbf{C}$
Fluid flow	Pressure	Velocity	Permeability/viscosity
Electrical conduction	Voltage	Current	Conductivity
Dielectric relaxation	Potential	Displacement	Dielectric permittivity
Diffusion (dispersion)	Concentration	Particle flux	Diffusion constant

corresponds to the diffusion flux, and $\mathbf{C}$ becomes the diffusivity. Table III summarizes the translation of P, $\mathbf{v}$, and $\mathbf{C}$ into various problems.

For a homogeneous and isotropic medium the transport coefficients $\mathbf{C}(\mathbf{r}) = C\mathbf{1}$, where $\mathbf{1}$ denotes the identify, are independent of $\mathbf{r}$, and Eq. (5.1) reduces to a Laplace equation for the field P. For a random medium the transport coefficients are random functions of $\mathbf{r}$ and the solutions $P(\mathbf{r})$ and $\mathbf{v}(\mathbf{r})$ depend on the realization of $\mathbf{C}(\mathbf{r})$. The averaged solutions $\langle P(\mathbf{r}) \rangle$ and $\langle \mathbf{v}(\mathbf{r}) \rangle$ are therefore of primary interest. The tensor of *effective transport coefficients* is $\bar{\mathbf{C}}$ defined as

$$\langle \mathbf{v}(\mathbf{r}) \rangle = -\bar{\mathbf{C}} \boldsymbol{\nabla} \langle P(\mathbf{r}) \rangle \qquad (5.3)$$

and it provides a relation between the average fields. The ensemble averages $\langle f(\mathbf{r}) \rangle$ in the definition can be replaced with spatial averages defined by

$$\bar{f} = \frac{1}{V(\mathbb{S})} \int f(\mathbf{r}) \chi_{\mathbb{S}}(\mathbf{r}) \, d^3\mathbf{r} \qquad (5.4)$$

where f stands for P or $\mathbf{v}$. Both the ensemble and the spatial average depend on the averaging region $\mathbb{S}$, and a residual variation of $\bar{f}$ or $\langle f \rangle$ is possible on scales larger than the size of $\mathbb{S}$. In the following it will always be assumed that $\bar{f} = \langle f \rangle$ if $\mathbb{S}$ is sufficiently large. The ensemble average notation will be preferred because it is notationally more convenient.

The purpose of introducing effective macroscopic transport coefficients is to replace the heterogeneous medium described by $\mathbf{C}(\mathbf{r})$ with an equivalent homogeneous medium described by $\bar{\mathbf{C}}$. If $\bar{\mathbf{C}}$ is known, then all the knowledge accumulated for the homogeneous problem can be utilized immediately, and, for example, the average field $\mathbf{v}$ can be obtained simply from solving a Laplace equation for P.

2. Discretization and Networks

If the function $\mathbf{C}(\mathbf{r})$ is known, then Eq. (5.1) can be solved to any desired accuracy using standard finite difference approximation schemes. To this end the sample space $\mathbb{S}$ of linear extension $\mathscr{L}$ is partitioned into cubes $\mathbb{K}_j$. The cubes are centered on the sites $\mathbf{r}_i$ of a simple cubic lattice with lattice spacing L. Other lattices may also be employed. The lengths L and $\mathscr{L}$ obey $L \ll \mathscr{L}$. The total number of cubes is $N = (\mathscr{L}/L)^d$.

For a stationary and isotropic medium with $\mathbf{C}(\mathbf{r}) = c(\mathbf{r})\mathbf{1}$ the discretization of Eq. (5.1) gives a system of linear equations for the pressure variables at the cube centers $P_i = P(\mathbf{r}_i)$

$$\sum_j c_{ij}(P_i - P_j) = 0 \tag{5.5}$$

for cubes $\mathbf{r}_i$ not located at the sample boundary. The boundary conditions at the sample boundary give rise to a nonvanishing right-hand side of the linear system if $\mathbf{r}_i$ is the center of a cube located close to $\partial\mathbb{S}$. The local transport coefficients c_{ij} are given as

$$c_{ij}(L) = c((\mathbf{r}_i + \mathbf{r}_j)/2) \tag{5.6}$$

if $\mathbf{r}_i$ and $\mathbf{r}_j$ are nearest neighbors. If $\mathbf{r}_i$ and $\mathbf{r}_j$ are not nearest neighbors the local coefficient vanishes, $c_{ij} = 0$. Because the location of the cube centers $\mathbf{r}_i$ depends on the resolution L the coefficients c_{ij} in the network equations depend on L and on the shape of the measurement cells $\mathbb{K}_i$.

The numerical solution of the discretized equations (5.5) can be obtained by many methods including relaxation, successive overrelaxation, or conjugate gradient schemes, transfer matrix calculations, series expansions, or recursion methods [40, 248, 263–267]. If the function $c(\mathbf{r})$ is known, then the solution to Eq. (5.1) is recovered in the limit $L \to 0$ to any desired accuracy. Within a certain class of lattices the limit is known to be independent of the choice of the approximating discrete lattice. To actually perform this limit, however, the function $c(\mathbf{r})$ must be known to arbitrary accuracy.

In most experimental and practical problems the function $c(\mathbf{r})$ is either completely unknown or not known to arbitrary accuracy. Therefore, it is necessary to have a theory for the *local transport coefficients* $c_{ij}(L)$ as a function of the resolution L of the discretization. At present, the only resolution dependent theories seem to be local porosity theory [168–175] and homogenization theory [38, 268–271], which will be discussed in more detail below. The basic idea of local porosity theory is to use the local geometry distributions defined in Section III.A.5 and to express the local transport coefficients in terms of the geometrical quantities characterizing the local geometry. The basic idea of homogenization theory is a double scale asymptotic expansion in the small parameter $L/\mathcal{L}$.

The discretized equations (5.5) are network equations. This explains the great importance and popularity of network models. In the more conventional *network models* [155, 157, 187, 220, 223, 225–233] the resolution dependence is neglected altogether. Instead one assumes a specific model for the local transport coefficients c_{ij} such that the *global* geometric characteristics (porosity, etc.) are reproduced by the model. Three immediate problems arise from this assumption:

- The connection with the underlying *local* geometry is lost, although the local value of the transport property depends on it.
- In the absence of an independent measurement of the local transport coefficients they become free fit parameters. Popular stochastic network models assume log normal or binary distributions for the local transport coefficients.
- Without a model for the local geometry an independent experimental or calculational determination of the local transport coefficients for one transport problem (say fluid flow) cannot be used for another transport problem (say diffusion) although the equations of motion [Eq. (5.1)] have the same mathematical form for both cases.

All of these problems are alleviated in local porosity theory or homogenization theory, which attempt to keep the connection with the underlying local geometry.

3. Simple Expressions for Effective Transport Coefficients

While a numerical solution of the network equations (5.5) is of great practical interest, its value for a scientific understanding of heterogeneous media is limited. Analytical expressions, be they exact or approximate, are better suited for developing the theory because they allow to extract the general model independent aspects. Unfortunately, only very few exact analytical results are available [272–275]. The one-dimensional case can be solved exactly by a change of variable. The exact result is the harmonic average

$$\bar{c} = \langle c^{-1} \rangle^{-1} \tag{5.7}$$

where the average denotes either an average with respect to $w(c)$, the probability density of local transport coefficients, or a spatial average as defined in Eq. (5.4). In two dimensions the geometric average

$$\bar{c} = \exp\langle \log c \rangle = \frac{\langle c \rangle^{1/2}}{\langle c^{-1} \rangle^{1/2}} \tag{5.8}$$

has been obtained exactly using duality in harmonic function theory [272] if the microstructure is homogeneous, isotropic, and symmetric. It was later rederived under less stringent conditions [273] and generalized to isomorphisms between associated microstructures [274].

Most analytical expressions for effective transport properties are approximate. In general dimensions, approximation formulas such as [275–277]

$$\bar{c} = \frac{\langle c \rangle^{(d-1)/d}}{\langle c^{-1} \rangle^{1/d}} \tag{5.9}$$

or [278, 279]

$$\bar{c} = \langle c^{(1-2/d)} \rangle^{1/(1-2/d)} \tag{5.10}$$

have been suggested, which reduce to the exact results for $d = 1$ and $d = \infty$. Various mean-field theories also provide approximate estimates for the effective permeabilities. The simplest mean-field theory

$$\bar{c} = \langle c \rangle \tag{5.11}$$

is obtained from Eqs. (5.9) or (5.10) by letting $d \to \infty$. Another very important approximation is the self-consistent effective medium approximation which reads

$$\left\langle \frac{c - \bar{c}}{c + (d-1)\bar{c}} \right\rangle = 0 \tag{5.12}$$

for a d-dimensional hypercubic lattice. For other regular lattices the factor $d - 1$ in the denominator has to be replaced with $(z/2) - 1$, where z is the coordination number of the lattice. Note that for $d = 1$ and $d \to \infty$ the effective medium approximation reproduces the exact result.

To distinguish the quality of these approximations it is instructive to consider a probability density $w(c)$ of local transport coefficients which has a finite fraction $p = 1 - \lim_{\varepsilon \to 0} \int_0^\varepsilon w(c)\, dc$ of blocking bonds. In dimension $d > 1$ this implies the existence of a percolation threshold $0 < p_c < 1$ below which $\bar{c} = 0$ vanishes identically (see Table II for values of p_c). Among the expressions (5.7)–(5.12) only the effective medium approximation (5.12) is able to predict the existence of a transition. The predicted critical value $p_c = 1/d$, however, is not exact as seen by comparison with Table II.

Another method for calculating the effective transport coefficient $\bar{c}$ will be discussed in homogenization theory in Section V.C.4. The resulting expression appears in Eq. (5.87) if one sets $\mathbf{K} = c\mathbf{1}$. It is given as a correction to the simplest mean-field expression (5.11). The correction involves the fundamental solution of the local transport problem [Eq. (5.88)]. In practice, the use of Eq. (5.87) is restricted to simple periodic microstructures [268, 280]. If the microstructure is periodic it suffices to obtain the fundamental solutions within the basic period, and to extend the average in Eq. (5.87) over that period. If the microstructure is not periodic then the solution of Eq. (5.88) and averaging in Eq. (5.87) quickly become as impractical as solving the original problem, because $c(\mathbf{r})$ is then unknown.

B. Dielectric Relaxation

1. Maxwell Equations in the Quasistatic Approximation

Consider a two component medium $\mathbb{S} = \mathbb{P} \cup \mathbb{M}$. The substances filling the sets $\mathbb{P}$ and $\mathbb{M}$ are assumed to be electrically homogeneous and characterized by their real frequency dependent conductivity $\sigma'(\omega)$ and dielectric function $\varepsilon'(\omega)$. The real dielectric function $\varepsilon'(\mathbf{r}, \omega)$ and conductivity $\sigma'(\mathbf{r}, \omega)$ of the composite are then given as

$$\varepsilon'(\mathbf{r}, \omega) = \varepsilon'_\mathbb{P}(\omega)\chi_\mathbb{P}(\mathbf{r}) + \varepsilon'_\mathbb{M}(\omega)\chi_\mathbb{M}(\mathbf{r}) \qquad (5.13)$$

$$\sigma'(\mathbf{r}, \omega) = \sigma'_\mathbb{P}(\omega)\chi_\mathbb{P}(\mathbf{r}) + \sigma'_\mathbb{M}(\omega)\chi_\mathbb{M}(\mathbf{r}) \qquad (5.14)$$

in terms of the functions $\varepsilon'_\mathbb{P}(\omega)$, $\varepsilon'_\mathbb{M}(\varepsilon)$ and $\sigma'_\mathbb{P}(\omega)$, $\sigma'_\mathbb{M}(\omega)$ characterizing the dielectric response of the constituents.

The propagation of electromagnetic waves in the composite medium is described by the macroscopic Maxwell equations (4.4)–(4.8). In the following, the magnetic permeabilities are assumed to be unity to simplify the analysis. The time variation of the fields is taken to be proportional to $\exp(-i\omega t)$. Fourier transforming and inserting Eq. (4.8) into Eq. (4.4) yields $\nabla \cdot \mathbf{D}(\mathbf{r}, \omega) = 0$, where the frequency-dependent displacement field

$$\mathbf{D}(\mathbf{r}, \omega) = \left(\varepsilon'(\mathbf{r}, \omega)\varepsilon_0 + \frac{i\sigma'(\mathbf{r}, \omega)}{\omega} \right)\mathbf{E}(\mathbf{r}, \omega) \qquad (5.15)$$

combines the free current density and the polarization current. In the quasistatic approximation one assumes that the frequency is small enough such that the inductive term on the right-hand side of Faradays law (4.6) can be neglected. By introducing the complex frequency dependent dielectric function

$$\varepsilon(\mathbf{r}, \omega) = \varepsilon'(\mathbf{r}, \omega) + i\varepsilon''(\mathbf{r}, \omega) = \varepsilon'(\mathbf{r}, \omega)\varepsilon_0 + i\frac{\sigma'(\mathbf{r}, \omega)}{\omega} \qquad (5.16)$$

the electric field and the displacement are found to satisfy the equations

$$\nabla \cdot \mathbf{D}(\mathbf{r}, \omega) = 0 \qquad (5.17)$$

$$\nabla \times \mathbf{E}(\mathbf{r}, \omega) = 0 \qquad (5.18)$$

$$\mathbf{D}(\mathbf{r}, \omega) = \varepsilon(\mathbf{r}, \omega)\mathbf{E}(\mathbf{r}, \omega) \qquad (5.19)$$

in the quasistatic approximation. If the electric field is replaced by the potential these equations assume the same form as Eq. (5.2), and hence the methods discussed in Section V.A can be employed in their analysis.

The neglect of the induced electromagnetic force is justified if the wavelength or penetration depth of the radiation is large compared to the typical linear dimension of the scatterers. If the scattering is caused by heterogeneities on the micrometer scale, as in many examples of interest, the approximation will be valid well into the infrared region.

2. Experimental Observations for Rocks

The electrical conductivity of rocks fully or partially saturated with brine is an important quantity for the reconstruction of subsurface geology from borehole logs [281, 282]. The main contribution to the total conductivity $\sigma'_{\mathbb{S}}$ of a sample $\mathbb{S} = \mathbb{P} \cup \mathbb{M}$ of brine-filled rock comes from the electrolyte. The contribution $\sigma'_{\mathbb{M}}$ from the rock matrix is usually negligible.[4] The electrolyte filling the pore space contributes through its intrinsic electrolytic conductivity $\sigma_{\mathbb{P}}$ as well as through electrochemical interactions at the interface. The total dc conductivity of the sample is written as

$$\sigma'_{\mathbb{S}}(0) = \frac{1}{F}\sigma'_{\mathbb{P}}(0) + \sigma'_{\partial\mathbb{P}}(0) \qquad (5.20)$$

where $\sigma'_{\mathbb{P}}(0)$ is the dc conductivity of the electrolyte (usually salt water) filling the pore space $\mathbb{P}$ and $\sigma'_{\partial\mathbb{P}}(0)$ denotes the conductivity resulting from the electrochemical boundary layer at the internal surface [283]. The surface conductivity $\sigma'_{\partial\mathbb{P}}(0)$ correlates well with the specific internal surface S and indirectly with other quantities related to it. The factor F is

---

[4] Nonvanishing matrix conductivity does, however, occur in veinlike ores.

called *electrical formation factor*. If the salinity of the pore water is high or electrochemical effects are absent, the second term in Eq. (5.20) can be neglected and the formation factor becomes identical with the dimensionless resistivity of the sample normalized by the water resistivity. In the following, the formation factor will be used synonymously for the dimensionless inverse dc conductivity $F = [\sigma'(0)]^{-1}$.

The formation factor is usually correlated with the bulk porosity $\bar{\phi}$ in a relation known as "Archie's first equation" [284]

$$F = \bar{\phi}^{-m} \tag{5.21}$$

where the so-called cementation index m scatters widely and often obeys $1.3 \le m \le 2.5$ [281, 282]. Smaller values of m are associated empirically with loosely packed media, while higher values are associated with more consolidated and compacted media. Equation (5.21) implies not only an algebraic correlation between a purely geometric quantity $\bar{\phi}$ and a transport coefficient F, but it also states that porous rocks do not show a conductor–insulator transition at any finite porosity.

The experimental evidence for the postulated algebraic correlation (5.21) between conductivity and porosity is weak. The available range of the porosity rarely spans more than a decade. The corresponding conductivity data scatter widely for measurements on porous rocks and other media [188, 254, 281, 285–288]. The most reliable tests of Archies law have been performed on artificial porous media made from sintering glass beads [254, 285, 287]. These media have a microstructure very similar to sandstone and are at the same time free from electrochemical effects. A typical experimental result for glass beads is shown in Fig. 18 [287]. Note the small range of porosities in the figure. The existence of nontrivial power law relations in such samples is better demonstrated by correlating the conductivity with the permeability [170, 284]. In other experiments on artificial media a mixture of rubber balls and water is successively compressed while monitoring its conductivity [217, 289]. These experiments show deviations from the pure algebraic behavior postulated by Eq. (5.21). If the cementation "exponent" in Eq. (5.21) is assumed to depend on ϕ, then it increases at low porosities in agreement with the general trend that higher values correspond to a higher degree of compaction.

A much better confirmed observation on natural and artificial porous rocks is dielectric enhancement caused by the disorder in the micro-structure [89, 287, 290–295]. Dielectric enhancement due to disorder has been studied extensively in percolation theory and experiment [40, 296,

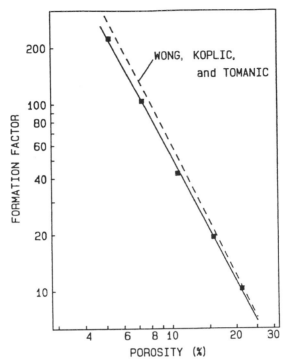

Figure 18. Log–log plot of formation factor versus porosity for sintered glass beads with 200–300-μm diameter. The solid line is a fit to $F = a\phi^{-m}$ with $a = 0.33$ and a cementation exponent $m = 2.2$ while the dashed line represents the results $a = 0.30$ and $m = 2.3$ on the same system in [254]. (Reproduced with permission from [287].)

297]. An example is shown in Fig. 19 for the sintered glass bead media containing thin-glass plates. In these media interfacial conductivity and other electrochemical effects can be neglected [287]. The frequency is plotted in units of $\omega_W = \sigma'_W/(\varepsilon_0 \varepsilon'_W)$ the relaxation frequency of water. Although salt water and glass are essentially dispersion free in the frequency range shown in Fig. 19 their mixture shows a pronounced dispersion, which exceeds the values of the dielectric constants of both components. Similar results can be found in [287]. In [295] the dielectric response of a large number of sandstones and carbonates is given in terms of the empirical Cole–Cole formula [298]. Interestingly, the corresponding temporal relaxation function appears within the recent theory of nonequilibrium systems [64], which is the same theory on which the macroscopic local porosity distributions in Section III.A.5 were based.

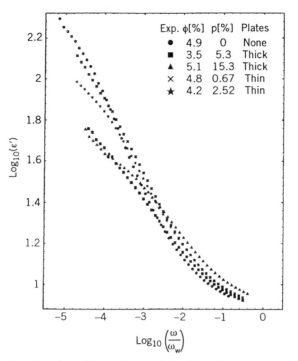

Figure 19. Log–log plot of the real part of the dielectric function versus normalized frequency ω/ω_w where $\omega_w = \sigma_w'/(\varepsilon_0 \varepsilon_w')$ is the relaxation frequency of water for samples with porosities close to 5%. The different symbols denote different concentrations of thin glass plates p defined as the ratio of the mass of the glass plates divided by the total mass of glass in the sample [362].

3. Theoretical Mixing Laws

a. Spectral Theories. Dielectric mixing laws express the frequency dependent dielectric function or conductivity of a two-component mixture in terms of the dielectric functions of the constituents [31, 35, 40, 46]. Spectral theories express the effective dielectric function in terms of an abstract pole spectrum which is independent of the dielectric functions ε_p and ε_M of the two constituents filling the pore and matrix space [293, 299–309]. Theoretically, the effective dielectric function may be written as

$$\bar{\varepsilon} = \varepsilon_M \left(1 - \sum_{n=1}^{\infty} \frac{a_n}{s - b_n} \right) \tag{5.22}$$

where

$$s = \left(1 - \frac{\varepsilon_{\mathbb{P}}}{\varepsilon_{\mathbb{M}}}\right)^{-1} \tag{5.23}$$

The constants a_n and b_n are the strength and location of the poles and they reflect the influence of the microgeometry. Unfortunately, these parameters do not have a direct geometrical interpretation, although under the assumption of stationarity and isotropy two sum rules are known that connect integrals of the pole spectrum with the bulk porosity [40].

b. Geometric Theories. The simplest geometric theories for the effective dielectric function $\bar{\varepsilon}$ are mean-field theories. In these approximations, a small spherical cell with a randomly valued dielectric constant ε is embedded into a homogeneous host medium of dielectric constant ε_h. Then the electrical analogue of Eq. (5.3) becomes

$$\bar{\varepsilon} = \varepsilon_h \left(1 + 2\left\langle \frac{\varepsilon - \varepsilon_h}{\varepsilon + 2\varepsilon_h} \right\rangle\right)\left(1 - \left\langle \frac{\varepsilon - \varepsilon_h}{\varepsilon + 2\varepsilon_h} \right\rangle\right)^{-1} \tag{5.24}$$

where the average denotes an ensemble average using the probability density $w(\varepsilon)$ of ε. A two-component medium $\mathbb{S} = \mathbb{P} \cup \mathbb{M}$ can be represented by the binary probability density

$$w(\varepsilon) = \bar{\phi}\delta(\varepsilon - \varepsilon_{\mathbb{P}}) + (1 - \bar{\phi})\delta(\varepsilon - \varepsilon_{\mathbb{M}}) \tag{5.25}$$

containing $\bar{\phi}$ as the only geometrical input parameter. The *Clausius–Mossotti approximation* [46, 310] for a two-component medium is obtained by setting $\varepsilon_h = \varepsilon_{\mathbb{M}}$ in the limit $\bar{\phi} \to 0$ or $\varepsilon_h = \varepsilon_{\mathbb{P}}$ in the limit $\bar{\phi} \to 1$ in Eq. (5.24). In the former case one obtains

$$\frac{\bar{\varepsilon} - \varepsilon_{\mathbb{M}}}{\bar{\varepsilon} + 2\varepsilon_{\mathbb{M}}} = \bar{\phi} \frac{\varepsilon_{\mathbb{P}} - \varepsilon_{\mathbb{M}}}{\varepsilon_{\mathbb{P}} + 2\varepsilon_{\mathbb{M}}} \tag{5.26}$$

which will be a good approximation at low porosities. Note that the Clausius–Mossotti approximation is not symmetrical under exchanging pore and matrix. A symmetrical and also self-consistent approximation is obtained from Eq. (5.24) by setting $\varepsilon_h = \bar{\varepsilon}$. This leads to the symmetrical *effective medium approximation* for a two-component medium

$$\bar{\phi} \frac{\varepsilon_{\mathbb{P}} - \bar{\varepsilon}}{\varepsilon_{\mathbb{P}} + 2\bar{\varepsilon}} + (1 - \bar{\phi}) \frac{\varepsilon_{\mathbb{M}} - \bar{\varepsilon}}{\varepsilon_{\mathbb{M}} + 2\bar{\varepsilon}} = 0 \tag{5.27}$$

which also could have been derived from using Eq. (5.25) in Eq. (5.12). The effective medium approximation is a very good approximation for microstructures consisting of a small concentration of nonoverlapping spherical grains embedded in a host. Recently, much effort has been expended to show that the effective medium approximation (EMA) becomes exact for certain pathological microstructures [311]. The so-called *asymmetrical* or *differential effective medium approximation* is obtained by iterating the Clausius–Mossotti equation, which gives the effective conductivity to lowest order in $\bar{\phi}$ [285, 312, 313]. One finds the result

$$\left(\frac{\bar{\varepsilon} - \varepsilon_M}{\varepsilon_P - \varepsilon_M} \left(\frac{\varepsilon_P}{\bar{\varepsilon}} \right)^{1/3} = \bar{\phi} \right) \tag{5.28}$$

for spherically shaped inclusions.

The symmetric and asymmetric effective medium approximations can be generalized to ellipsoidal inclusions because the electric field and polarization inside the ellipsoid remain uniform in an applied external field [40, 310, 312]. For aligned oblate spheroids whose quadratic form is $(x/b_2)^2 + (y/b_2)^2 + (z/b_1)^2 = 1$ with $b_1 \leq b_2$ the effective medium theory for a two-component composite results in two coupled equations

$$\bar{\phi} \frac{\varepsilon_{P_i} - \bar{\varepsilon}_i}{1 + \Lambda_i(\varepsilon_{P_i} - \bar{\varepsilon}_i)} + (1 - \bar{\phi}) \frac{\varepsilon_{M_i} - \bar{\varepsilon}_i}{1 + \Lambda_i(\varepsilon_{M_i} - \bar{\varepsilon}_i)} = 0 \tag{5.29}$$

where the index $i = 1$ denotes vertical conductivities and the index $i = 2$ denotes horizontal conductivities. The two equations in Eq. (5.29) are coupled through

$$\Lambda = \frac{L_i}{\bar{\varepsilon}_i} \tag{5.30}$$

$$2L_2 = (-L_1) \tag{5.31}$$

$$L_1 = (e^{-3} + e^{-1})(e - \arctan e) \tag{5.32}$$

$$e = \sqrt{\frac{b_2^2 \bar{\varepsilon}_1}{b_1^2 \bar{\varepsilon}_2} - 1} \tag{5.33}$$

with $\frac{1}{3} \leq L_1 \leq 1$. The generalization of the asymmetric effective medium theory [Eq. (5.28)] to aligned spheroids with depolarization factor L was given in [285] as

$$\frac{\bar{\varepsilon} - \varepsilon_M}{\varepsilon_p - \varepsilon_M} \left(\frac{\varepsilon_p}{\bar{\varepsilon}}\right)^L = \bar{\phi} \tag{5.34}$$

For spheroids with identical shape but isotropically distributed orientations

$$\frac{\bar{\varepsilon} - \varepsilon_M}{\varepsilon_p - \varepsilon_M} \left(\frac{\varepsilon_p}{\bar{\varepsilon}}\right)^{[3L(1-L)/(1+3L)]}$$

$$\times \left(\frac{(5-3L)\varepsilon_p + (1+3L)\varepsilon_M}{(5-3L)\bar{\varepsilon} + (1+3L)\varepsilon_M}\right)^{[2(1-3L)^2/(1+3L)(5-3L)]} = \bar{\phi} \tag{5.35}$$

was obtained in [205, 292, 314]. Equation (5.34) will be referred to as the Sen–Scala–Cohen model (SSC) and Eq. (5.35) will be called the uniform spheroid model (USM).

Recently, local porosity theory has been proposed as an alternative generalization of effective medium theories [168–175]. The simplest mean-field theories [Eqs. (5.26)–(5.28)] are based on the simplest geometric characterization theories of Section III.A.1. The theories are usually interpreted geometrically in terms of grain models (see Section III.B.2) with spherical grains embedded into a homogeneous host material. The generalizations (5.29), (5.34), and (5.35) are obtained by generalizing the interpretation to more general grain models. Local porosity theory on the other hand is based on generalizing the geometric characterization by using local geometry distributions (see Section III.A.5) rather than simply porosity or specific surface area alone. In Section III.A.5 two different types of local geometry distributions were introduced: Macroscopic distributions with infinitely large measurement cells defined in Eq. (3.52), and mesoscopic distributions with measurement cells of finite volume defined in Eq. (3.33). For a mesoscopic partitioning $\mathcal{K}$ of the sample using a simple cubic lattice with cubic unit cell $\mathbb{K}$ the self-consistency equation of local porosity theory for $\bar{\varepsilon}$ reads

$$\int_0^1 \left[\frac{\varepsilon_C(\phi) - \bar{\varepsilon}}{\varepsilon_C(\phi) + 2\bar{\varepsilon}} \lambda(\phi; \mathbb{K}) + \frac{\varepsilon_B(\phi) - \bar{\varepsilon}}{\varepsilon_B(\phi) + 2\bar{\varepsilon}} (1 - \lambda(\phi; \mathbb{K}))\right] \mu(\phi; \mathbb{K}) \, d\phi = 0 \tag{5.36}$$

where $\lambda(\phi; \mathbb{K})$ is the local percolation probability defined in Section III.A.5.d, and $\varepsilon_C(\phi)$ and $\varepsilon_B(\phi)$ are the local dielectric functions of percolating or conducting (index C) and nonpercolating or blocking (index B) measurement cells. In Eq. (5.40) it is assumed that the local dielectric response depends only on the porosity, but this may be generalized to include other geometrical characteristics.

Equation (5.36) has two interesting special cases. For a cubic measurement lattice ($z = 6$) in the limit $L \to 0$ in which the side length L of the cubic cells is small the one-cell local porosity distribution is given by Eq. (3.31) or (3.34) if the medium is mixing. Inserting Eq. (3.31) or (3.34) into Eq. (5.36) and using $\lambda(0) = 0$, $\lambda(1) = 1$, $\varepsilon_C(0) = \varepsilon_B(0) = \varepsilon_M$, and $\varepsilon_C(1) = \varepsilon_B(1) = \varepsilon_P$ yields Eq. (5.27) for traditional effective medium theory. Note, however, that in the limit $L \to 0$ the local porosities become *highly correlated* rendering a description of the geometry in terms of the one-cell function $\mu(\phi; \mathbb{K})$ more and more inadequate. This argument does not apply in the opposite limit $L \to \infty$ in which the measurement cells $\mathbb{K}$ become very large. For stationary media, the local porosities in nonoverlapping measurement cells are uncorrelated in the limit $L \to \infty$. For stationary and mixing media the local porosity distribution

$$\lim_{b \to \infty} \mu(\phi; b\mathbb{K}) = \delta(\phi - \bar{\phi}) \tag{5.37}$$

becomes concentrated at a single point according to Eq. (3.32) or (3.35). Assuming, as before, that the limit is independent of the shape of $\mathbb{K}$ Equation (5.36) reduces to

$$\lambda(\bar{\phi}) \frac{\varepsilon_C(\bar{\phi}) - \bar{\varepsilon}}{\varepsilon_C(\bar{\phi}) + 2\bar{\varepsilon}} + (1 - \lambda(\bar{\phi})) \frac{\varepsilon_B(\bar{\phi}) - \bar{\varepsilon}}{\varepsilon_B(\bar{\phi}) + 2\bar{\varepsilon}} = 0 \tag{5.38}$$

which is identical to Eq. (5.27) except for the replacement of $\bar{\phi}$ by $\lambda(\bar{\phi})$, ε_M by $\varepsilon_B(\bar{\phi})$, and ε_P by $\varepsilon_C(\bar{\phi})$. Repeating the same differential replacement arguments [285] that lead to Eq. (5.28) for $\lambda(\bar{\phi})$ instead of $\bar{\phi}$ gives a differential version of mesoscopic local porosity theory in the limit $L \to \infty$

$$\frac{\bar{\varepsilon} - \varepsilon_B(\bar{\phi})}{\varepsilon_C(\bar{\phi}) - \varepsilon_B(\bar{\phi})} \left(\frac{\varepsilon_C(\bar{\phi})}{\bar{\varepsilon}} \right)^{1/3} = \lambda(\bar{\phi}) \tag{5.39}$$

Note that the limiting equations (5.38) and (5.39) for $L \to \infty$ contain geometric information about the pore space that goes beyond the bulk porosity and is contained in the function $\lambda(\bar{\phi})$.

As discussed in Section III.A.5.e the δ distribution is not the only possible macroscopic limit. Macroscopically heterogeneous media are described by the limiting macroscopic local porosity densities

$\mu(\phi; \varpi, C, D)$ defined in Eq. (3.52). The δ distribution $\delta(\phi - \bar{\phi})$ is obtained in the limit $\varpi \rightarrow 1$ or in the degenerate case arising for $D = 0$. The macroscopic form of local porosity theory for the effective dielectric constant is given by the integral equation

$$\int_0^1 \left[\frac{\varepsilon_C(\phi) - \bar{\varepsilon}}{\varepsilon_C(\phi) + 2\bar{\varepsilon}} \lambda(\phi) + \frac{\varepsilon_B(\phi) - \bar{\varepsilon}}{\varepsilon_B(\phi) + 2\bar{\varepsilon}} (1 - \lambda(\phi)) \right]$$
$$\times \mu(\phi; \varpi, C, D) \, d\phi = 0 \tag{5.40}$$

with $\mu(\phi; \varpi, C, D)$ defined in Eq. (3.52) and $\varpi \neq 1$ and $D \neq 0$. If the parameters $\varpi = \varpi(\bar{\phi}, \phi^2, \phi^3)$, $D = D(\bar{\phi}, \phi^2, \phi^3)$, and $C = C(\bar{\phi}, \phi^2, \phi^3)$, are expressed in terms of the moments according to Eq. (3.53) the resulting effective dielectric function $\bar{\varepsilon} = \bar{\varepsilon}(\bar{\phi}, \phi^2, \phi^3)$ is found to be a function of the bulk porosity and its fluctuations. This observation indicates the possibility to study Archie's law within the framework of local porosity theory.

4. Archie's Law

Archie's law [Eq. (5.21)] concerns the effective dc conductivity, $\sigma'(0)$, and can be studied by replacing ε with $\sigma'(0)$ in all the formulas of Section V.B.3. For notational convenience the shorthand notation $\sigma'(0) = \sigma$ will be employed in this section. Archie's law can then be discussed by replacing ε with σ throughout and setting $\sigma_M = 0$. From the Clausius–Mossotti formula (5.26) one obtains the relation

$$\bar{\sigma} = \sigma_{\mathbb{P}} \frac{2\bar{\phi}}{3 - \bar{\phi}} \tag{5.41}$$

which reproduces Archie's law (5.21) with a cementation exponent $m = 1$. The symmetrical effective medium approximation (5.27) gives

$$\bar{\sigma} = \frac{\sigma_{\mathbb{P}}}{2} (3\bar{\phi} - 1) \tag{5.42}$$

for $\bar{\phi} > \frac{1}{3}$ and $\bar{\sigma} = 0$ for $\bar{\phi} \leq \frac{1}{3}$. Thus the symmetrical effective medium theory predicts a percolation transition at $\bar{\phi}_c = \frac{1}{3}$ and does not agree with Archie's law (5.21) in this respect. The same conclusion holds for the anisotropic generalization of the symmetric theory to nonspherical inclusions given in Eq. (5.29).

For the asymmetric effective medium theory in its simplest form (5.28) one finds

$$\bar{\sigma} = \sigma_p \bar{\phi}^{3/2} \tag{5.43}$$

consistent with Archies law (5.21) with cementation exponent $m = \frac{3}{2}$. This expression has found much attention because it yields $m \neq 1$ [205, 285, 312, 314, 315]. There are, however, several problems with Eq. (5.43). Its derivation implies that the solid component is not connected [312]. The experimentally observed behavior is often not algebraic, and if a power law is nevertheless assumed its exponent is often very different from $\frac{3}{2}$. Most importantly, the frequency dependent theory does not predict sufficient dielectric enhancement. The first problem can be circumvented by generalizing to a three-component medium [313, 316], the second can be overcome by considering nonspherical inclusions [205, 285, 312, 314, 315]. As an example the generalization to nonspherical grains [Eq. (5.34)] gives

$$\bar{\sigma} = \sigma_p \bar{\phi}^{1/(1-L)} \tag{5.44}$$

with $\frac{1}{3} \leq L \leq 1$, and the uniform spheroid model gives a similar result. The most serious problem, however, is the fact pointed out in [285, 292] that Eq. (5.28) cannot reproduce the frequency dependence of $\bar{\sigma}$ and the observed dielectric enhancement. This will be discussed further in Section V.B.5.

Local porosity theory contains geometrical information above and beyond the average porosity $\bar{\phi}$. Consequently, it predicts more general relationships between porosity and conductivity. In its simplest form Eq. (5.38) leads to

$$\bar{\sigma} = \frac{\sigma_C(\bar{\phi})}{2} (3\lambda(\bar{\phi}) - 1) \tag{5.45}$$

which may or may not have a percolation transition depending on whether the equation

$$\lambda(\bar{\phi}_c) = \frac{1}{3} \tag{5.46}$$

has a solution $0 < \bar{\phi}_c < 1$ that can be interpreted as a critical porosity. Therefore, the percolation threshold can arise at any porosity including $\bar{\phi}_c = 0$. This fact reconciles percolation theory with Archie's law. Note also that the behavior is nonuniversal[5] and depends on the local percolation probability function $\lambda(\phi)$, and the local response $\sigma_C(\bar{\phi})$. Similarly, the differential form (5.39) of local porosity theory yields

$$\bar{\sigma} = \sigma_C(\bar{\phi})\lambda(\bar{\phi})^{3/2} \tag{5.47}$$

which is more versatile than Eq. (5.44). The preceding results hold for large measurement cells when $\mu(\phi) \approx \delta(\phi - \bar{\phi})$. For general local porosity distributions equation (5.36) gives the result [168]

$$\bar{\sigma} \approx \sigma_0(p - p_c) \tag{5.48}$$

where $p_c = \frac{1}{3}$,

$$\frac{1}{\sigma_0} = \int_0^1 \frac{\lambda(\phi; \mathbb{K})\mu(\phi; \mathbb{K})}{\sigma_C(\phi)} \, d\phi \tag{5.49}$$

and p is the control parameter of the percolation transition

$$p = \int_0^1 \lambda(\phi; \mathbb{K})\mu(\phi; \mathbb{K}) \, d\phi \tag{5.50}$$

giving the total fraction of percolating local geometries. The result [Eq. (5.48)] applies if $\bar{\phi} \to 0$ for all values of p, and also if $p \to p_c$ at arbitrary $\bar{\phi}$. It holds universally as long as

$$\int_0^1 \frac{1}{\phi} \lambda(\phi; \mathbb{K})\mu(\phi; \mathbb{K}) \, d\phi < \infty \tag{5.51}$$

the inverse first moment is finite [317]. This condition is violated for the macroscopic distributions $\mu(\phi; \varpi, C, D)$ if all cells are percolating, $\lambda(\phi) = 1$. In such a case if $\lambda(\phi)\mu(\phi) \propto \phi^{-\alpha}$ as $\phi \to 0$, then Eq. (5.48) is replaced with [317]

$$\bar{\sigma} \propto (p - p_c)^{1/[1 - \alpha(\bar{\phi}, \overline{\phi^3}, \overline{\phi^3})]} \tag{5.52}$$

where α depends on the moments $\overline{\phi^n}$ because ϖ depends on them through Eq. (3.53).

Compaction and consolidation processes will in general change the local porosity distribution $\mu(\phi)$ where its dependence on $\mathbb{K}$ or the parameters ϖ, C, D has been suppressed. Assume that it is possible to

---

[5] The statement in [41] that local porosity theory predicts Archies law with a universal exponent is incorrect.

describe the consolidation process as a one parameter family $\mu_q(\phi)$ of local porosity distributions depending on a parameter q, which characterizes the compaction process. Then the total fraction p, the bulk porosity $\bar{\phi}$, and the integral (5.49) become functions of q. If it is possible to invert the relation $\bar{\phi} = \bar{\phi}(q)$, then the fraction p becomes $p = p(\bar{\phi})$, and equally $\sigma_0 = \sigma_0(\bar{\phi})$. Therefore, Eqs. (5.48) and (5.52) become porosity–conductivity relations which depend on the consolidation process.[6] If the condition (5.51) and the asymptotic expansion $\sigma_0(\bar{\phi}) = \bar{\phi}^\beta(1 + \cdots)$ and $p(\bar{\phi}) = p_c + \bar{\phi}^\gamma(p_1 + \cdots)$ hold, then these equations yield Archies law (5.21) with a nonuniversal cementation index $m = \beta + \gamma$. If the condition (5.51) does not hold and $\lambda(\phi)\mu(\phi) \propto \phi^{-\alpha}$, then

$$m = \beta + \gamma\left(1 + \frac{\alpha}{1-\alpha}\right) \tag{5.53}$$

which is even less universal. The validity of the expansion $p(\bar{\phi}) \propto p_c + \bar{\phi}^\gamma(p_1 + \cdots)$ has been tested by experiment [173, 174]. Figure 20 shows the function $p(\bar{\phi})$ obtained for sintering of glass beads. The measured data are the points, the solid curve represents a fit $p(\bar{\phi}) = 1.51\bar{\phi}^{0.45}$ through the data. This fit was chosen to indicate that the consolidation process of sintering glass beads is expected to show a percolation transition at small but finite threshold $\bar{\phi}_c$ [173, 174]. Note, however, that the data of Fig. 20 are consistent with the form $p(\bar{\phi}) \propto p_c + \bar{\phi}(p_1 + \cdots)$ corresponding to $\gamma = 1$.

5. Dielectric Dispersion and Enhancement

The theoretical mixing laws for the frequency dependent dielectric function discussed in Section V.B.3 can be compared with experiment. Spectral theories generally give good fits to the experimental data [287, 293] but do not allow a geometrical interpretation. Geometrical theories on the other hand contain independently observable geometric characteristics, and can be falsified by experiment.

The single parameter mean-field theories Eqs. (5.12), (5.26) and (5.28) contain only the bulk porosity as a geometrical quantity. They are generally unable to reproduce the observed dielectric dispersion and enhancement. This is illustrated in Fig. 21, which shows the experimental measurements of the real part of the frequency dependent dielectric function as solid circles [175]. The results were obtained for a brine

---

[6] The statement in [41] that local porosity theory predicts Archies law with a universal cementation index is incorrect.

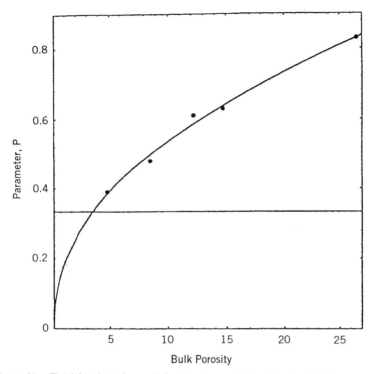

Figure 20. Total fraction of percolating local geometries p for the sintering process of 90 μm glass beads as parametrized by the bulk porosity $\bar{\phi}$ [173].

saturated sample of sintered 250-μm glass spheres. The porosity of the specimen was 10.7%, and the water conductivity was 12.4 mS m^{-1}. A cross-sectional image of the pore space has been displayed in Fig. 13. The frequency in Fig. 21 is dimensionless and measured in units of the relaxation frequency of water $\omega_W = \sigma'_W/(\varepsilon_0\varepsilon'_W)$ where σ'_W and ε'_W are the conductivity and dielectric constant of water, which are constant over the frequency range of interest.[7] With the porosity known from independent measurements the simple mean-field mixing laws can be tested without adjustable parameters. The prediction of the Clausius–Mossotti approximation (5.26) is shown as the solid line with water as the uniform background, and as the dashed line with glass as background. The prediction of the symmetrical effective medium theory [Eq. (5.27)] is shown as the dotted line. The asymmetrical (differential) effective

---

[7] For water with $\sigma'_W = 12.4$ mS m^{-1} the relaxation frequency is $\omega_W = 2.82$ MHz.

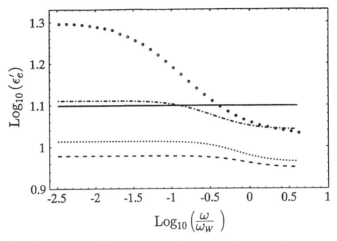

Figure 21. Comparison of simple mean-field theory predictions with the experimentally observed dielectric response of sintered glass beads as a function of frequency. The solid circles are the experimental results for a 10.7% sample of sintered 250-μm glass beads filled with water having $\sigma'_w = 12.4$ mS m^{-1}. The solid line is the Clausius–Mossotti prediction [Eq. (5.26)] with water background, the dashed line corresponds to the same with glass background. The dotted line is the symmetrical effective medium theory [Eq. (5.27)], the dash–dotted line is the asymmetrical effective medium theory [Eq. (5.28)] [175].

medium scheme (5.28), shown as the dash–dotted curve, appears to reproduce the high-frequency behavior correctly, but does not give the low-frequency enhancement.

To compare the experimental observations with the Sen–Scala–Cohen model [Eq. (5.34)] or with the uniform spheroid model [Eq. (5.35)] the depolarization factor L of the ellipsoids has to be treated as a free-fit parameter [175]. In the case of local porosity theory the local porosity distributions $\mu(\phi; \mathbb{K})$ have been measured independently from cross sections through the pore space using image processing techniques. The resulting distributions have been displayed in Fig. 15 for different sizes L of the cubic measurement cells.[8] The local percolation probability function $\lambda(\phi)$ on the other hand has not yet been measured directly from pore-space reconstruction. Instead, the result for $p(\bar{\phi})$ displayed as the power law fit in Fig. 20 was combined with the fact that $\lambda(\bar{\phi}) = p(\bar{\phi})$ in the limit of large measurement cells in which $\mu(\phi)$ becomes a δ distribution concentrated at $\bar{\phi}$ [see Eq. (3.32) or (3.35)]. These observa-

---

[8] The side length of the measurement cell and the depolarization factor have been denoted by the same symbol L. Their distinction should be clear from the context.

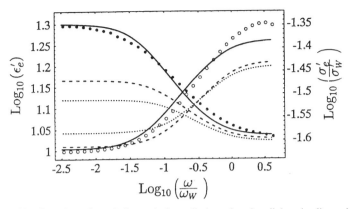

Figure 22. Log–log plot of theoretical predictions for the dielectric dispersion with experimental observations of the effect. Solid circles represent experimental measurements of the effective real dielectric constant and the effective real conductivity as a function of frequency. The dielectric function decreases from left to right, and its values are read from the labels on the left axis. The conductivity increases from left to right and is indicated on the right axis. The dotted line is obtained as a best fit to Eq. (5.34) with $L = 0.387$ for the depolarization factor. The dashed line corresponds to the uniform spheroid model [Eq. (5.34)] with $L = 0.597$ corresponding to an aspect ratio of 2.58 for the oblate spheroid. The solid lines are obtained from local porosity theory (5.36) using $\zeta = 0.2035$ [Eq. (5.28)] [175].

tions and measurements motivate the Ansatz $\lambda(\bar{\phi}) = \bar{\phi}^{\zeta}$ treating ζ as a single free-fit parameter. Fig. 22 shows fits for the frequency dependent real dielectric function $\bar{\varepsilon}'(\omega)$ and inverse formation factor $F^{-1}(\omega)$. In Fig. 22 the circles are the experimental results. All curves represent one parameter fits to the experimental data. The solid curves are obtained from local porosity theory [Eq. (5.36)] using $\zeta = 0.2035$ as the best value of the fit parameter. The local porosity distributions were those of Fig. 15 for measurement cells of side length 50 pixels. The dashed curves correspond to the uniform spheroid model [Eq. (5.35)] with a depolarization factor of 0.597 as the best value of its fit parameter. The dotted curves represent the Sen–Scala–Cohen model with a best value of 0.387 for the depolarization.

Similar experimental results for the dielectric dispersion have been observed in natural rock samples [292]. Figure 8 in [292] compares the measurements only to the uniform spheroid model. Similar to the results of [175] on sintered glass beads the uniform spheroid model did not reproduce the dielectric enhancement, and required too high aspect ratios to be realistic for the observed microstructure.

Local porosity theory has also been used to estimate the broadening of the dielectric relaxation of polymers blends [174].

C. Single-Phase Fluid Flow

1. Permeability and Darcy's Law

The permeability is the most important physical property of a porous medium in much the same way as the porosity is its most important geometrical property. Some authors define porous media as media with a nonvanishing permeability [2]. Permeability measures quantitatively the ability of a porous medium to conduct fluid flow. The permeability tensor $\mathbf{K}$ relates the macroscopic flow density $\bar{\mathbf{v}}$ to the applied pressure gradient ∇P or external field $\mathbf{F}$ through

$$\bar{\mathbf{v}} = \frac{\mathbf{K}}{\eta}(\mathbf{F} - \nabla P) \tag{5.54}$$

where η is the dynamic viscosity of the fluid. The parameter $\bar{\mathbf{v}}$ is the volumetric flow rate per unit area of cross section. Equation (5.54) is known as *Darcy's law*.

The permeability has dimensions of an area, and it is measured in units of Darcy (d). If the pressure is measured in physical atmospheres one has $1\text{d} = 0.9869\ \mu\text{m}^2$ while $1\text{d} = 1.0197\ \mu\text{m}^2$ if the pressure is measured in technical atmospheres. To within practical measuring accuracy one may often assume $1\text{d} = 10^{-12}\ \text{m}^2$. An important question arising from the fact that $\mathbf{K}$ is dimensionally an area concerns the interpretation of this area or length scale in terms of the underlying geometry. This fundamental question has recently found renewed interest [4, 43, 170, 172, 318–320]. Unfortunately, most answers proposed in these discussions [4, 318–320] give a dynamic rather than geometrical interpretation of this length scale. The traditional answer to this basic problem is provided by hydraulic radius theory [2, 3]. It gives a geometrical interpretation that is based on the capillary models of Section III.B.1, and it will be discussed in Section V.C.2.

The permeability does not appear in the microscopic Stokes or Navier–Stokes equations. Darcy's law and with it the permeability concept can be derived from microscopic Stokes flow equations using homogenization techniques [38, 268–271, 321], which are asymptotic expansions in the ratio of microscopic to macroscopic length scales. The derivation will be given in Section V.C.3 below.

The linear Darcy law holds for flows at low Reynolds numbers in which the driving forces are small and balanced only by the viscous forces. Various nonlinear generalizations of Darcy's law have also been derived using homogenization or volume averaging methods [1, 38, 268, 269, 271, 321–325]. If a nonlinear Darcy law governs the flow in a given experiment this would appear in the measurement as if the permeability becomes velocity dependent. The linear Darcy law breaks down also if the flow becomes too slow. In this case interactions between the fluid and the pore walls become important. Examples occur during the slow movement of polar liquids or electrolytes in finely porous materials with high specific internal surface.

2. Hydraulic Radius Theory

The hydraulic radius theory or Carman–Kozeny model is based on the geometrical models of capillary tubes discussed above in Section III.B.1. In such capillary models the permeability can be obtained exactly from the solution of the Navier–Stokes Eq. (4.9) in the capillary. Consider a cylindrical capillary tube of length L and radius a directed along the x direction. The velocity field $\mathbf{v}(\mathbf{r})$ for creeping laminar flow is of the form $\mathbf{v}(\mathbf{r}) = v(r)\mathbf{e}_x$, where $\mathbf{e}_x$ denotes a unit vector along the pipe, and r measures the distance from the center of the pipe. The pressure has the form $P(\mathbf{r}) = P(x)\mathbf{e}_x$. Assuming "no-slip" boundary conditions, $v(a) = 0$, at the tube walls one obtains for $v(r)$ the familiar Hagen–Poiseuille result [326]

$$P(x) = P(0) - (P(0) - P(L))\frac{x}{L} \tag{5.55}$$

$$v(r) = \frac{P(0) - P(L)}{4\eta L}(a^2 - r^2) \tag{5.56}$$

with a parabolic velocity and linear pressure profile. The volume flow rate Q is obtained through integration as

$$Q = \int_0^a v(r)2\pi r\, dr = \frac{\pi a^4}{8\eta}\frac{P(0) - P(L)}{L} \tag{5.57}$$

Consider now the capillary tube model of Section III.B.1 with a cubic sample space $\mathbb{S}$ of side length L. The pore space $\mathbb{P}$ consists of N nonintersecting capillary tubes of radii a_i and lengths L_i distributed according to a joint probability density $\Pi(a, L)$. The pressure drop must then be calculated over the length L_i and thus the right-hand side of Eq. (5.57) is multiplied by a factor L/L_i. Because the tubes are nonintersecting, the volume flow Q_i through each of the tubes can be added to give the macroscopic volume flow rate per unit area $\bar{v} = (1/L^2) \sum_{i=1}^N Q_i$. Thus the permeability of the capillary tube model is simply additive, and it reads

$$k = \frac{\pi}{8L} \sum_{i=1}^{N} \frac{a_i^4}{L_i} \qquad (5.58)$$

Dimensional analysis of Eqs. (3.58), (3.59), and (5.58) shows that kS^2/ϕ^3 is dimensionless. Averaging Eq. (5.58) as well as Eqs. (3.58) and (3.59) for the porosity and specific internal surface of the capillary tube model yields the relation

$$\langle k \rangle = \frac{C}{2} \frac{\langle \phi \rangle^3}{\langle S \rangle^2} \qquad (5.59)$$

where the mixed-moment ratio

$$C = L^2 \left\langle \frac{a^4}{L} \right\rangle \frac{\langle aL \rangle^2}{\langle a^2 L \rangle^3} \qquad (5.60)$$

is a dimensionless number, and the angular brackets denote as usual the average with respect to $\Pi(a, L)$.

The *hydraulic radius theory* or Carman–Kozeny model is obtained from a mean-field approximation which assumes $\langle f(x) \rangle \approx f(\langle x \rangle)$. The approximation becomes exact if the distribution is sharply peaked or if $L_i = L$ and $a_i = a$ for all N. With this approximation the average permeability $\langle k \rangle$ may be rewritten in terms of the average hydraulic radius $\langle R_H \rangle$ defined in Eq. (3.66) as

$$\langle k \rangle \approx \frac{\langle \phi \rangle}{2 \langle \mathcal{T} \rangle^2} \frac{\langle \phi \rangle^2}{\langle S \rangle^2} \approx \frac{\langle \phi \rangle}{2 \langle \mathcal{T} \rangle^2} \left\langle \frac{\phi^2}{S^2} \right\rangle \approx \frac{\langle \phi \rangle \langle R_H \rangle^2}{2 \langle \mathcal{T} \rangle^2} \approx \frac{\langle \phi \rangle \langle a \rangle^2}{8 \langle \mathcal{T} \rangle^2}$$

$$(5.61)$$

where $\langle \mathcal{T} \rangle = \langle L \rangle / L$ is the average of the tortuosity defined above in Eq. (3.62). Equation (5.61) is one of the main results of hydraulic radius theory. The permeability is expressed as the square of an average hydraulic radius $\langle R_H \rangle$, which is related to the average "pore width" as $\langle R_H \rangle = \langle a \rangle / 2$.

It must be stressed that hydraulic radius theory is not exact even for the simple capillary tube model because in general $\langle R_H \rangle \neq \langle \phi \rangle / \langle S \rangle$ and $C \neq \langle \mathcal{T} \rangle^2$. However, interesting exact relations for the average permeability can be obtained from Eqs. (5.59) and (5.60) in various special cases without employing the mean-field approximation of hydraulic radius theory. If the tube radii and lengths are independent, then the distribution factorizes as $\Pi(a, L) = \Pi_a(a)\Pi_L(L)$. In this case the permeability may be written as

$$\langle k \rangle = \frac{1}{2} \frac{\langle 1/\mathcal{T} \rangle}{\langle \mathcal{T} \rangle} \frac{\langle a^4 \rangle \langle a \rangle^2}{\langle a^2 \rangle^3} \frac{\langle \phi \rangle^3}{\langle S \rangle^2} = \frac{\langle \phi \rangle}{8} \frac{\langle 1/\mathcal{T} \rangle}{\langle \mathcal{T} \rangle} \frac{\langle a^4 \rangle}{\langle a^2 \rangle} \quad (5.62)$$

where $\langle \mathcal{T} \rangle$ is the average of the tortuosity factor defined in Eq. (3.62). The last equality interprets $\langle k \rangle$ in terms of the microscopic effective cross section $\langle a^4 \rangle / \langle a^2 \rangle$ determined by the variance and curtosis of the distribution of tube radii. Further specialization to the cases $L_i = l$ or $a_i = a$ is readily carried out from these results.

Finally, it is of interest to consider also the capillary slit model of Section III.B.1. The model assumes again a cubic sample of side length L containing a pore space consisting of parallel slits with random widths governed by a probability density $\Pi(b)$. For flat planes without undulations the analogue of tortuosity is absent. The average permeability is obtained in this case as

$$\langle k \rangle = \frac{1}{3} \frac{\langle b^3 \rangle}{\langle b \rangle^3} \frac{\langle \phi \rangle^3}{\langle S \rangle^2} \quad (5.63)$$

which has the same form as Eq. (5.59) with a constant $C = \langle b^3 \rangle / \langle b \rangle^3$. The prefactor one-third is due to the different shape of the capillaries, which are planes rather than tubes.

3. Derivation of Darcy's Law from Stokes Equation

The previous section V.C.2 has shown that Darcy's law arises in the capillary models. This raises the question of whether it can be derived more generally. This section shows that Darcy's law can be obtained from Stokes equation for a slow flow. It arises to lowest order in an asymptotic expansion whose small parameter is the ratio of microscopic to macroscopic length scales.

Consider the stationary and creeping (low Reynolds number) flow of a Newtonian incompressible fluid through a porous medium whose matrix is assumed to be rigid. The microscopic flow through the pore space $\mathbb{P}$ is governed by the stationary Stokes equations for the velocity $\mathbf{v}(\mathbf{r})$ and pressure $P(\mathbf{r})$

$$\eta \, \Delta\mathbf{v}(\mathbf{r}) + \mathbf{F} - \nabla P(\mathbf{r}) = 0 \quad (5.64)$$

$$\nabla^T \cdot \mathbf{v}(\mathbf{r}) = 0 \quad (5.65)$$

inside the pore space, $\mathbb{P} \ni \mathbf{r}$, with no slip boundary condition

$$\mathbf{v}(\mathbf{r}) = \mathbf{0} \qquad (5.66)$$

for $\mathbf{r} \in \partial\mathbb{P}$. The body force $\mathbf{F}$ and the dynamic viscosity η are assumed to be constant.

The derivation of Darcy's law assumes that the pore space $\mathbb{P}$ has a characteristic length scale ℓ, which is small compared to some macroscopic scale L. The microscopic scale ℓ could be the diameter of grains, the macroscale L could be the diameter of the sample $\mathbb{S}$ or some other macroscopic length such as the diameter of a measurement cell or the wavelength of a seismic wave. The small ratio $\varepsilon = l/L$ provides a small parameter for an asymptotic expansion. The expansion is constructed by assuming that all properties and fields can be written as functions of two new space variables $\mathbf{x}, \mathbf{y}$, which are related to the original space variable $\mathbf{r}$ as $\mathbf{x} = \mathbf{r}$ and $\mathbf{y} = \mathbf{r}/\varepsilon$. All functions $f(\mathbf{r})$ are now replaced with functions $f(\mathbf{x}, \mathbf{y})$ and the slowly varying variable $\mathbf{x}$ is allowed to vary independently of the rapidly varying variable $\mathbf{y}$. This requires to replace the gradient according to

$$\nabla f(\mathbf{r}) = \nabla f\left(\mathbf{r}, \frac{\mathbf{r}}{\varepsilon}\right) = \nabla_\mathbf{x} f(\mathbf{x}, \mathbf{y}) + \frac{1}{\varepsilon}\nabla_\mathbf{y} f(\mathbf{x}, \mathbf{y}) \qquad (5.67)$$

and the Laplacian is replaced similarly. The velocity and pressure are now expanded in ε where the leading orders are chosen such that the solution is not reduced to the trivial zero solution and the problem remains physically meaningful. In the present case this leads to the expansions [268, 271, 280]

$$\mathbf{v}(\mathbf{r}) = \varepsilon^2 \mathbf{v}_0(\mathbf{x}, \mathbf{y}) + \varepsilon^3 \mathbf{v}_1(\mathbf{x}, \mathbf{y}) + \cdots \qquad (5.68)$$

$$P(\mathbf{r}) = P_0(\mathbf{x}, \mathbf{y}) + \varepsilon P_1(\mathbf{x}, \mathbf{y}) + \cdots \qquad (5.69)$$

where $\mathbf{x} = \mathbf{r}$ and $\mathbf{y} = \mathbf{r}/\varepsilon$. Inserting into Eqs. (5.64)–(5.66) yields to lowest order in ε the system of equations

$$\nabla_\mathbf{y} P_0(\mathbf{x}, \mathbf{y}) = 0 \quad \text{in } \mathbb{P} \qquad (5.70)$$

$$\nabla_\mathbf{y}^T \cdot \mathbf{v}_0 = 0 \quad \text{in } \mathbb{P} \qquad (5.71)$$

$$\eta \, \Delta_\mathbf{y} \mathbf{v}_0 - \nabla_\mathbf{y} P_1 - \nabla_\mathbf{x} P_0 + \mathbf{F} = 0 \quad \text{in } \mathbb{P} \qquad (5.72)$$

$$\nabla_\mathbf{x}^T \cdot \mathbf{v}_0 + \nabla_\mathbf{y}^T \cdot \mathbf{v}_1 = 0 \quad \text{in } \mathbb{P} \qquad (5.73)$$

$$\mathbf{v}_0 = \mathbf{0} \quad \text{on } \partial \mathbb{P} \tag{5.74}$$

in the fast variable $\mathbf{y}$. It follows from the first equation that $P_0(\mathbf{x}, \mathbf{y})$ depends only on the slow variable $\mathbf{x}$, and thus it appears as an additional external force for the determination of the dependence of $\mathbf{v}_0(\mathbf{x}, \mathbf{y})$ on $\mathbf{y}$ from the remaining equations. Because the equations are linear the solution $\mathbf{v}_0(\mathbf{x}, \mathbf{y})$ has the form

$$\mathbf{v}_0(\mathbf{x}, \mathbf{y}) = \sum_{i=1}^{3} \left(F_i - \frac{\partial P_0}{\partial x_i} \right) \mathbf{u}_i(\mathbf{x}, \mathbf{y}) \tag{5.75}$$

where the three vectors $\mathbf{u}_i(\mathbf{x}, \mathbf{y})$ (and the scalars $Q_i(\mathbf{x}, \mathbf{y})$) are the solutions of the three systems $(i = 1, 2, 3)$

$$\mathbf{\nabla}_\mathbf{y}^T \cdot \mathbf{u}_i = 0 \quad \text{in } \mathbb{P} \tag{5.76}$$

$$\eta \, \Delta_\mathbf{y} \mathbf{u}_i - \mathbf{\nabla}_\mathbf{y} Q_i - \mathbf{e}_{y_i} = 0 \quad \text{in } \mathbb{P} \tag{5.77}$$

$$\mathbf{u}_i = \mathbf{0} \quad \text{on } \partial \mathbb{P} \tag{5.78}$$

and $\mathbf{e}_{y_i}$ is a unit vector in the direction of the y_i-axis.

It is now possible to average $\mathbf{v}_0$ over the fast variable $\mathbf{y}$. The spatial average over a convex set $\mathbb{K}$ is defined as

$$\bar{\mathbf{v}}_0(\mathbf{x}; \mathbb{K}) = \frac{1}{V(\mathbb{K})} \int \mathbf{v}_0(\mathbf{x}, \mathbf{y}) \chi_\mathbb{K}(\mathbf{x}, \mathbf{y}) \, d^3\mathbf{y} \tag{5.79}$$

where $\mathbb{K}$ is centered at $\mathbf{x}$ and $\chi_\mathbb{K}(\mathbf{x}, \mathbf{y}) = \chi_\mathbb{K}(\mathbf{r}, \mathbf{r}/\varepsilon)$ equals 1 or 0 depending on whether $\mathbf{r} \in \mathbb{K}$ or not. The dependence on the averaging region $\mathbb{K}$ has been indicated explicitly. By using the notation of Eq. (2.20) the average over all space is obtained as the limit $\lim_{s \to \infty} \bar{\mathbf{v}}_0(\mathbf{x}; s\mathbb{K}) = \bar{\mathbf{v}}_0(\mathbf{x})$. The function P_0 need not be averaged as it depends only on the slow variable $\mathbf{x}$. If $\mathbf{v}_0$ is constant then $\overline{\mathbf{v}}_0(\mathbf{x}) = \mathbf{v}_0 \bar{\phi}(\mathbf{x})$, which is known as the law of Dupuit–Forchheimer [1]. Averaging Eq. (5.75) gives Darcy's law (5.54) in the form

$$\bar{\mathbf{v}}_0(\mathbf{x}; \mathbb{K}) = \frac{\mathbf{K}(\mathbf{x}; \mathbb{K})}{\eta} [\mathbf{F} - \mathbf{\nabla}_\mathbf{x} P_0(\mathbf{x})] \tag{5.80}$$

where the components $k_{ij}(\mathbf{x}; \mathbb{K}) = (\mathbf{K}(\mathbf{x}; \mathbb{K}))_{ij}$ of the permeability tensor $\mathbf{K}$ are expressed in terms of the solutions $\mathbf{u}_j(\mathbf{x}; \mathbb{K})$ to Eqs. (5.76)–(5.78) within the region $\mathbb{K}$ as

$$(\mathbf{K}(\mathbf{x}; \mathbb{K}))_{ij} = (\overline{\mathbf{u}_j}(\mathbf{x}; \mathbb{K}))_i \tag{5.81}$$

The permeability tensor is symmetric and positive definite [268]. Its dependence on the configuration of the pore space $\mathbb{P}$ and the averaging region $\mathbb{K}$ have been made explicit because they will play an important role below. For isotropic and strictly periodic or stationary media the permeability tensor reduces to a constant independent of $\mathbf{x}$. For (quasi-) periodic microgeometries or (quasi-)stationary random media averaging Eq. (5.73) leads to the additional macroscopic relation

$$\nabla_{\mathbf{x}}^{T} \cdot \bar{\mathbf{v}}_0(\mathbf{x}; \mathbb{K}) = 0 \qquad (5.82)$$

Equations (5.80) and (5.82) are the macroscopic laws governing the microscopic Stokes flow obeying Eqs. (5.64)–(5.66) to leading order in $\varepsilon = \ell/L$.

The importance of the homogenization technique illustrated here in a simple example lies in the fact that it provides a systematic method to obtain the reference problem for an effective medium treatment.

Many of the examples for transport and relaxation in porous media listed in Section IV can be homogenized using a similar technique [268]. The heterogeneous elliptic equation (4.2) is of particular interest. The linear Darcy flow derived in this section can be cast into the form of Eq. (4.2) for the pressure field. The permeability tensor may still depend on the slow variable $\mathbf{x}$, and it is therefore of interest to iterate the homogenization procedure in order to see whether Darcy's law becomes again modified on larger scales. This question is discussed next.

4. Iterated Homogenization

The permeability $\mathbf{K}(\mathbf{x})$ for the macroscopic Darcy flow was obtained from homogenizing the Stokes equation by averaging the fast variable $\mathbf{y}$ over a region $\mathbb{K}$. The dependence on the slow variable $\mathbf{x}$ allows for macroscopic inhomogeneities of the permeability. This raises the question whether the homogenization may be repeated to arrive at an averaged description for a much larger megascopic scale.

If Eq. (5.80) is inserted into Eq. (5.82) and $\mathbf{F} = 0$ is assumed the equation for the macroscopic pressure field becomes

$$\nabla^{T} \cdot (\mathbf{K}(\mathbf{x})\nabla P(\mathbf{x})) = 0 \qquad (5.83)$$

which is identical with Eq. (4.2). The equation must be supplemented with boundary conditions that can be obtained from the requirements of mass and momentum conservation at the boundary of the region for which Eq. (5.83) was derived. If the boundary marks a transition to a region with different permeability the boundary conditions require continuity of pressure and normal component of the velocity.

Equation (5.83) holds at length scales L much larger than the pore scale ℓ, and much larger than the diameter of the averaging region $\mathbb{K}$. To homogenize it one must therefore consider length scales $\mathscr{L}$ much larger than ℓ such that

$$\ell \ll L \ll \mathscr{L} \tag{5.84}$$

is fulfilled. The ratio $\delta = L/\mathscr{L}$ is then a small parameter in terms of which the homogenization procedure of Section V.C.3 can be iterated. The pressure is expanded in terms of δ as

$$P(\mathbf{x}) = P_0(\mathbf{s}, \mathbf{z}) + \delta P_1(\mathbf{s}, \mathbf{z}) + \cdots \tag{5.85}$$

where now $\mathbf{s} = \mathbf{x}$ is the slow variable, and $\mathbf{z} = \mathbf{s}/\delta$ is the rapidly varying variable. Assuming that the medium is stationary, that is, $\mathbf{K}(\mathbf{z})$ does not depend on the slow variable $\mathbf{s}$, the result becomes [268, 271, 280]

$$\nabla^T \cdot (\bar{\mathbf{K}} \nabla P_0(\mathbf{s})) = 0 \tag{5.86}$$

where $P_0(\mathbf{s})$ is the first term in the expansion of the pressure that is independent of $\mathbf{z}$, and the tensor $\bar{\mathbf{K}}$ has components

$$(\bar{\mathbf{K}})_{ij} = \overline{k_{ij}(\mathbf{z}) + \sum_{l=1}^{3} k_{il}(\mathbf{z}) \frac{\partial Q_j(\mathbf{z})}{\partial z_l}} \tag{5.87}$$

given in terms of three scalar fields $Q_j (j = 1, 2, 3)$, which are obtained from solving an equation of the form

$$-\sum_{i,j} \frac{\partial}{\partial z_i} \left(k_{ij}(\mathbf{z}) \frac{\partial Q_k(\mathbf{z})}{\partial z_j} \right) = \sum_i \frac{\partial k_{ik}(\mathbf{z})}{\partial z_i} \tag{5.88}$$

analogous to Eqs. (5.76)–(5.78) in the homogenization of Stokes equation.

If the assumption of strict stationarity is relaxed the averaged permeability depends in general on the slow variable, and the homogenized equation (5.86) then has the same form as the original Eq. (5.83). This shows that the form of the macroscopic equation does not change under further averaging. This highlights the importance of the averaged permeability as a key element of every macroscopically homogeneous description. Note, however, that the averaged tensor $\bar{\mathbf{K}}$ may have a different symmetry than the original permeability. If $\mathbf{K}(\mathbf{x}) = k(\mathbf{x})\mathbf{1}$ is isotropic (**1** denotes the unit matrix) then $\bar{\mathbf{K}}$ may become anisotropic because of the second term appearing in Eq. (5.87).

5. Network Model

Consider a porous medium described by Eq. (5.83) for Darcy flow with a stationary and isotropic local permeability function $\mathbf{K}(\mathbf{x}) = k(\mathbf{x})\mathbf{1}$. The expressions (5.87) and (5.88) for the effective permeability tensor $\bar{\mathbf{K}}$ are difficult to use for general random microstructures. Therefore, it remains necessary to follow the strategy outlined in Section V.A.2 and to discretize Eq. (5.83) using a finite difference scheme with lattice constant L. As before, it is assumed that $\ell \ll L \ll \mathscr{L}$, where ℓ is the pore scale and $\mathscr{L}$ is the system size. The discretization results in the linear network equations (5.5) for a regular lattice with lattice constant L.

To make further progress it is necessary to specify the local permeabilities. A microscopic network model of tubes results from choosing the expression

$$k(a, \ell, L) = \frac{\pi}{8L} \frac{a^4}{\ell} \qquad (5.89)$$

for a cylindrical capillary tube of radius a and length ℓ in a region of size L. The parameters a and ℓ must obey the geometrical conditions $a \le L/2$ and $\ell \ge L$. In the resulting network model each bond represents a winding tube with circular cross section whose diameter and length fluctuate from bond to bond. The network model is completely specified by assuming that the local geometries specified by a and ℓ are independent and identically distributed random variables with joint probability density $\Pi(a, \ell)$. Note that the probability density $\Pi(a, \ell)$ depends also on the discretization length through the constraints $a \le L/2$ and $\ell \ge L$.

By using the effective medium approximation to the network equations the effective permeability $\bar{k}$ for this network model is the solution of the self-consistency equation

$$\int_L^\infty \int_0^{L/2} \frac{\pi a^4 - 8L\ell\bar{k}}{\pi a^4 + 16L\ell\bar{k}} \Pi(a, \ell) \, da \, d\ell = 0 \qquad (5.90)$$

where the restrictions on a and ℓ are reflected in the limits of integration. In simple cases, as for binary or uniform distributions, this equation can be solved analytically, in other cases it is solved numerically. The effective medium prediction agrees well with an exact solution of the network equations [231]. The behavior of the effective permeability depends qualitatively on the fraction p of conducting tubes defined as

$$p = 1 - \lim_{\varepsilon \to 0} \int_0^\varepsilon \Pi(a) \, da \qquad (5.91)$$

where $\Pi(a) = \int_L^\infty \Pi(a, \ell) \, d\ell$. For $p > \frac{1}{3}$ the permeability is positive while for $p < \frac{1}{3}$ it vanishes. At $p = p_c = \frac{1}{3}$ the network has a percolation transition. Note that $p \neq \bar{\phi}$ is not related to the average porosity.

6. Local Porosity Theory

Consider, as in Section V.C.5, a porous medium described by Eq. (5.83) for Darcy flow with a stationary and isotropic local permeability function $\mathbf{K}(\mathbf{x}) = k(\mathbf{x})\mathbf{1}$. A glance at Section III shows that the one-cell local geometry distribution defined in Eq. (3.45) is particularly well adapted to the discretization of Eq. (5.83). As before, the discretization employs a cubic lattice with lattice constant L and cubic measurement cells $\mathbb{K}$ and yields a local geometry distribution $\mu(\phi, S; \mathbb{K})$. It is then natural to use the Carman equation (5.59) locally because it is often an accurate description as illustrated in Fig. 23. The straight line in Fig. 23 corresponds to Eq. (5.59). The local percolation probabilities defined in Section III.A.5.d complete the description. Each local geometry is characterized by its local porosity, specific internal surface, and a binary random variable indicating whether the geometry is percolating or not. The self-consistent effective medium equation now reads

$$\int_0^\infty \int_0^1 \frac{3C\phi^3 \lambda(\phi, S; \mathbb{K}) \mu(\phi, S; \mathbb{K})}{C\phi^3 + 4S^2 \bar{k}} \, d\phi \, dS = 1 \qquad (5.92)$$

for the effective permeability $\bar{k}$. The control parameter for the underlying percolation transition was given in Eq. (3.47) as

$$p(L) = \int_0^\infty \int_0^1 \lambda(\phi, S; \mathbb{K}) \mu(\phi, S; \mathbb{K}) \, d\phi \, dS \qquad (5.93)$$

and it gives the total fraction of percolating local geometries. If the quantity

$$k_0 = \left(\int_0^\infty \int_0^1 \frac{2S^2}{C\phi^3} \lambda(\phi, S; \mathbb{K}) \mu(\phi, S; \mathbb{K}) \, d\phi \, dS \right)^{-1} \qquad (5.94)$$

is finite then the solution to Eq. (5.92) is given approximately as

$$\bar{k} \approx k_0(p - p_c) \qquad (5.95)$$

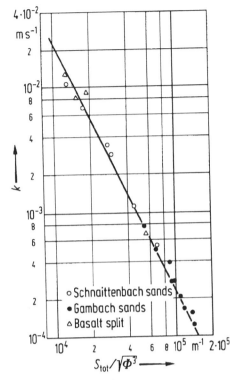

Figure 23. Log–log plot of the permeability coefficient $kg\rho/\eta$, where k is the permeability, g is the acceleration of gravity, ρ is the fluid density, and η is the fluid viscosity against the combination $S/(\phi)^{3/2}$ of porosity ϕ and specific surface S for sands and basalt split. (Reproduced with permission from J. Schopper, "Porosität und Permeabilität," in *Landolt–Börnstein: Physikalische Eigenschaften der Gesteine*, K.-H. Hellwege, Ed., Vol. V/1a, Springer, Berlin, 1982, p. 184. Copyright © Springer-Verlag, 1982.)

for $p > p_c = \frac{1}{3}$ and as $\bar{k} = 0$ for $p < p_c$. This result is analogous to Eq. (5.48) for the electrical conductivity. Note that the control parameter for the underlying percolation transition differs from the bulk porosity $p \neq \bar{\phi}$.

To study the implications of Eq. (5.92) it is necessary to supply explicit expressions for the local geometry distribution $\mu(\phi, S; L)$. Such an expression is provided by the local porosity reduction model reviewed in Section III.B.6. By writing the effective medium approximation for the number $\bar{n}$ defined in Eq. (3.87) and using Eqs. (3.86) and (3.88) it has been shown that the effective permeability may be written approximately as [170]

$$\bar{k} = \bar{\phi}^{\beta} \lambda(\bar{\phi}) \tag{5.96}$$

where the exponent β depends on the porosity reduction factor r and the type of consolidation model characterized by Eq. (3.88) as

$$\beta = (3 - 2\alpha)\frac{\ln r}{r - 1} \tag{5.97}$$

If all local geometries are percolating, that is, if $\lambda = 1$, then the effective permeability depends algebraically on the bulk porosity $\bar{\phi}$ with a *strongly nonuniversal exponent* β. This dependence will be modified if the local percolation probability $\lambda(\bar{\phi})$ is not constant. The large variability is consistent with experience from measuring permeabilities in experiment. Figure 24 demonstrates the large data scatter seen in experimental results. While in general small permeabilities correlate with small porosities the correlation is not very pronounced.

D. Permeability Length Scales

The fact that the effective permeability $\bar{k}$ has dimensions of area raises the question whether $\sqrt{\bar{k}}$ has an interpretation as a length scale. The traditional answer to this question is provided by hydraulic radius theory, which uses the approximate result of Eq. (5.61) for the capillary tube

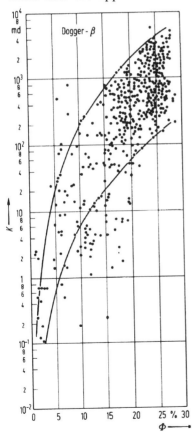

Figure 24. Logarithmic plot of permeability versus porosity for Dogger-β (Jurassic) sandstone. (Reproduced with permission from J. Schopper, "Porosität und Permeabilität," in *Landolt–Börnstein: Physikalische Eigenschaften der Gesteine*, K.-H. Hellwege, Ed., Vol. V/1a, Springer, Berlin, 1982, p. 184. Copyright © Springer-Verlag, 1982.)

model to postulate more generally the relation

$$\bar{k} \propto \frac{\bar{\phi} R_H^2}{2 \mathscr{T}^2} \qquad (5.98)$$

where $R_H = \phi(\mathbb{S})/S(\mathbb{S})$ is the hydraulic radius. This generalization has been modified by incorporating the formation factor to write [112, 327]

$$\bar{k} \propto \frac{\Lambda^2}{F} \qquad (5.99)$$

where the length scale $\Lambda = R_H$ is still given by the hydraulic radius, and the geometrical tortuosity $\mathscr{T}^2$ was replaced by the electrical tortuosity defined as $\mathscr{T}_{el}^2 = F\bar{\phi}$. Because the length scale is still given by the hydraulic radius this theory is still faced with the objection that the hydraulic radius R_H contains contributions from the dead ends that do not contribute to the transport.

An alternative was proposed in [43, 318]. It postulates $\Lambda = l_c$, where l_c is a length scale related to the breakthrough pressure in mercury injection experiments. The length scale l_c is well defined for network models with a broad distribution of cylindrical pores. A dynamical interpretation of Λ was proposed in [319, 320, 328] as

$$\frac{2}{\Lambda} = \frac{\int |\mathbf{E}(\mathbf{r})|^2 \chi_{\partial \mathbb{P}}(\mathbf{r}) \, d^2 \mathbf{r}}{\int |\mathbf{E}(\mathbf{r})|^2 \chi_{\mathbb{P}}(\mathbf{r}) \, d^3 \mathbf{r}} \qquad (5.100)$$

where $\mathbf{E}(\mathbf{r})$ is the unknown exact solution of the microscopic dielectric problem. This "electrical length" is expected to measure, somehow, the "dynamically connected pore size" [4, 319, 328]. The interpretation of Λ within local porosity theory is obtained by eliminating $(p - p_c)$ between the result [Eq. (5.48)] for the conductivity, and Eq. (5.95) for the permeability. This generally yields

$$\Lambda^2 \approx \frac{\displaystyle\int_0^\infty \int_0^1 \frac{\lambda(\phi, S; \mathbb{K}) \mu(\phi, S; \mathbb{K})}{\sigma_{loc}(\phi, S)} \, d\phi \, dS}{\displaystyle\int_0^\infty \int_0^1 \frac{\lambda(\phi, S; \mathbb{K}) \mu(\phi, S; \mathbb{K})}{k_{loc}(\phi, S)} \, d\phi \, dS} \qquad (5.101)$$

where $\sigma_{loc}(\phi, S)$ and $k_{loc}(\phi, S)$ are the local electrical conductivity and the local permeability. Thus Λ involves macroscopic geometrical information through μ and λ and microscopic dynamical and geometrical information through the local transport coefficients. If one assumes the hydraulic radius expressions $\sigma_{loc}(\phi, S) \propto \phi$ and $k_{loc}(\phi, S) \propto \phi^3/S^2$ locally and the expression $\mu(\phi, S; \mathbb{K}) \approx \delta(\phi - \bar{\phi})\delta(S - \bar{S})$ is valid for large measurement cells, then it follows that $\Lambda \propto \bar{\phi}/\bar{S}$ becomes the local hydraulic radius [170]. This expression is no longer proportional to the total internal surface but only to the average local internal surface. Thus the argument against hydraulic radius theories no longer apply.

VI.   IMMISCIBLE DISPLACEMENT

This section discusses the transition between microscopic and macroscopic length scales for the flow of two immiscible fluids through a porous medium. Contrary to the previous sections, the macroscopic equations of motion describing the immiscible displacement process are assumed to be known from averaging the microscopic equations. The upscaling problem is addressed by comparing the dimensional analysis of the given microscopic and macroscopic equations. The original dimensional analysis dates back to [49, 329–331], but continues to attract the attention of recent authors [332–336]. This fact indicates the presence of unresolved problems, which can be seen most clearly in capillary desaturation experiments where they appear as large unexplained discrepancies in the macroscopic balance of viscous and capillary forces. Recently, these problems were traced to a tacit assumption underlying the traditional dimensional analysis [47, 48]. This finding could lead to a resolution of the discrepancies and has additional implications for laboratory measurements of relative permeabilities [337], which will be discussed in Section VI.C.3. The dimensional analysis of immiscible displacement will therefore be reviewed in this section providing a basis for quantitative estimates of the relative importance of macroscopic viscous, capillary, and gravitational forces. Such estimates were distorted in the traditional analysis. The revised analysis allows one to predict segregation front widths or gravitational relaxation times for different porous media [47, 48].

A.   Experimental Observations

Consider the displacement of oil from an oil saturated porous medium through injecting water at constant velocity. After steady-state flow conditions are established a certain fraction S_{0r}[9] of oil remains microscopically trapped inside the medium. The trapped oil can be mobilized if the viscous forces overcome the capillary retention forces [333]. Displacement experiments in a variety of porous media including micromodels show a strong correlation between the residual oil saturation S_{0r} and the

---

[9] In all of Section VI the letter S denotes saturation, and should not be confused with the usage in previous sections.

capillary number Ca of the waterflood [2, 28, 203, 332, 338–341]. The *capillary number*, defined as $Ca = \mu u / \sigma$, is the dimensionless ratio of viscous/capillary forces. Here u denotes an average microscopic velocity, μ is the viscosity, and σ is the surface tension between the fluids.

The experimental curves $S_{Or}(Ca)$ are called capillary number correlations, recovery curves, or capillary desaturation curves, and they give the residual oil saturation as a function of the capillary number of the flood. All such capillary desaturation curves exhibit a critical capillary number Ca_c below which the residual oil saturation remains constant. This critical capillary number Ca_c marks the point where the viscous forces equal the capillary forces. Figure 25 shows a schematic drawing of the capillary desaturation curves for unconsolidated sand, sandstone, and limestone (after [28, 203]). Surprisingly, all experimentally observed values for Ca_c are much smaller than 1. For unconsolidated sand Ca_c is often reported to be $Ca_c \approx 10^{-4}$, while for sandstone $Ca_c \approx 3 \times 10^{-6}$ and for limestone $Ca_c \approx 2 \times 10^{-7}$ [28]. The exceedingly small values of Ca_c as well as their dependence on the type of porous medium strongly suggest that the microscopically defined capillary number Ca cannot be an adequate measure of the balance between macroscopic viscous and macroscopic capillary forces.

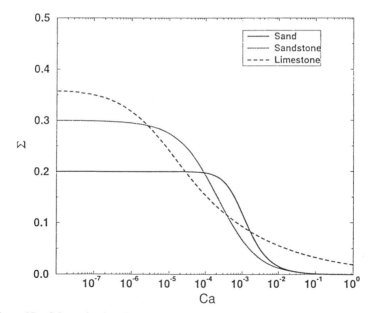

Figure 25. Schematic plot of residual oil saturation $S_{Or}(Ca)$ as function of microscopic capillary number Ca for unconsolidated sands, sandstones, and limestones.

The subsequent sections review recent work that relates the large discrepancy between the observed force balance and the force balance estimated from Ca to an implicit assumption in the traditional dimensional analysis [49, 329–331]. First, the microscopic equations of motion and their dimensional analysis are recalled. This leads to the familiar dimensionless numbers of fluid dynamics. Next, the accepted macroscopic equations of motion are analyzed. This leads to macroscopic dimensionless number, which are then related to the traditional microscopic dimensionless groups. The results are shown to be applicable to the quantitative estimation of residual oil saturation, gravitational relaxation times, and the width of the oil–water contact.

B. Microscopic Description

1. *Microscopic Equations of Motion*

Microscopic equations of motion for two-phase flow in porous media are commonly given as Stokes (or Navier–Stokes) equations for two incompressible Newtonian fluids with no-slip and stress–balance boundary conditions at the interfaces [270, 322, 342]. In the following, the wetting fluid (water) will be denoted by a subscript $\mathbb{W}$ while the nonwetting fluid (oil) is indexed with $\mathbb{O}$. The solid rock matrix, indexed as $\mathbb{M}$, is assumed to be porous and rigid. It fills a closed subset $\mathbb{M} \subset \mathbf{R}^3$ of three-dimensional space. The pore space $\mathbb{P}$ is filled with the two fluid phases described by the two closed subsets $\mathbb{W}(t)$, $\mathbb{O}(t) \subset \mathbf{R}^3$, which are in general time dependent, and related to each other through the condition $\mathbb{P} = \mathbb{W}(t) \cup \mathbb{O}(t)$. Note that $\mathbb{P}$ is independent of time because $\mathbb{M}$ is rigid while $\mathbb{O}(t)$ and $\mathbb{W}(t)$ are not. The rigid rock surface will be denoted as $\partial\mathbb{M}$, and the mobile oil–water interface as $\partial(\mathbb{OW})(t) = \mathbb{O}(t) \cap \mathbb{W}(t)$. A standard formulation of pore-scale equations of motion for two incompressible and immiscible fluids flowing through a porous medium are the Navier–Stokes equations

$$\rho_{\mathbb{W}} \frac{\partial \mathbf{v}_{\mathbb{W}}}{\partial t} + \rho_{\mathbb{W}}(\mathbf{v}_{\mathbb{W}}^T \cdot \boldsymbol{\nabla})\mathbf{v}_{\mathbb{W}} = \mu_{\mathbb{W}} \Delta \mathbf{v}_{\mathbb{W}} + \rho_{\mathbb{W}} g \boldsymbol{\nabla} z - \boldsymbol{\nabla} P_{\mathbb{W}}$$
$$\rho_{\mathbb{O}} \frac{\partial \mathbf{v}_{\mathbb{O}}}{\partial t} + \rho_{\mathbb{O}}(\mathbf{v}_{\mathbb{O}}^T \cdot \boldsymbol{\nabla})\mathbf{v}_{\mathbb{O}} = \mu_{\mathbb{O}} \Delta \mathbf{v}_{\mathbb{O}} + \rho_{\mathbb{O}} g \boldsymbol{\nabla} z - \boldsymbol{\nabla} P_{\mathbb{O}} \tag{6.1}$$

and the incompressibility conditions

$$\boldsymbol{\nabla}^T \cdot \mathbf{v}_{\mathbb{W}} = 0$$
$$\boldsymbol{\nabla}^T \cdot \mathbf{v}_{\mathbb{O}} = 0 \tag{6.2}$$

where $\mathbf{v}_{\mathbb{W}}(\mathbf{x}, t)$, $\mathbf{v}_{\mathbb{O}}(\mathbf{x}, t)$ are the velocity fields for water and oil; $P_{\mathbb{W}}(\mathbf{x}, t)$,

$P_O(\mathbf{x}, t)$ are the pressure fields in the two phases; ρ_W, ρ_O are the densities; μ_W, μ_O are the dynamic viscosities; and g is the gravitational constant. The vector $\mathbf{x}^T = (x, y, z)$ denotes the coordinate vector, t is the time, $\mathbf{\nabla}^T = (\partial/\partial x, \partial/\partial y, \partial/\partial z)$ is the gradient operator, Δ is the Laplacian, and the superscript T denotes transposition. The gravitational force is directed along the z axis and it represents an external body force. Although gravity effects are often small for pore-scale processes [see Eq. (6.37)], there has recently been a growing interest in modeling gravity effects at the pore scale [42, 245, 246, 343].

The microscopic formulation is completed by specifying an initial fluid distribution $\mathbb{W}(t = 0)$, $\mathbb{O}(t = 0)$ and boundary conditions. The latter are usually no-slip boundary conditions at solid–fluid interfaces,

$$
\begin{aligned}
\mathbf{v}_W &= 0 \quad \text{at} \quad \partial \mathbb{M} \\
\mathbf{v}_O &= 0 \quad \text{at} \quad \partial \mathbb{M}
\end{aligned}
\tag{6.3}
$$

as well as for the fluid–fluid interface,

$$
\mathbf{v}_W = \mathbf{v}_O \quad \text{at} \quad \partial(\mathbb{O}\mathbb{W})(t) \tag{6.4}
$$

combined with stress–balance across the fluid–fluid interface,

$$
\boldsymbol{\tau}_W \cdot \mathbf{n} = \boldsymbol{\tau}_O \cdot \mathbf{n} + 2\sigma_{OW}\kappa\mathbf{n} \quad \text{at} \quad \partial(\mathbb{O}\mathbb{W})(t) \tag{6.5}
$$

Here σ_{OW} denotes the water–oil interfacial tension, κ is the curvature of the oil–water interface, and $\mathbf{n}$ is a unit normal to it. The stress tensor $\boldsymbol{\tau}(\mathbf{x}, t)$ for the two fluids is given in terms of $\mathbf{v}$ and P as

$$
\boldsymbol{\tau} = -P\mathbf{1} + \mu\mathscr{S}\mathbf{\nabla}\mathbf{v}^T \tag{6.6}
$$

where the symmetrization operator $\mathscr{S}$ acts as

$$
\mathscr{S}\mathbf{A} = \tfrac{1}{2}(\mathbf{A} + \mathbf{A}^T - \tfrac{2}{3}\operatorname{tr}\mathbf{A}\mathbf{1}) \tag{6.7}
$$

on the matrix $\mathbf{A}$, and $\mathbf{1}$ is the identity matrix.

The pore-space boundary $\partial\mathbb{M}$ is given and fixed while the fluid–fluid interface $\partial(\mathbb{O}\mathbb{W})(t)$ has to be determined self-consistently as part of the solution. For $\mathbb{W} = \emptyset$ or $\mathbb{O} = \emptyset$ the above formulation of two-phase flow at the pore scale reduces to the standard formulation of single-phase flow of water or oil at the pore scale.

2.   The Contact Line Problem

The pore-scale equations of motion given in Section VI.B.1 contain a self-contradiction. The problem arises from the system of contact lines defined as

$$\partial(\mathbb{M}\mathbb{O}\mathbb{W})(t) = \partial\mathbb{M} \cup \partial(\mathbb{O}\mathbb{W})(t) \tag{6.8}$$

on the inner surface of the porous medium. The contact lines must in general slip across the surface of the rock in direct contradiction to the no-slip boundary condition Eq. (6.3). This self-contradiction is not specific for flow in porous media but exists also for immiscible two-phase flow in a tube or in other containers [344–346].

There exist several ways out of this classical dilemma depending on the wetting properties of the fluids. For complete and uniform wetting a microscopic precursor film of water wets the entire rock surface [344]. In that case $\mathbb{M} \cap \mathbb{O}(t) = \emptyset$ and thus

$$\partial(\mathbb{M}\mathbb{O}\mathbb{W})(t) = \{[\mathbb{M} \cap \mathbb{W}(t)] \cup [\mathbb{M} \cap \mathbb{O}(t)]\} \cap [\mathbb{O}(t) \cap \mathbb{W}(t)] = \emptyset \tag{6.9}$$

the problem does not appear.

For other wetting properties a phenomenological slipping model for the manner in which the slipping occurs at the contact line is needed to complete the pore-scale description of two-phase flow. The phenomenological slipping models describe the region around the contact line microscopically. The typical size of this region, called the "slipping length," is around 10^{-9} m. Therefore, the problem of contact lines is particularly acute for immiscible displacement in microporous media, and the Navier–Stokes description of Section VI.B.1 does not apply for such media.

3. Microscopic Dimensional Analysis

Given a microscopic model for contact line slipping the next step is to evaluate the relative importance of the different terms in the equations of motion at the pore scale. This is done by casting them into dimensionless form using the definitions

$$\mathbf{x} = \ell\hat{\mathbf{x}} \tag{6.10}$$

$$\mathbf{\nabla} = \frac{\hat{\mathbf{\nabla}}}{\ell} \tag{6.11}$$

$$\mathbf{v} = u\hat{\mathbf{v}} \tag{6.12}$$

$$t = \frac{\ell\hat{t}}{u} \tag{6.13}$$

$$\kappa = \frac{\hat{\kappa}}{\ell} \tag{6.14}$$

$$P = \frac{\sigma_{OW}}{\ell} \hat{P} \tag{6.15}$$

where ℓ is a microscopic length, u is a microscopic velocity, and $\hat{A}$ denotes the dimensionless equivalent of the quantity A.

With these definitions the dimensionless equations of motion on the pore scale can be written as

$$\frac{\partial \hat{\mathbf{v}}_W}{\partial \hat{t}} + (\hat{\mathbf{v}}_W^T \cdot \hat{\mathbf{\nabla}})\hat{\mathbf{v}}_W = \frac{1}{Re_W} \hat{\Delta}\hat{\mathbf{v}}_W + \frac{1}{Fr^2} \hat{\mathbf{\nabla}}\hat{z} - \frac{1}{We_W} \hat{\mathbf{\nabla}}\hat{P}_W$$

$$\frac{\partial \hat{\mathbf{v}}_O}{\partial \hat{t}} + (\hat{\mathbf{v}}_O^T \cdot \hat{\mathbf{\nabla}})\hat{\mathbf{v}}_O = \frac{1}{Re_O} \hat{\Delta}\hat{\mathbf{v}}_O + \frac{1}{Fr^2} \hat{\mathbf{\nabla}}\hat{z} - \frac{1}{We_O} \hat{\mathbf{\nabla}}\hat{P}_O \tag{6.16}$$

$$\hat{\mathbf{\nabla}}^T \cdot \hat{\mathbf{v}}_W = 0$$

$$\hat{\mathbf{\nabla}}^T \cdot \hat{\mathbf{v}}_O = 0 \tag{6.17}$$

with dimensionless boundary conditions

$$\hat{\mathbf{v}}_W = \hat{\mathbf{v}}_O = 0 \qquad \text{at} \qquad \partial_M \tag{6.18}$$

$$\hat{\mathbf{v}}_W = \hat{\mathbf{v}}_O \qquad \text{at} \qquad \partial_{OW}(t) \tag{6.19}$$

$$(\hat{P}_O - \hat{P}_W)\mathbf{n} = \left(\frac{We_W}{Re_W} \mathscr{S}\hat{\mathbf{\nabla}}\hat{\mathbf{v}}_W - \frac{We_O}{Re_O} \mathscr{S}\hat{\mathbf{\nabla}}\hat{\mathbf{v}}_O \right) \cdot \mathbf{n}$$

$$+ 2\hat{\kappa}\mathbf{n} \qquad \text{at} \ \partial_{OW}(t) \tag{6.20}$$

In these equations the microscopic dimensionless ratio

$$Re_W = \frac{\text{inertial forces}}{\text{viscous forces}} = \frac{\rho_W u\ell}{\mu_W} = \frac{u\ell}{\nu_W^*} \tag{6.21}$$

is the Reynolds number, and

$$\nu_W^* = \frac{\mu_W}{\rho_W} \tag{6.22}$$

is the kinematic viscosity, which may be interpreted as a specific action or a specific momentum transfer. The other fluid dynamic numbers are defined as

$$Fr = \sqrt{\frac{u^2}{g\ell}} = \sqrt{\frac{\text{inertial forces}}{\text{gravitational forces}}} \qquad (6.23)$$

for the Froude number, and

$$We_w = \frac{\rho_w u^2 \ell}{\sigma_{ow}} = \frac{\text{inertial forces}}{\text{capillary forces}} \qquad (6.24)$$

for the Weber number. The corresponding dimensionless ratios for the oil phase are related to those for the water phase as

$$Re_O = Re_w \frac{\rho_O}{\rho_w} \frac{\mu_w}{\mu_O} \qquad (6.25)$$

$$We_O = We_w \frac{\rho_O}{\rho_w} \qquad (6.26)$$

by viscosity and density ratios.

Table IV gives approximate values for densities, viscosities, and surface tensions under reservoir conditions [47, 48]. In the following, these values will be used to make order of magnitude estimates. Typical pore sizes in an oil reservoir are of order $\ell \approx 10^{-4}$ m and microscopic fluid velocities for reservoir floods range around $u \approx 3 \times 10^{-6} \, \text{ms}^{-1}$. Combining these estimates with those of Table IV shows that the dimensionless ratios obey Re_O, Re_w, Fr^2, We_O, $We_w \ll 1$. Therefore, the pore-scale Eqs. (6.16) reduce to the simpler Stokes form

$$0 = \hat{\Delta}\hat{v}_w + \frac{1}{Gr_w} \hat{\nabla}\hat{z} - \frac{1}{Ca_w} \hat{\nabla}\hat{P}_w$$

$$0 = \hat{\Delta}\hat{v}_O + \frac{1}{Gr_O} \hat{\nabla}\hat{z} - \frac{1}{Ca_O} \hat{\nabla}\hat{P}_O \qquad (6.27)$$

TABLE IV

Order of Magnitude Estimates for Densities, Viscosities, and Surface Tension of Oil and Water under Reservoir Conditions

ρ_O	ρ_w	μ_O	μ_w	σ_{ow}
800 kg m^{-3}	1000 kg m^{-3}	0.0018 N m^{-2} s	0.0009 N m^{-2} s	0.035 N m^{-1}

where

$$Ca_W = \frac{We_W}{Re_W} = \frac{\text{viscous forces}}{\text{capillary forces}} = \frac{\mu_W u}{\sigma_{OW}} = \frac{u}{u_W^*} \qquad (6.28)$$

is the microscopic capillary number of water, and

$$Gr_W = \frac{Fr^2}{Re_W} = \frac{\text{viscous forces}}{\text{gravity forces}} = \frac{\mu_W u}{\rho_W g \ell^2} \qquad (6.29)$$

is the microscopic "gravity number" of water. The capillary number is a measure of velocity in units of

$$u_W^* = \frac{\sigma_{OW}}{\mu_W} \qquad (6.30)$$

a characteristic velocity at which the coherence of the oil–water interface is destroyed by viscous forces. The capillary and gravity numbers for the oil phase can again be expressed through density and viscosity ratios as

$$Ca_O = Ca_W \frac{\mu_O}{\mu_W} \qquad (6.31)$$

$$Gr_O = Gr_W \frac{\rho_W}{\rho_O} \frac{\mu_O}{\mu_W} \qquad (6.32)$$

Many other dimensionless ratios may be defined. Of general interest are dimensionless space and time variables. Such ratios are formed as

$$Gl_W = \frac{Ca_W}{Gr_W} = \frac{We_W}{Fr^2} = \frac{\text{gravity forces}}{\text{capillary forces}} = \frac{\rho_W g \ell^2}{\sigma_{OW}} = \frac{\ell^2}{\ell_W^{*\,2}} \qquad (6.33)$$

which has been called the "gravillary number" [47, 48]. The gravillary number becomes the better known bond number if the density ρ_W is replaced with the density difference $\rho_W - \rho_O$. The corresponding length

$$\ell_W^* = \sqrt{\frac{\sigma_{OW}}{\rho_W g}} \qquad (6.34)$$

separates capillary waves with wavelengths below ℓ_W^* from gravity waves with wavelengths above ℓ_W^*. A dimensionless time variable is formed from the gravillary and capillary numbers as

$$\frac{\sqrt{Gl_W}}{Ca_W} = \frac{Re_W}{Fr\sqrt{We_W}} = \frac{(\text{gravity f})^{3/2}}{(\text{capillary f})^{1/2} \times \text{viscous f}}$$

$$= \frac{\sqrt{\rho_W \sigma_{OW} g} t}{\mu_W} = \frac{t}{t_W^*} \tag{6.35}$$

where

$$t_W^* = \frac{\ell_W^*}{u_W^*} = \frac{\mu_W}{\sqrt{\sigma_{OW} \rho_W g}} \tag{6.36}$$

is a characteristic time after which the influence of gravity dominates viscous and capillary effects. The reader is cautioned not to misinterpret the value of t_W^* in Table V as an indication that gravity forces dominate on the pore scale.

Table V collects definitions and estimates for the dimensionless groups and the numbers ℓ^*, u^*, and ν^* characterizing the oil–water system. For these estimates the values in Table IV together with the above estimates of ℓ and u have been used. Table V shows that

$$\text{viscous forces} \ll \text{gravity forces} \ll \text{capillary forces} \tag{6.37}$$

and hence capillary forces dominate on the pore scale [2, 47, 48, 333].

From the Stokes equation (6.27) it follows immediately that for low-capillary number floods ($Ca \ll 1$) the viscous term as well as the shear term in the boundary condition (6.20) become negligible. Therefore, the velocity field drops out, and the problem reduces to finding the equilibrium capillary pressure field. The equilibrium configuration of the oil–water interface then defines time-independent pathways for the flow of oil and water. Hence, for flows with microscopic capillary numbers $Ca \ll 1$ an improved methodology for a quantitative description of immiscible displacement from pore-scale physics requires improved calculations of capillary pressures from the pore scale, and much research is devoted to this topic [246a, 347, 348].

C. Macroscopic Description

1. Macroscopic Equations of Motion

The accepted large-scale equations of motion for two-phase flow involve a generalization of Darcy's law to relative permeabilities including off-diagonal viscous coupling terms [270, 321, 322, 349–352]. The importance

TABLE V
Overview of Definitions and Estimates for Charac-
teristic Microscopic Numbers Describing Oil and
Water Flow under Reservoir Conditions

Quantity	Definition	Estimate
$\mathrm{Re_W}$	$\dfrac{\rho_{\mathrm{W}} u \ell}{\mu_{\mathrm{W}}}$	3.3×10^{-4}
$\mathrm{Ca_W}$	$\dfrac{\mu_{\mathrm{W}} u}{\sigma_{\mathrm{OW}}}$	7.7×10^{-8}
$\mathrm{Gr_W}$	$\dfrac{\mu_{\mathrm{W}} u}{\rho_{\mathrm{W}} g \ell^2}$	2.8×10^{-5}
$\mathrm{Gl_W}$	$\dfrac{\rho_{\mathrm{W}} g \ell^2}{\sigma_{\mathrm{OW}}}$	2.8×10^{-3}
ν_{W}^*	$\dfrac{\mu_{\mathrm{W}}}{\rho_{\mathrm{W}}}$	$9 \times 10^{-7}\,\mathrm{m^2\,s^{-1}}$
u_{W}^*	$\dfrac{\sigma_{\mathrm{OW}}}{\mu_{\mathrm{W}}}$	$38.9\,\mathrm{m\,s^{-1}}$
ℓ_{W}^*	$\sqrt{\dfrac{\sigma_{\mathrm{OW}}}{\rho_{\mathrm{W}} g}}$	$1.9\,\mathrm{cm}$
t_{W}^*	$\dfrac{\mu_{\mathrm{W}}}{\sqrt{\sigma_{\mathrm{OW}} \rho_{\mathrm{W}} g}}$	$4.9 \times 10^{-4}\,\mathrm{s}$

of viscous coupling terms has been recognized relatively late [353–357]. The equations that are generally believed to describe multiphase flow on the reservoir scale as well as on the laboratory scale may be written as [322, 350]

$$\bar{\phi}\,\frac{\partial \bar{S}_{\mathrm{W}}}{\partial \bar{t}} = \bar{\boldsymbol{\nabla}} \cdot \bar{\mathbf{v}}_{\mathrm{W}}$$
$$\bar{\phi}\,\frac{\partial \bar{S}_{\mathrm{O}}}{\partial \bar{t}} = \bar{\boldsymbol{\nabla}} \cdot \bar{\mathbf{v}}_{\mathrm{O}} \tag{6.38}$$

$$\bar{\mathbf{v}}_{\mathrm{W}} = -\left[\mathbf{K}_{\mathrm{WW}}^r\,\frac{\mathbf{K}}{\mu_{\mathrm{W}}}\,(\bar{\boldsymbol{\nabla}}\bar{P}_{\mathrm{W}} - \rho_{\mathrm{W}} g \bar{\boldsymbol{\nabla}}\bar{z}) + \mathbf{K}_{\mathrm{WO}}^r\,\frac{\mathbf{K}}{\mu_{\mathrm{O}}}\,(\bar{\boldsymbol{\nabla}}\bar{P}_{\mathrm{O}} - \rho_{\mathrm{O}} g \bar{\boldsymbol{\nabla}}\bar{z}) \right]$$
$$\bar{\mathbf{v}}_{\mathrm{O}} = -\left[\mathbf{K}_{\mathrm{OW}}^r\,\frac{\mathbf{K}}{\mu_{\mathrm{W}}}\,(\bar{\boldsymbol{\nabla}}\bar{P}_{\mathrm{W}} - \rho_{\mathrm{W}} g \bar{\boldsymbol{\nabla}}\bar{z}) + \mathbf{K}_{\mathrm{OO}}^r\,\frac{\mathbf{K}}{\mu_{\mathrm{O}}}\,(\bar{\boldsymbol{\nabla}}\bar{P}_{\mathrm{O}} - \rho_{\mathrm{O}} g \bar{\boldsymbol{\nabla}}\bar{z}) \right] \tag{6.39}$$

$$\bar{S}_W + \bar{S}_O = 1 \tag{6.40}$$

$$\bar{P}_O - \bar{P}_W = \bar{P}_c(\bar{S}_W) \tag{6.41}$$

where $\bar{A}$ denotes the macroscopic volume averaged equivalent of the pore-scale quantity A. In the equations above $\mathbf{K}$ stands for the absolute (single-phase flow) permeability tensor, $\mathbf{K}^r_{WW}$ is the relative permeability tensor for water, $\mathbf{K}^r_{OO}$ the oil relative permeability tensor, and $\mathbf{K}^r_{WO}$, $\mathbf{K}^r_{OW}$ denote the possibly anisotropic coupling terms. The relative permeabilities are matrix-valued functions of saturation. The saturations are denoted as $\bar{S}_W$, $\bar{S}_O$, and they depend on the macroscopic space and time variables $(\bar{\mathbf{x}}, \bar{t})$. The capillary pressure curve $\bar{P}_c(\bar{S}_W)$ and the relative permeability tensors $\mathbf{K}^r_{ij}(\bar{S}_W)$, i, $j = W$, O must be known either from solving the pore-scale equations of motion, or from experiment. The parameters $\mathbf{K}^r_{ij}(\bar{S}_W)$ and $\bar{P}_c(\bar{S}_W)$ are conventionally assumed to be independent of $\bar{\mathbf{v}}$ and $\bar{P}$ and this convention is followed here, although it is conceivable that this is not generally correct [354].

Eliminating $\bar{\mathbf{v}}$ and choosing $\bar{P}_W(\bar{\mathbf{x}}, \bar{t})$ and $\bar{S}_W(\bar{\mathbf{x}}, \bar{t})$ as the principal unknowns one arrives at the large-scale two-phase flow equations

$$\bar{\phi}\,\frac{\partial \bar{S}_W}{\partial \bar{t}} = \bar{\nabla} \cdot \left\{ \mathbf{K}^r_{WW}(\bar{S}_W)\,\frac{\mathbf{K}}{\mu_W}\,(\bar{\nabla}\bar{P}_W - \rho_W g \bar{\nabla} \bar{z}) \right.$$

$$\left. + \mathbf{K}^r_{WO}(\bar{S}_W)\,\frac{\mathbf{K}}{\mu_O}\,[(\bar{\nabla}\bar{P}_W - \rho_W g \bar{\nabla} \bar{z}) + \bar{\nabla}\bar{P}_c(\bar{S}_W) + (\rho_W - \rho_O)g\bar{\nabla}\bar{z}] \right\} \tag{6.42}$$

$$\bar{\phi}\,\frac{\partial (1 - \bar{S}_W)}{\partial \bar{t}} = \bar{\nabla} \cdot \left\{ \mathbf{K}^r_{OW}(\bar{S}_W)\,\frac{\mathbf{K}}{\mu_W}\,(\bar{\nabla}\bar{P}_W - \rho_W g \bar{\nabla} \bar{z}) \right.$$

$$+ \mathbf{K}^r_{OO}(\bar{S}_W)\,\frac{\mathbf{K}}{\mu_O}\,[(\bar{\nabla}\bar{P}_W - \rho_W g \bar{\nabla} \bar{z}) + \bar{\nabla}\bar{P}_c(\bar{S}_W)$$

$$\left. + (\rho_W - \rho_O)g\bar{\nabla}\bar{z}] \right\} \tag{6.43}$$

for these two unknowns. Equations (6.42) and (6.43) are coupled nonlinear partial differential equations for the large-scale pressure and saturation field of the water phase.

These equations must be complemented with large-scale boundary conditions. For core experiments these are typically given by a surface source on one side of the core, a surface sink on the opposite face, and impermeable walls on the other faces. For a reservoir the boundary conditions depend on the drive configuration and the geological modeling of the reservoir environment, so that Dirichlet as well as von Neumann problems arise in practice [339, 351, 352].

2. Macroscopic Dimensional Analysis

The large-scale equations of motion can be cast in dimensionless form using the definitions

$$\bar{\mathbf{x}} = \bar{\ell}\hat{\bar{\mathbf{x}}} \tag{6.44}$$

$$\bar{\boldsymbol{\nabla}} = \frac{\hat{\bar{\boldsymbol{\nabla}}}}{\bar{\ell}} \tag{6.45}$$

$$\bar{\mathbf{v}} = \bar{u}\hat{\bar{\mathbf{v}}} \tag{6.46}$$

$$\bar{t} = \frac{\bar{\ell}\hat{\bar{t}}}{\bar{u}} \tag{6.47}$$

$$\bar{P} = \bar{P}_b\hat{\bar{P}} \tag{6.48}$$

where as before $\hat{\bar{A}}$ denotes the dimensionless equivalent of the macroscopic quantity $\bar{A}$. The length $\bar{\ell}$ is now a macroscopic length, and $\bar{u}$ a macroscopic (seepage or Darcy) velocity. The pressure $\bar{P}_b$ denotes the "breakthrough" pressure from the capillary pressure curve $\bar{P}_c(\bar{S}_W)$. It is defined as

$$\bar{P}_b = \bar{P}_c(\bar{S}_b) \tag{6.49}$$

where $\bar{S}_b$ is the breakthrough saturation defined as the solution of the equation

$$\frac{d^2\bar{P}_c(\bar{S}_W)}{d\bar{S}_W^2} = 0 \tag{6.50}$$

Thus the dimensionless pressure is defined in terms of the inflection point $(\bar{P}_b, \bar{S}_b)$ on the capillary pressure curve, and it gives a measure of the macroscopic capillary pressure. Note that $\bar{P}_b$ is process dependent, that is, it will in general differ between imbibition and drainage. This dependence reflects the influence of microscopic wetting properties [348] and flow mechanisms on the macroscale [358].

The definition (6.48) differs from the traditional analysis [49, 329–

331]. In the traditional analysis the normalized pressure field is defined as

$$\bar{P} = \frac{\mu_{\mathrm{W}} \bar{u} \bar{\ell}}{k} \, \hat{\bar{P}} \tag{6.51}$$

which immediately gives rise to three problems. First, the permeability is a tensor, and thus a certain nonuniqueness results in anisotropic situations [339]. Second, Eq. (6.51) neglects the importance of microscopic wetting and saturation history dependence. The main problem, however, is that Eq. (6.51) is not based on macroscopic capillary pressures but on Darcy's law, which describes macroscopic viscous pressure effects. On the other hand, the normalization Eq. (6.48) is free from these problems and it includes macroscopic capillarity in the same was as the microscopic normalization Eq. (6.15) includes microscopic capillarity.

With the normalizations introduced above the dimensionless form of the macroscopic two-phase flow equations (6.42) and (6.43) becomes

$$\bar{\phi} \, \frac{\partial \bar{S}_{\mathrm{W}}}{\partial \hat{\bar{t}}} = \hat{\bar{\nabla}} \cdot \left\{ \mathbf{K}^{r}_{\mathrm{WW}}(\bar{S}_{\mathrm{W}})(\overline{\mathbf{Ca}}_{\mathrm{W}}^{-1} \hat{\bar{\nabla}} \hat{\bar{P}}_{\mathrm{W}} - \overline{\mathbf{Gr}}_{\mathrm{W}}^{-1} \hat{\bar{\nabla}} \hat{z}) \right.$$

$$+ \mathbf{K}^{r}_{\mathrm{WO}}(\bar{S}_{\mathrm{W}}) \frac{\mu_{\mathrm{W}}}{\mu_{\mathrm{O}}} \left[(\overline{\mathbf{Ca}}_{\mathrm{W}}^{-1} \hat{\bar{\nabla}} \hat{\bar{P}}_{\mathrm{W}} - \overline{\mathbf{Gr}}_{\mathrm{W}}^{-1} \bar{\nabla} \hat{z}) \right.$$

$$\left. \left. + \overline{\mathbf{Ca}}_{\mathrm{W}}^{-1} \hat{\bar{\nabla}} \hat{\bar{P}}_{c}(\bar{S}_{\mathrm{W}}) + \left(1 - \frac{\rho_{\mathrm{O}}}{\rho_{\mathrm{W}}} \right) \overline{\mathbf{Gr}}_{\mathrm{W}}^{-1} \hat{\bar{\nabla}} \hat{z} \right] \right\} \tag{6.52}$$

$$\bar{\phi} \, \frac{\partial (1 - \bar{S}_{\mathrm{W}})}{\partial \hat{\bar{t}}} = \hat{\bar{\nabla}} \cdot \left\{ \mathbf{K}^{r}_{\mathrm{OW}}(\bar{S}_{\mathrm{W}})(\overline{\mathbf{Ca}}_{\mathrm{W}}^{-1} \hat{\bar{\nabla}} \hat{\bar{P}}_{\mathrm{W}} - \overline{\mathbf{Gr}}_{\mathrm{W}}^{-1} \hat{\bar{\nabla}} \hat{z}) \right.$$

$$+ \mathbf{K}^{r}_{\mathrm{OO}}(\bar{S}_{\mathrm{W}}) \frac{\mu_{\mathrm{W}}}{\mu_{\mathrm{O}}} \left[(\overline{\mathbf{Ca}}_{\mathrm{W}}^{-1} \hat{\bar{\nabla}} \hat{\bar{P}}_{\mathrm{W}} - \overline{\mathbf{Gr}}_{\mathrm{W}}^{-1} \bar{\nabla} \hat{z}) \right.$$

$$\left. \left. + \overline{\mathbf{Ca}}_{\mathrm{W}}^{-1} \hat{\bar{\nabla}} \hat{\bar{P}}_{c}(\bar{S}_{\mathrm{W}}) + \left(1 - \frac{\rho_{\mathrm{O}}}{\rho_{\mathrm{W}}} \right) \overline{\mathbf{Gr}}_{\mathrm{W}}^{-1} \hat{\bar{\nabla}} \hat{z} \right] \right\} \tag{6.53}$$

In these equations the dimensionless tensor

$$\overline{\mathbf{Ca}}_{\mathrm{W}} = \frac{\mu_{\mathrm{W}} \bar{u} \bar{\ell}}{\bar{P}_{b}} \, \mathbf{K}^{-1} = \frac{\text{macroscopic viscous pressure drop}}{\text{macroscopic capillary pressure}} \tag{6.54}$$

plays the role of a macroscopic or large-scale capillary number. Similarly, the tensor

$$\overline{\mathbf{Gr}}_W = \frac{\mu_W \bar{u}}{\rho_W g} \mathbf{K}^{-1} = \frac{\text{macroscopic viscous pressure drop}}{\text{macroscopic gravitational pressure}} \quad (6.55)$$

corresponds to the macroscopic gravity number.

If the traditional normalization [Eq. (6.51)] is used instead of the normalization Eq. (6.48), and isotropy is assumed, then the same dimensionless equations are obtained with

$$\overline{Ca}_W = 1 \qquad (6.56)$$

where $\overline{Ca}_W$ is the macroscopic capillary number. Thus the traditional normalization is equivalent to the assumption that the macroscopic viscous pressure drop always equals the macroscopic capillary pressure. While this assumption is not generally valid, it sometimes is a reasonable approximation as illustrated below. First, however, the consequences of the traditional assumption (6.56) for the measurement of relative permeabilities will be discussed.

3. Measurement of Relative Permeabilities

For simplicity only the isotropic case will be considered from now on, that is, let $\mathbf{K} = k\mathbf{1}$ where $\mathbf{1}$ is the identity matrix. The tensors $\overline{\mathbf{Ca}}_W$ and $\overline{\mathbf{Gr}}_W$ then become $\overline{\mathbf{Ca}}_W = \overline{Ca}_W \mathbf{1}$ and $\overline{\mathbf{Gr}}_W = \overline{Gr}_W \mathbf{1}$, where $\overline{Ca}_W$ and $\overline{Gr}_W$ are the macroscopic capillary and gravity numbers.

The unsteady state or displacement method of measuring relative permeabilities consists of monitoring the production history and pressure drop across the sample during a laboratory displacement process [2, 337, 359]. The relative permeability is obtained as the solution of an inverse problem. The inverse problem consists in matching the measured production history and pressure drop to the solutions of the multiphase flow equations (6.52) and (7.53) using the Buckley–Leverett approximation.

In the present formulation the Buckley–Leverett approximation comprises several independent assumptions. First, it is assumed that gravity effects are absent, which amounts to the assumption

$$\overline{Ca}_W \ll \overline{Gr}_W \qquad (6.57)$$

Second, the viscous coupling terms are neglected, that is,

$$k^r_{WO} \frac{\mu_W}{\mu_O} \ll \overline{Ca}_W \qquad \text{and} \qquad k^r_{OW} \ll \overline{Ca}_W \qquad (6.58)$$

Finally, the resulting equations

$$\bar{\phi}\,\frac{\partial \bar{S}_{\mathrm{w}}}{\partial \hat{t}} = \hat{\bar{\nabla}} \cdot \left\{ k^r_{\mathrm{ww}}(\bar{S}_{\mathrm{w}})\,\frac{\hat{\bar{\nabla}}\hat{\bar{P}}_{\mathrm{w}}}{\overline{\mathrm{Ca}}_{\mathrm{w}}} \right\} \tag{6.59}$$

$$\bar{\phi}\,\frac{\partial (1 - \bar{S}_{\mathrm{w}})}{\partial \hat{t}} = \hat{\bar{\nabla}} \cdot \left\{ k^r_{\mathrm{oo}}(\bar{S}_{\mathrm{w}})\,\frac{\mu_{\mathrm{w}}}{\mu_{\mathrm{o}}}\left[\frac{\hat{\bar{\nabla}}\hat{\bar{P}}_{\mathrm{w}}}{\overline{\mathrm{Ca}}_{\mathrm{w}}} + \frac{\hat{\bar{\nabla}}\hat{\bar{P}}_{c}(\bar{S}_{\mathrm{w}})}{\overline{\mathrm{Ca}}_{\mathrm{w}}} \right] \right\} \tag{6.60}$$

are further simplified by assuming that the term involving $\hat{\bar{P}}_{c}(\bar{S}_{\mathrm{w}})$ in Eq. (6.60) may be neglected [332].

Combining Eq. (6.57) with the traditional normalization Eq. (6.56) yields the consistency condition

$$\overline{\mathrm{Gr}}_{\mathrm{w}} \gg 1 \tag{6.61}$$

for the application of Buckley–Leverett theory in the determination of relative permeabilities. It is now clear from the definition of the macroscopic gravity number [see Eq. (6.55)] that the consistent use of Buckley–Leverett theory for the unsteady-state measurement of relative permeabilities depends strongly on the flow regime. This is valid whether or not the capillary pressure term $\hat{\bar{P}}_{c}(\bar{S}_{\mathrm{w}})$ in Eq. (6.60) is neglected. In addition to these consistency problems the Buckley–Leverett theory is also plagued with stability problems [360].

4. Pore-Scale to Large-Scale Comparison

The comparison between the macroscopic and the microscopic dimensional analysis is carried out by relating the microscopic and macroscopic velocities and length scales. The macroscopic velocity is taken to be a Darcy velocity defined as [see discussion following Eq. (5.79)]

$$\bar{u} = \bar{\phi}u \tag{6.62}$$

where $\bar{\phi}$ is the bulk porosity and u denotes the average microscopic flow velocity introduced in the microscopic analysis [Eq. (6.12)]. The length scales ℓ and $\bar{\ell}$ are identical ($\bar{\ell} = \ell$).

Using these relations between microscopic and macroscopic length and time scales together with the assumption of isotropy yields

$$\overline{\mathrm{Ca}}_{\mathrm{w}} = \frac{\mu_{\mathrm{w}}\bar{\phi}u\ell}{k\bar{P}_{b}} = \frac{u\ell}{\nu_{\mathrm{w}}^{*}} = \frac{\sigma_{\mathrm{ow}}\bar{\phi}\ell}{k\bar{P}_{b}}\,\mathrm{Ca}_{\mathrm{w}} \tag{6.63}$$

as the relationship between microscopic and macroscopic capillary numbers. Similarly, one obtains

$$\overline{\mathrm{Gr}}_{\mathrm{w}} = \frac{\mu_{\mathrm{w}} \bar{\phi} u}{\rho_{\mathrm{w}} g k} = \frac{u}{\overline{u}_{\mathrm{w}}^{*}} = \frac{\bar{\phi} \ell^2}{k} \mathrm{Gr}_{\mathrm{w}} \qquad (6.64)$$

for the gravity numbers. Taking the quotient gives

$$\overline{\mathrm{Gl}}_{\mathrm{w}} = \frac{\overline{\mathrm{Ca}}_{\mathrm{w}}}{\overline{\mathrm{Gr}}_{\mathrm{w}}} = \frac{\rho_{\mathrm{w}} g \ell}{\bar{P}_b} = \frac{\ell}{\ell_{\mathrm{w}}^{*}} = \frac{\sigma_{\mathrm{ow}}}{\ell \bar{P}_b} \mathrm{Gl}_{\mathrm{w}} \qquad (6.65)$$

for the macroscopic gravillary number. Note that the ratio $\sigma_{\mathrm{ow}}/(\ell \bar{P}_b)$ is the ratio of the microscopic to the macroscopic capillary pressures. The characteristic numbers

$$\overline{\nu}_{\mathrm{w}}^{*} = \frac{k \bar{P}_b}{\bar{\phi} \mu_{\mathrm{w}}} \qquad (6.66)$$

$$\overline{u}_{\mathrm{w}}^{*} = \frac{\rho_{\mathrm{w}} g k}{\mu_{\mathrm{w}} \bar{\phi}} \qquad (6.67)$$

$$\overline{\ell}_{\mathrm{w}}^{*} = \frac{\bar{P}_b}{\rho_{\mathrm{w}} g} \qquad (6.68)$$

are the macroscopic counterparts of the microscopic numbers defined in Eqs. (6.22), (6.30), and (6.34).

An interesting way of rewriting these relationships arises from interpreting the permeability as an effective microscopic cross-sectional area of flow, combined with the Leverett J function. More precisely, let

$$\Lambda = \sqrt{\frac{k}{\phi}} \qquad (6.69)$$

denote a microscopic length that is characteristic for the pore-space transport properties. Then Eqs. (6.63)–(6.65) may be rewritten as

$$\overline{\mathrm{Ca}}_{\mathrm{w}} = \frac{\bar{\ell}}{\Lambda} \frac{\mathrm{Ca}_{\mathrm{w}}}{J(\bar{S}_b) \cos \theta} \qquad (6.70)$$

$$\overline{\mathrm{Gr}}_{\mathrm{w}} = \frac{\bar{\ell}^2}{\Lambda^2} \mathrm{Gr}_{\mathrm{w}} \qquad (6.71)$$

$$\overline{\text{Gl}}_{\text{W}} = \frac{\Lambda}{\bar{\ell}} \frac{\text{Gl}_{\text{W}}}{\bar{J}(\bar{S}_b) \cos \theta} \qquad (6.72)$$

where $J(\bar{S}_b) = (\bar{P}_b \sqrt{k/\bar{\phi}})/((\sigma_{\text{OW}} \cos \theta)$ is the value of the Leverett–J function [2, 28] at the saturation corresponding to breakthrough, and θ is the wetting angle.

The capillary number scales as $(\bar{\ell}/\Lambda)$ while the gravity number scales as $(\bar{\ell}/\Lambda)^2$. Inserting Eqs. (6.71) and (6.72) into Eq. (6.57) implies that the Buckley–Leverett approximation (6.57) becomes invalid whenever $\bar{\ell} < \Lambda \text{Gl}_{\text{W}} / [J(\bar{S}_b) \cos \theta]$.

5. Macroscopic Estimates

This section gives order of magnitude estimates for the relative importance of capillary, viscous, and gravity effects at different scales in representative categories of porous media. These estimates illustrate the usefulness of the macroscopic dimensionless ratios for the problem of upscaling.

Three types of porous media are considered: high-permeability unconsolidated sand, intermediate permeability sandstone, and low-permeability limestone. Representative values for $\bar{\phi}$, k, and $\bar{P}_b$ are shown in Table VI.

To estimate the dimensionless numbers, the same microscopic velocity $u \approx 3 \times 10^{-6} \, \text{m s}^{-1}$ as for the microscopic estimates will be used. The length scale ℓ, however, differs between a laboratory displacement and a reservoir process. The parameters $\ell_{\text{lab}} \approx 0.1 \, \text{m}$ and $\ell_{\text{res}} \approx 100 \, \text{m}$ are used as representative values. Combining these values with those in Tables IV and VI yields the results shown in Table VII.

The first row in Table VII can be used to check the consistency of the Buckley–Leverett approximation with the traditional normalization. The consistency condition [Eq. (6.61)] is violated for unconsolidated sand and

TABLE VI
Representative Values for Porosity, Permeability, and Breakthrough Capillary Pressure in Unconsolidated Sand, Sandstone, and Low-Permeability Limestone

Quantity	Sand	Sandstone	Limestone
$\bar{\phi}$	0.36	0.22	0.20
k	10,000 mD	400 mD	3 mD
$\bar{P}_b$	2,000 Pa	10^4 Pa	10^5 Pa

TABLE VII

Definition and Representative Values for Macroscopic Dimensionless Numbers in Different Porous Media on Laboratory ($\ell_{lab} \approx 0.1$ m) and Reservoir Scale ($\ell_{res} \approx 100$ m) under Uniform Flow Conditions ($u \approx 3 \times 10^{-6}$ m s^{-1})

Quantity	Definition	Unconsolidated Sand		Sandstone		Limestone	
		Laboratory	Reservoir	Laboratory	Reservoir	Laboratory	Reservoir
$\overline{Gr}_w$	$\dfrac{\mu_w \bar{\phi} u}{\rho_w g k}$	0.01	0.01	0.13	0.13	18.6	18.6
$\overline{Ca}_w$	$\dfrac{\mu_w \bar{\phi} u \ell}{k \bar{P}_b}$	0.005	4.9	0.015	15.0	0.19	187.5
$\overline{Gl}_w$	$\dfrac{\rho_w g \ell}{\bar{P}_b}$	0.5	492	0.1	115	0.01	10
Λ	$\sqrt{\dfrac{k}{\phi}}$	5.2 μm	5.2 μm	1.3 μm	1.3 μm	0.1 μm	0.1 μm
$Ca_w/\overline{Ca}_w$	$\dfrac{\Lambda}{\ell} J(\bar{S}_b) \cos\theta$	1.5×10^{-4}	1.5×10^{-7}	4.8×10^{-6}	4.8×10^{-9}	2.8×10^{-7}	2.8×10^{-10}
$\dfrac{Ca_c \overline{Ca}_w}{Ca_w}$	$\dfrac{\Lambda}{\ell} J(\bar{S}_b) \cos\theta$	0.67	–	0.63	–	0.71	–

sandstones. Such a conclusion, of course, assumes that the values given in Table VI are representative for these media.

The fifth row in Table VII gives the ratio between macroscopic and microscopic capillary numbers, which according to Eq. (6.63) is length scale dependent. The last row in Table VII compares this ratio to the typical critical capillary number Ca_c reported for laboratory desaturation curves in the different porous media. Using the $Ca_c \approx 10^{-4}$ for sand, $Ca_c \approx 3 \times 10^{-6}$ for sandstone, and $Ca_c \approx 2 \times 10^{-7}$ for limestone [28] as before one finds that the corresponding critical macroscopic capillary number is close to 1. This indicates that the macroscopic capillary number is indeed an appropriate measure of the relative strength of viscous and capillary forces.

Consequently, one expects differences between residual oil saturation S_{0r} in laboratory and reservoir floods. Given a laboratory measured capillary desaturation curve $S_{0r}(Ca_w)$ as a function of the microscopic capillary number Ca_w the analysis predicts that the residual oil saturation in a reservoir flood can be estimated from the laboratory curve as $S_{0r}(Ca_c \cdot \overline{Ca}_w)$ [47, 48]. For $\overline{Ca}_w > 1$ the S_{0r} value based on macroscopic capillary numbers will in general be lower than the value $S_{0r}(Ca_w)$

expected from using microscopic capillary numbers. Such differences have been frequently observed, and Morrow [361] recently raised the question why field recoveries are sometimes significantly higher than those observed in the laboratory. The revised macroscopic analysis of [47, 48] suggests a possible answer to this question.

The values of the dimensionless numbers in Table VII allow an assessment of the relative importance of the different forces for a displacement. To illustrate this consider the values $\overline{Gr}_w = 0.01$, $\overline{Ca}_w = 0.005$, and $\overline{Gl}_w = 0.5$ for unconsolidated sand on the laboratory scale. A moment's reflection shows that this implies $V \ll G \approx C$, where V stands for macroscopic viscous forces, C for macroscopic capillary forces, and G for gravity forces. The notation $A \ll B$ indicates that $A/B < 10^{-2}$, while $A < B$ means $10^{-2} < A/B < 0.5$ and $A \approx B$ stands for $0.5 < A/B < 2$. Repeating this for all cases in Table VII yields the results shown in Table VIII. Table VIII also contains the results from the microscopic dimensional analysis, as well as the results one would obtain from a traditional macroscopic dimensional analysis that assumes $\overline{Ca} = 1$ [see Eq. (6.56)].

Obviously, the relative importance of the different forces may change depending on the type of medium, the characteristic fluid velocities, and the length scale. Perhaps this explains part of the general difficulty of scaling up from the laboratory to the reservoir scale for immiscible displacement.

6. Applications

The characteristic macroscopic velocities, length scales, and kinematic viscosities defined, respectively, in Eqs. (6.66)–(6.68) are intrinsic physical characteristics of the porous media and the fluid displacement processes. These characteristics can be useful in applications such as

TABLE VIII

Relative Importance of Viscous (V), Gravity (G), and Capillary (C) Forces in Unconsolidated Sand, Sandstone, and Limestone[a]

			Sand	Sandstone	Limestone
Pore scale			$V \ll G \ll C$		
Large scale		Traditional analysis	$V = C \ll G$	$V = C < G$	$G < V = C$
	[47]	Laboratory scale	$V \ll G \approx C$	$V < G < C$	$G < V < C$
	[48]	Field scale	$C < V \ll G$	$C < V < G$	$C < G < V$

The notation $A \ll B$ (with A, $B \in \{V, G, C\}$) indicates that $A/B < 10^{-2}$, while $A < B$ means $10^{-2} < A/B < 0.5$ and $A \approx B$ stands for $0.5 < A/B < 2$.

estimating the width of a gravitational segregation front, the energy input required to mobilize residual oil, or gravitational relaxation times.

The macroscopic gravillary number $\overline{\mathrm{Gl}}_\mathrm{w}$ defines an intrinsic length scale ℓ_w^* [see Eq. (6.65)]. Because $\overline{\mathrm{Gl}}_\mathrm{w}$ gives the ratio of the gravity to the capillary forces the length ℓ_w^* directly gives the width of a gravitational segregation front when the fluids are at rest and in gravitational equilibrium, that is when viscous forces are negligible or absent. By using the same estimates for $\bar{\phi}$, k, and $\bar{P}_b$ as those used for Table VII one obtains a characteristic front width of 20 cm for unconsolidated sand, 1 m for sandstone, and roughly 10 m for a low-permeability limestone.

Similarly, the macroscopic capillary number defines an intrinsic specific action (or energy input) ν_w^* via Eq. (6.63), which is the energy input required to mobilize residual oil if gravity forces may be considered negligible or absent. Representative estimates are given in Table IX.

The gravitational relaxation time is the time needed to return to gravitational equilibrium after its disturbance. This may be defined from the balance of gravitational forces versus the combined effect of viscous and capillary forces. Analogous to Eq. (6.35) for the microscopic case the dimensionless ratio becomes

$$\frac{\overline{\mathrm{Gl}}_\mathrm{w}}{\overline{\mathrm{Gr}}_\mathrm{w}} = \frac{(\text{macr gravitational pressure})^2}{(\text{macr capillary pressure}) \times (\text{macr viscous pressure drop})}$$

$$= \frac{\overline{\mathrm{Ca}}_\mathrm{w}}{\overline{\mathrm{Gr}}_\mathrm{w}^2} = \frac{\rho_\mathrm{w}^2 g^2 k \ell}{\mu_\mathrm{w} \bar{\phi} \bar{P}_b u} = \frac{t}{\bar{t}_\mathrm{w}^*} \tag{6.73}$$

which defines the gravitational relaxation time $\bar{t}_\mathrm{w}^*$ as

$$\bar{t}_\mathrm{w}^* = \frac{\overline{\ell_\mathrm{w}^*}}{\overline{u_\mathrm{w}^*}} = \frac{\mu_\mathrm{w} \bar{\phi} \bar{P}_b}{\rho_\mathrm{w}^2 g^2 k} \tag{6.74}$$

TABLE IX
Characteristic Macroscopic Energies, Velocities, Length Scales, Time Scales, and Volumetric Flow Rates for Oil–Water Flow under Reservoir Conditions in Unconsolidated Sand, Sandstone, and Low-Permeability Limestone

Quantity	Sand	Sandstone	Limestone
$\overline{\nu_\mathrm{w}^*}$	$6.1 \times 10^{-5}\,\mathrm{m^2\,s^{-1}}$	$2.0 \times 10^{-5}\,\mathrm{m^2\,s^{-1}}$	$1.6 \times 10^{-6}\,\mathrm{m^2\,s^{-1}}$
$\overline{u_\mathrm{w}^*}$	$2.99 \times 10^{-4}\,\mathrm{m\,s^{-1}}$	$2.17 \times 10^{-5}\,\mathrm{m\,s^{-1}}$	$1.61 \times 10^{-7}\,\mathrm{m\,s^{-1}}$
$\overline{\ell_\mathrm{w}^*}$	0.2 m	1.02 m	10.2 m
$\overline{t_\mathrm{w}^*}$	669 s	$4.7 \times 10^4\,\mathrm{s}$	$6.36 \times 10^7\,\mathrm{s}$
$\overline{Q^*}$	$1.22 \times 10^{-6}\,\mathrm{m^3\,s^{-1}}$	$2.04 \times 10^{-5}\,\mathrm{m^3\,s^{-1}}$	$1.63 \times 10^{-5}\,\mathrm{m^3\,s^{-1}}$

Estimated values are given in Table IX. They correspond to gravitational relaxation times of roughly 10 min for unconsolidated sand, 13 h for a sandstone, and 736 days for a low-permeability limestone.

Another interesting intrinsic number arises from comparing the strength of macroscopic capillary forces versus the combined effect of viscous and gravity forces

$$(\overline{Gl}_W \overline{Ca}_W)^{-1} = \frac{(\text{macr capillary pressure})^2}{(\text{macr grav pressure}) \times (\text{macr viscous pressure drop})}$$

$$= \frac{\overline{Gr}_W}{\overline{Ca}_W^2} = \frac{k\bar{P}_b^2}{\phi\mu_W \rho_W g u \ell^2} = \frac{\overline{Q_W^*}}{Q} \tag{6.75}$$

where Q denotes the volumetric flow rate. Thus $\overline{Q_W^*}$ defined as

$$\overline{Q_W^*} = \overline{\ell_W^{*2}} \overline{u_W^*} = \frac{k\bar{P}_b^2}{\phi\mu_W \rho_W g} \tag{6.76}$$

is an intrinsic system specific characteristic flow rate. The estimates for v_W^*, u_W^*, ℓ_W^*, t_W^*, and $\overline{Q_W^*}$ are summarized in Table IX.

In summary, the dimensional analysis of the upscaling problem for two-phase immiscible displacement suggests normalizing the macroscopic pressure field in a way that differs from the traditional normalization. This gives rise to a macroscopic capillary number $\overline{Ca}$ that differs from the traditional microscopic capillary number Ca in that it depends on length scale and the breakthrough capillary pressure $\bar{P}_b$. The traditional normalization corresponds to the tacit assumption that viscous and capillary forces are of equal magnitude. With the new macroscopic capillary number $\overline{Ca}$ the breakpoint Ca_c in capillary desaturation curves seems to occur at $\overline{Ca} \approx 1$ for all types of porous media. Representative estimates of $\overline{Ca}$ for unconsolidated sand, sandstones, and limestones suggest that the residual oil saturation after a field flood will in general differ from that after a laboratory flood performed under the same conditions. Order of magnitude estimates of gravitational relaxation times and segregation front widths for different media are consistent with experiment.

ACKNOWLEDGMENTS

The author is grateful to Thor Engøy, Karl-Sigurd Årland, and Christian Ostertag-Henning for providing him with Figs. 5, 8, and 12, and especially to Espen Haslund for allowing the use of Fig. 19 prior to publication. He thanks B. Virgin for technical assistance, Professor B. Nøst, Professor T. Jøssang, and Professor J. Feder for their hospitality in Oslo, and Professor D. H. Welte and Dr. U. Mann for their interest and support. He gratefully

414   R. HILFER

acknowledges Norges Forskningsråd and Forschungszentrum Jülich for partial financial support.

REFERENCES

1. J. Schopper, Permeabilist und Porosität, in *Landolt–Börnstein: Physikalische Eigenschaften de Gesteine*, K.-H. Hellwege, Ed., Springer, Berlin, 1982, Vol. V/1a, p. 184.
2. F. Dullien, *Porous Media—Fluid Transport and Pore Structure*, San Diego, Academic, 1992.
3. A. Scheidegger, *The Physics of Flow through Porous Media*, University of Toronto Press, Toronto, 1974.
4. L. Schwartz, F. Auzerais, J. Dunsmuir, N. Martys, D. Bentz, and S. Torquato, *Physica A*, **207**, 28 (1994).
5. J. Bear and Y. Bachmat, *Introduction to Modeling of Transport Phenomena in Porous Media*, Kluwer Academic, Dordrecht, 1990.
6. G. Stell, in *The Wonderful World of Stochastics*, M. Shlesinger and G. Weiss, Eds., Elsevier, Amsterdam, 1985, p. 127.
7. G. Stell and P. Rikvold, *Chem. Eng. Commun.*, **87**, 233 (1987).
8. H. Reiss, *J. Phys. Chem.*, **96**, 4736 (1992).
9. E. Harding and D. Kendall, Eds., *Stochastic Geometry*, London, Wiley, 1974.
10. G. Matheron, *Random Sets and Integral Geometry*, New York, Wiley, 1975.
11. J. Klafter, R. Rubin, and M. Shlesinger, Eds., *Transport and Relaxation in Random Materials*, World Scientific, Singapore, 1986.
12. V. Lehmann and U. Gösele, *Appl. Phys. Lett.*, **58**, 856 (1991).
13. A. Cullis and L. Canham, *Nature (London)*, **353**, 335 (1991).
14. J. Drake, J. Klafter, R. Kopelman, and D. Awschalom, Eds., *Dynamics in Small Confining Systems*, Vol. 290, Materials Research Society, Pittsburgh, 1993.
15. M. Hoffmann, *MRS Bulletin*, **20**, 28 (1995).
16. K. Yoshida, *J. Phys. Soc. Jpn.*, **59**, 4087 (1990).
17. U. Mann, in *Geofluids: Origin, Migration and Evolution of Fluids in Sedimentary Basins*, J. Parnell, Ed., Geological Society, 1994, p. 233, Vol. 78.
18. J. Bear and M. Corapcioglu, *Advances in Transport in Porous Media*, Martinus Nijhoff, Dordrecht, 1987.
19. I. Goldsmith and P. King, in *Diagenesis of Sedimentary Sequences*, J. Marshall, Ed., Geological Society, 1987, p. 1.
20. G. Marsily, *Quantitative Hydrogeology—Groundwater Hydrology for Engineers*, Academic, San Diego, 1986.
21. Z. Kabala and A. Hunt, *Stochastic Hydrology Hydraulics*, **7**, 255 (1993).
22. D. Yale, *Geophysics*, **50**, 2480 (1985).
23. W. England, A. Mackenzie, D. Mann, and T. Quigley, *J. Geophys. Soc. (London)*, **144**, 327 (1987).
24. Y. Tang and M. Aral, *Water Resources Res.*, **28**, 1389 (1992).
25. R. Bales, S. Li, K. Maguire, M. Yahya, and C. Gerba, *Water Resources Res.*, **29**, 957 (1993).
26. Y. Jang, N. Sitar, and A. Kiureghian, *Water Resources Res.*, **30**, 2435 (1994).

27. U. Ahmed, S. Crary, and G. Coates, *J. Pet. Technol.*, 578, May 1991.

28. L. Lake, *Enhanced Oil Recovery*, Prentice Hall, Englewood Cliffs, NJ, 1989.

29. M. Honarpour, L. Koederitz, and A. Harvey, *Relative Permeability of Petroleum Reservoirs*, CRC Press, Boca Raton, FL, 1986.

30. M. Yenkie and G. Natarajan, *Separation Sci. Technol.*, **28**, 1177 (1993).

31. W. Mochan and R. Barrera, Eds., *ETOPIM 3, Proceedings of the Third International Conference on Electrical Transport and Optical Properties of Inhomogeneous Media*, Vol. Physica A 207, North-Holland, Amsterdam, 1994.

32. J. Cushman, Ed., *Dynamics of Fluids in Hierarchical Porous Media*, Academic, London, 1990.

33. P. Adler, *Porous Media*, Butterworth-Heinemann, Boston, 1992.

34. D. Stoyan and H. Stoyan, *Fractals, Random Shapes and Point Fields*, Wiley, Chichester, 1994.

35. J. Lafait and D. Tanner, Eds., *ETOPIM 2, Proceedings of the Second International Conference on Electrical Transport and Optical Properties of Inhomogeneous Media*, Vol. Physica A 157, North-Holland, Amsterdam, 1989.

36. M. Allen, G. Behie, and J. Trangenstein, *Multiphase Flow in Porous Media*, Vol. 34 of *Lecture Notes in Engineering*, Springer Verlag, Berlin, 1988.

37. D. Stoyan, W. Kendall, and J. Mecke, *Stochastic Geometry and its Applications*, Akademie-Verlag/Wiley, Berlin/Chichester, 1987.

38. E. Sanchez-Palencia and A. Zaoui, *Homogenization Techniques for Composite Media*, Vol. 272 of *Lecture Notes in Physics*, Springer-Verlag, Berlin, 1987.

38a. M. Sahimi, *Flow and Transport in Porous Media and Fractured Rock.* Weinheim: VCH Verlagsgesellschaft mbH, 1995.

39. J. Parker, *Rev. Geophys.*, **27**, 311 (1989).

40. D. Bergman and D. Stroud, in *Solid State Physics*, H. Ehrenreich and D. Turnbull, Eds., Academic, New York, 1992, p. 147.

41. M. Sahimi, *Rev. Mod. Phys.*, **65**, 1393 (1993).

42. J. Feder and T. Jøssang, in *Fractals in Petroleum Geology and Earth Processes*, C. Barton and P. L. Pointe, Eds., Plenum, New York, 1995, p. 179.

43. A. Thompson, A. Katz, and C. Krohn, *Adv. Phys.*, **36**, 625 (1987).

44. K. Meyer, P. Lorenz, B. Böhl-Kuhn, and P. Klobes, *Cryst. Res. Tech.*, **29**, 903 (1994).

45. S. Torquato, *Appl. Mech. Rev.*, **47**, S29 (1994).

46. R. Landauer, in *Electrical Transport and Optical Properties of Inhomogeneous Materials*, J. Garland and D. Tanner, Eds., American Institute of Physics, New York, 1978, p. 2.

47. R. Hilfer and P. Øren, 1993, Statoil Publ. Nr. F&U-LoU-94001.

48. R. Hilfer and P. Øren, *Transport Porous Media*, in press, 1995.

49. M. Leverett, W. Lewis, and M. True, *Trans. AIME*, **146**, 175 (1942).

50. M. Cole, J. Harvey, R. Lux, D. Eckart, and R. Tsu, *Appl. Phys. Lett.*, **60**, 2800 (1992).

51. S. Gardelis, U. Bangert, and B. Hamilton, *Thin Solid Films*, **255**, 167 (1995).

52. E. Takasuka and K. Kamei, *Appl. Phys. Lett.*, **65**, 484 (1994).

53. A. Cullis, L. Canham, G. Williams, P. Smith, and O. Dosser, *J. Appl. Phys.*, **75**, 493 (1994).

54. H. Lee, Y. Seo, D. Oh, K. Nahm, E. Suh, Y. Lee, H. Lee, Y. Hwang, K. Park, S. Chang, and E. Lee, *Appl. Phys. Lett.*, **62**, 855 (1993).

55. P. Smith, in *Phase Transitions in Soft Condensed Matter*, T. Riste and D. Sherrington, Eds., Plenum, New York, 1989, p. 353.

56. E. Weibel, in *Geometrical Probability and Biological Structures: Buffon's 200th Anniversary*, R. Miles and J. Serra, Eds., Springer, Berlin, 1978, p. 171.

57. P. Philippi, P. R. Yunes, C. Fernandes, and F. Magnani, *Transport Porous Media*, **14**, 219 (1994).

58. B. Mandelbrot, *The Fractal Geometry of Nature*, San Francisco, Freeman, 1982.

59. M. Zähle, *Math. Nachr.*, **108**, 49 (1982).

60. M. Zähle, *Math. Nachr.*, **110**, 179 (1983).

61. K. Falconer, *The Geometry of Fractal Sets*, Cambridge University Press, Cambridge, MA, 1985.

62. J. Wohlenberg, Dichte der Minerale, in *Landolt–Börnstein: Physikalische Eigenschaften de Gesteine*, K.-H. Hellwege, Ed., Vol. V/1a, Springer, Berlin, 1982, p. 66.

63. R. Hilfer, *Int. J. Mod. Phys. B*, **7**, 4371 (1993).

64. R. Hilfer, *Phys. Rev. E*, **48**, 2466 (1993).

65. R. Hilfer, in *Random Magnetism and High-Temperature Superconductivity*, W. Beyermann, N. Huang-Liu, and D. MacLaughlin, Eds., World Scientific Publishers, Singapore, 1994, p. 85.

66. R. Hilfer and L. Anton, *Phys. Rev. E, Rapid Commun.*, **51**, 848 (1995).

67. R. Hilfer, *Fractals*, **3(1)**, 211, 1995.

68. R. Hilfer, *Chaos, Solitons, Fractals*, Vol. 5, 1995, p. 1475.

69. R. Hilfer, *Physica A* 221 (1995), p. 9.

70. D. Kendall, in *Stochastic Geometry*, E. Harding and D. Kendall, Eds., Wiley, London, 1974, p. 322.

71. N. Cressie and G. Lassett, *SIAM Rev.*, **29**, 577 (1987).

72. G. Choquet, *Ann. Inst. Fourier*, V, 131 (1953).

73. W. Feller, *An Introduction to Probability Theory and Its Applications*, Vol. I, Wiley, New York, 1968.

74. W. Feller, *An Intoduction to Probability Theory and Its Applications*, Vol. II, Wiley, New York, 1971.

75. W. Rudin, *Real and Complex Analysis*, McGraw-Hill, New York, 1974.

76. Y. Bachmat and J. Bear, *Transport Porous Media*, **1**, 213 (1986).

77. R. S. Mikhail and E. Robens, *Microstructure and Thermal Analysis of Solid Surfaces*, Wiley, Chichester, UK, 1983.

78. P. Wong, J. Howard, and J. Lin, *Phys. Rev. Lett.*, **57**, 637 (1986).

79. P. Wong, *Phys. Today*, **24**, December, 1988.

80. J. Hansen and A. Skjeltorp, *Phys. Rev. B*, **38**, 2635 (1988).

81. C. Harris, in *Reservoir Characterization II*, L. Lake, H. Carroll, and T. Wesson, Eds., Academic, San Diego, 1991, p. 2.

82. C. Jacquin and P. Adler, *Transport Porous Media*, **2**, 28 (1987).

83. J. Fripiat, in *Fractal Approach to Heterogeneous Chemistry*, D. Avnir, Ed., Wiley, Chichester, UK, 1989, p. 331.

84. P. Adler, in *Fractal Approach to Heterogenous Chemistry*, D. Avnir, Ed., Wiley, Chichester, UK, 1989, p. 341.

85. O. Dinariev, *Fluid Dynamics*, **27**, 682 (1992).

86. J. Feder, *Fractals*, Plenum, New York, 1988.

87. R. Lenormand, *Physica D*, **38**, 230 (1989).

88. R. Lenormand, *Proc. R. Soc. London*, **423**, 159 (1989).

89. C. Ruffet, Y. Gueguen, and M. Darot, *Geophys.*, **56**, 758 (1991).

90. H. Davis, R. Novy, L. Scriven, and P. Toledo, *J. Phys. C*, **2**, SA457 (1990).

91. C. Barton and P. L. Pointe, Eds., *Fractals in Petroleum Geology and Earth Processes*, Plenum, New York, 1995.

92. K. Gaida, W. Rühl, and W. Zimmerle, *Erdöl Erdgas Z.*, **89**, 336 (1973).

93. S. Brunauer, P. Emmer, and E. Teller, *J. Am. Chem. Soc.*, **60**, 309 (1938).

94. R. Haul and G. Dümbgen, *Chemie-Ing.-Tech.*, **35**, 586 (1963).

95. C. Meng and Y. Wang, *J. Non-Cryst. Solids*, **122**, 41 (1990).

96. P. Debye and A. Bueche, *J. Appl. Phys.*, **20**, 518 (1949).

97. P. Debye, H. Anderson, and H. Brumberger, *J. Appl. Phys.*, **28**, 679 (1957).

98. S. Prager, *Phys. Fluids*, **4**, 1477 (1961).

99. S. Prager, *Physica*, **29**, 129 (1963).

100. H. Weissberg and S. Prager, *Phys. Fluids*, **5**, 1390 (1962).

101. H. Weissberg, *J. Appl. Phys.*, **34**, 2636 (1963).

102. W. Haller, *J. Chem. Phys.*, **42**, 686 (1965).

103. R. Reck and S. Prager, *J. Chem. Phys.*, **42**, 3027 (1965).

104. M. Doi, *J. Phys. Soc. Jpn.*, **40**, 567 (1976).

105. S. Torquato and G. Stell, *J. Chem. Phys.*, **77**, 2071 (1982).

106. S. Torquato and G. Stell, *J. Chem. Phys.*, **78**, 3262 (1983).

107. S. Torquato and G. Stell, *J. Chem. Phys.*, **79**, 1505 (1983).

108. P. Rikvold and G. Stell, *J. Chem. Phys.*, **82**, 1014 (1985).

109. P. Rikvold and G. Stell, *J. Colloid Interface Sci.*, **108**, 158 (1985).

110. G. Stell and P. Rikvold, *Int. J. Thermophys.*, **7**, 863 (1986).

111. J. Berryman, *J. Appl. Phys.*, **57**, 2374 (1985).

112. J. Berryman and S. Blair, *J. Appl. Phys.*, **60**, 1930 (1986).

113. M. Yanuka, F. Dullien, and D. Elrick, *J. Colloid Interface Sci.*, **112**, 24 (1986).

114. M. Kwiecen, I. MacDonald, and F. Dullien, *J. Microsc.*, **159**, 343 (1990).

115. F. Dullien, *Transport Porous Media*, **6**, 581 (1991).

116. S. Torquato and M. Avellaneda, *J. Chem. Phys.*, **95**, 6477 (1991).

117. C. Lin and M. Cohen, *J. Appl. Phys.*, **53**, 4152 (1982).

118. J. Thovert, J. Salles, and P. Adler, *J. Microsc.*, **170**, 65 (1993).

119. P. Spanne, J. Thovert, C. Jacquin, W. Lundquist, K. Jones, and P. Adler, *Phys. Rev. Lett.*, **73**, 2001 (1994).

120. J. Harlan, D. Picot, and P. Loll, *Anal. Biochem.*, **224**, 557 (1995).

121. K. Hosoya, K. Kimata, and N. Tanaka, *J. Liq. Chromatog.*, **16**, 3059 (1993).

122. S. Sakai, *J. Membrane Sci.*, **96**, 91 (1994).

123. S. Mochizuki and A. Zydney, *J. Membrane Sci.*, **82**, 211 (1993).

124. E. Grosgogeat, J. Fried, and R. Jenkins, *J. Membrane Sci.*, **57**, 237 (1991).

418 R. HILFER

125. M. Sasthav, W. P. Raj, and M. Cheung, *J. Colloid Interface Sci.*, **152**, 376 (1992).
126. H. Kamiya, K. Isomura, and T. Jun-ichiro, *J. Am. Chem. Soc.*, **78**, 49 (1995).
127. W. W. Chen and B. Dunn, *J. Am. Ceramic Soc.*, **76**, 2086 (1993).
128. L. Garrido and J. L. Ackerman, *Ceramic Eng. Sci. Proc.*, **12**, 2042 (1992).
129. N. Naito, L. De Jonghe, and M. Rahaman, *J. Materials Sci.*, **25**, 1686 (1990).
130. R. Murdey and W. Machin, *Langmuir*, **10**, 3842 (1994).
131. S. Zeng, A. Hunt, and R. Greif, *J. Heat Transfer*, **116**, 756 (1994).
132. H. Naono, M. Hakuman, and K. Nakai, *J. Coll. Interf. Sci.*, **165**, 532 (1994).
133. B. Russell and M. LeVan, *Carbon*, **32**, 845 (1994).
134. C. Lastoskie, K. E. Gubbins, and N. Quirke, *J. Phys. Chem.*, **97**, 4786 (1993).
135. P. Gu, P. Xie, and Y. Fu, *Cement Concrete Res.*, **24**, 86 (1994).
136. L. Konecny and S. Naqvi, *Cement Concrete Res.*, **23**, 1223 (1993).
137. L. Tang and L.-O. Nilsson, *Cement Concrete Res.*, **22**, 541 (1992).
138. E. Smith, W. Powers, and P. Shea, *Soil Sci.*, **159**, 23 (1995).
139. Y. Nagarajarao, Z. *Pflanzenern. Bodenkd.*, **157**, 81 (1994).
140. H. Yamaguchi, Y. Hashizume, and H. Ikenaga, *Soils Foundations*, **32**, 1 (1992).
141. A. Netto, *Aapg Bull.*, **77**, 1101 (1993).
142. J. Howard and W. Kenyon, *Marine Petroleum Geol.*, **9**, 139 (1992).
143. F. J. Griffiths and R. C. Joshi, *Can. Geotech. J.*, **28**, 20 (1991).
144. A. Watkinson, Y. Xu, and Y. Koga, *Fuel*, **73**, 1797 (1994).
145. J. Kloubek, *J. Adhesion Sci. Technol.*, **6**, 667 (1992).
146. V. Karathanos and G. Saravacos, *J. Food Eng.*, **18**, 259 (1993).
147. W. D. Machin, *Langmuir*, **10**, 1235 (1994).
148. S. Sato, *J. Chem. Eng. Jpn.*, **21**, 534 (1988).
149. D. Milburn, B. Adkins, and B. Davis, *Appl. Catal.*, **119**, 205 (1994).
150. G. Zgrablich, S. Mendioroz, L. Daza, J. Pajares, V. Mayagoita, F. Rojas, and W. Conner, *Langmuir*, **7**, 779 (1991).
151. H. Ritter and L. Drake, *Ind. Eng. Chem. (Anal. Ed.)*, **17**, 782 (1945).
152. W. Purcell, *Am. Inst. Min. Metall. Petrol. Eng.*, **186**, 39 (1949).
153. A. Thompson, A. Katz, and R. Raschke, *Phys. Rev. Lett.*, **58**, 29 (1987).
154. A. Lane, N. Shah, and W. Conner, *J. Colloid Interface Sci.*, **137**, 315 (1991).
155. N. Wardlaw, Y. Li, and D. Forbes, *Transport Porous Media*, **2**, 597 (1987).
156. M. Ioannidis and I. Chatzis, *J. Colloid Interface Sci.*, **161**, 278 (1993).
157. M. Ioannidis, I. Chatzis, and A. Payatakes, *J. Colloid Interface Sci.*, **143**, 22 (1991).
158. C. D. Tsakiroglou and A. C. Payatakes, *J. Colloid Interface Sci.*, **146**, 479 (1991).
159. M. Spearing and P. Matthews, *Transport Porous Media*, **6**, 71 (1991).
160. C. D. Tsakiroglou and A. Payatakes, *J. Colloid Interface Sci.*, **137**, 315 (1990).
161. A. Rosenfeld and A. Kak, *Digital Picture Processing*, Academic, New York, 1982.
162. R. DeHoff, E. Aigeltinger, and K. Craig, *J. Microscopy*, **95**, 69 (1972).
163. L. Muche and D. Stoyan, *J. Appl. Probability*, **29**, 467 (1992).
164. Lu, Binglin. S. Torquato, *J. Chem. Phys.*, **98**, 6472 (1993).

165. S. Torquato and B. Lu, *Phys. Rev. E*, **47**, 2950 (1993).

166. T. B. Borak, *Radiation Res.*, **137**, 346 (1994).

167. P. Levitz and D. Tchoubar, *J. Phys. I France*, **2**, 771 (1992).

168. R. Hilfer, *Phys. Rev. B*, **44**, 60 (1991).

169. R. Hilfer, *Phys. Scr.*, **T44**, 51 (1992).

170. R. Hilfer, *Phys. Rev. B*, **45**, 7115 (1992).

171. F. Boger, J. Feder, R. Hilfer, and T. Jøssang, *Physica A*, **187**, 55 (1992).

172. R. Hilfer, *Physica A*, **194**, 406 (1993).

173. B. Hansen, E. Haslund, R. Hilfer, and B. Nøst, *Mat. Res. Soc. Proc.*, **290**, 185 (1993).

174. R. Hilfer, B. Nøst, E. Haslund, T. Kautzsch, B. Virgin, and B. Hansen, *Physica A*, **207**, 19 (1994).

175. E. Haslund, B. Hansen, R. Hilfer, and B. Nøst, *J. Appl. Phys.*, **76**, 5473 (1994).

176. J. Cardy, Ed., *Finite-Size Scaling*, North-Holland, Amsterdam, 1988.

177. V. Privman, Ed., *Finite-Size Scaling and Numerical Simulation of Statistical Systems*, World Scientific, Singapore, 1990.

178. R. Hilfer, *Z. Physik B*, **96**, 63 (1994).

179. E. O'Neill, *Introduction to Statistical Optics*, Addison-Wesley, Reading, 1963.

180. B. Lu and S. Torquato, *J. Chem. Phys.*, **93**, 3452 (1990).

181. B. Lu and S. Torquato, *J. Opt. Soc. Am. A*, **7**, 7171 (1990).

182. R. Guyer, *Phys. Rev. B*, **37**, 5713 (1988).

183. C. Ostertag-Henning, B. Virgin, Th. Rage, R. Hilfer, R. Koch, and U. Mann, 1995, to be published.

184. C. Andraud, B. Virgin, E. Haslund, R. Hilfer, A. Beghdadi, and J. Lafait, unpublished results.

185. B. Gnedenko, *The Theory of Probability*, Chelsea, New York, 1962.

186. B. Nøst, B. Hansen, and E. Haslund, *Phys. Scr.*, **T44**, 67 (1992).

187. J. Koplik, C. Lin, and M. Vermette, *J. Appl. Phys.*, **56**, 3127 (1984).

188. P. Doyen, *J. Geophys. Res.*, **93**, 7729 (1988).

188a. J. Fredrich, B. Menendez, and T. Wong, *Science*, **268**, 276 (1995).

189. C. Andraud, A. Beghdadi, and J. Lafait, *Physica A*, **207**, 208 (1994).

190. R. Hilfer, *Mod. Phys. Lett. B*, **6**, 773 (1992).

191. A. Prudnikov, Y. Brychkov, and O. Marichev, *Integrals and Series*, Gordon and Breach, New York, 1990, Vol. 3.

192. D. Ruelle, *Statistical Mechanics*, Benjamin, London, 1969.

193. P. Nutting, *London, Edinburgh Dublin Philos. Mag. Ser. 6*, **26**, 423 (1913).

194. G. Porod, *Koll. Z.*, **124**, 83 (1951).

195. G. Porod, *Koll. Z.*, **125**, 51 (1952).

196. B. Widom and J. Rowlinson, *J. Chem. Phys.*, **52**, 1670 (1970).

197. J. Ziman, *Models of Disorder*, Cambridge University Press, Cambridge, 1982.

198. J. Finney, *Proc. R. Soc.*, **319A**, 479 (1970).

199. L. Schwartz, J. Banavar, and B. Halperin, *Phys. Rev. B*, **40**, 9155 (1989).

200. S. Bryant, D. Mellor, and C. Cade, *AIChE J.*, **39**, 387 (1993).

201. L. Schwartz and J. Banavar, *Phys. Rev. B*, **39**, 11965 (1989).

202. R. Jullien, A. Pavlovich, and P. Meakin, *J. Phys. A*, **25**, 4103 (1992).

203. N. Morrow, I. Chatzis, and J. Taber, *SPE Proceedings*, Vol. 60th SPE Conference, Las Vegas, 1985.

204. S. Bryant and M. Blunt, *Phys. Rev. A*, **46**, 2004 (1992).

205. P. Sen, *Geophysics*, **49**, 586 (1984).

206. N. Martys, S. Torquato, and D. Bentz, *Phys. Rev. E*, **50**, 403 (1994).

207. I. C. Kim and S. Turquato, *J. Appl. Phys.*, **74**, 1844 (1993).

208. B. U. Felderhof, *Physica A*, **207**, 13 (1994).

209. B. U. Felderhof and P. Iske, *Phys. Rev. A*, **45**, 611 (1992).

210. R. Coleman, technical report, Department of Theoretical Statistics, University of Aarhus, Denmark, 1979.

211. S. Wicksell, *Biometrika*, **17**, 84 (1925).

212. S. Wicksell, *Biometrika*, **18**, 152 (1926).

213. D. Stauffer and A. Aharony, Taylor and Francis, London, 1992.

214. P. Sen, J. Roberts, and B. Halperin, *Phys. Rev. B*, **32**, 3306 (1985).

215. S. Feng, B. Halperin, and P. Sen, *Phys. Rev. B*, **35**, 197 (1987).

216. E. Cinlar and S. Torquato, *J. Stat. Phys.*, **78**, 827 (1995).

217. S. Miyazima, K. Maruyama, and K. Okumura, *J. Phys. Soc. Jpn.*, **9**, 2805 (1991).

218. J. A. Given, I. C. Kim, and S. Torquto, *J. Chem. Phys.*, **93**, 5128 (1990).

219. W. Elam, A. Kerstein, and J. Rehr, *Phys. Rev. Lett.*, **52**, 1516 (1984).

220. I. Fatt, *AIME Pet. Trans.*, **207**, 144 (1956).

221. I. Fatt, *AIME Pet. Trans.*, **207**, 160 (1956).

222. I. Fatt, *AIME Pet. Trans.*, **207**, 164 (1956).

223. R. Ehrlich and F. Crane, *Trans. AIME*, **246**, 221 (1969).

224. I. Chatzis and F. Dullien, *J. Can. Pet. Technol.*, 97 (January–March 1977).

225. M. Dias and A. Payatakes, *J. Fluid Mech.*, **164**, 305 (1986).

226. C. Diaz, I. Chatzis, and F. Dullien, *Transport Porous Media*, **2**, 215 (1987).

227. J. Koplik and T. Lasseter, *Chem. Eng. Commun.*, **26**, 285 (1984).

228. K. McCall, D. Johnson, and R. Guyer, *Phys. Rev. B*, **44**, 7344 (1991).

229. M. Blunt and P. King, *Phys. Rev. A*, **42**, 4780 (1990).

230. J. Bear, C. Braester, and P. Menier, *Transport Porous Media*, **2**, 301 (1987).

231. C. O'Carroll and K. Sorbie, *Phys. Rev. E*, **47**, 3467 (1993).

232. P. Øren, J. Billiotte, and W. Pinczewski, *SPE Formation Evaluation*, March, 1992, p. 70.

233. G. Matthews and M. Spearing, *Marine Petrol. Geol.*, **9**, 146 (1992).

234. U. Oxaal, *Phys. Rev. A*, **44**, 5038 (1991).

235. U. Oxaal, F. Boger, J. Feder, T. Jøssang, P. Meakin, and A. Aharony, *Phys. Rev. A*, **44**, 6564 (1991).

236. R. Lenormand, *J. Phys. C*, **2**, 79 (1990).

237. R. Lenormand, E. Touboul, and C. Zarcone, *J. Fluid Mech.*, **189**, 165 (1988).

238. M. McKellar and N. Wardlaw, *J. Can. Pet. Technol.*, **21**, 39 (1982).

239. R. Larson, L. Scriven, and H. Davis, *Chem. Eng. Sci.*, **36**, 57 (1981).

240. N. Seaton, *Chem. Eng. Sci.*, **46**, 1895 (1991).

241. M. Yanuka, *J. Colloid Interface Sci.*, **127**, 35 (1989).

242. M. Yanuka, *J. Colloid Interface Sci.*, **127**, 48 (1989).

243. M. Yanuka, *Transport Porous Media*, **7**, 265 (1992).

244. R. Chandler, J. Koplik, K. Lerman, and J. Willemsen, *J. Fluid Mech.*, **119**, 249 (1982).

245. P. Meakin, J. Feder, V. Frette, and T. Jøssang, *Phys. Rev. A*, **46**, 3357 (1992).

246. A. H. Hirsch, Lee M. Thompson, *Phys. Rev. E*, **50**, 2069 (1994).

246a. M. Blunt, M. King, and H. Scher, *Phys. Rev. A*, **46**, 7680 (1992).

247. J. Essam, *Rep. Prog. Phys.*, **43**, 835 (1980).

248. A. Aharony, in *Directions in Condensed Matter Physics*, G. Grinstein and G. Mazenko, Eds., World Scientific, Singapore, 1986, p. 1.

249. J. Quiblier, *J. Colloid Interface Sci.*, **98**, 84 (1984).

250. P. Crossley, L. Schwartz, and J. Banavar, *Appl. Phys. Lett.*, **59**, 3553 (1991).

251. R. Blumenfeld and S. Torquato, *Phys. Rev. E*, **48**, 4492 (1993).

252. J. Salles, J. Thovert, and P. Adler, *J. Contaminant Hydrology*, **13**, 3 (1993).

253. E. Guyon, L. Oger, and T. Plona, *J. Phys. D*, **20**, 1637 (1987).

254. P. Wong, J. Koplik, and J. Tomanic, *Phys. Rev. B*, **30**, 6606 (1984).

255. J. Roberts and L. Schwartz, *Phys. Rev. B*, **31**, 5990 (1985).

256. L. Schwartz and S. Kimminau, *Geophysics*, **52**, 1402 (1987).

257. H. Scher and R. Zallen, *J. Chem. Phys.*, **53**, 3759 (1970).

258. V. Shante and S. Kirkpatrick, *Adv. Phys.*, **20**, 325 (1971).

259. I. Gel'fand and G. Shilov, *Generalized Functions*, Academic, New York, 1964, Vol. I.

260. F. Leij, T. Skaggs, and M. Genuchten, *Water Resources Res.*, **27**, 2719 (1991).

261. M. Quintard and S. Whitaker, *Adv. Water Res.*, **17**, 221 (1994).

262. P. Germann, *J. Contaminant Hydrol.*, **7**, 39 (1991).

263. S. Kirkpatrick, *Rev. Mod. Phys.*, **45**, 574 (1973).

264. I. Webman, J. Jortner, and M. Cohen, *Phys. Rev. B*, **15**, 5712 (1977).

265. J. Bernasconi, *Phys. Rev. B*, **18**, 2185 (1978).

266. J. Yeomans and R. Stinchcombe, *J. Phys. C*, **11**, 4095 (1978).

267. R. Hilfer, *Renormierungsansätze in der Theorie ungeordneter Systeme*, Verlag Harri Deutsch, Frankfurt, 1986.

268. E. Sanchez-Palencia and A. Zaoui, *Non-Homogenous Media and Vibration Theory*, Vol. 127 of *Lecture Notes in Physics*, Springer Verlag, Berlin, 1980.

269. R. Burridge and J. Keller, *J. Acoust. Soc. Am.*, **70**, 1140 (1981).

270. S. Whitaker, *Transport in Porous Media*, **1**, 3 (1986).

271. H. Ene, in *Dynamics of Fluids in Hierarchical Porous Media*, J. Cushman, Ed., Academic, London, 1990, p. 223.

272. J. Keller, *J. Math. Phys.*, **5**, 548 (1964).

273. A. Dykhne, *Sov. Phys. JETP*, **32**, 63 (1971).

274. G. Milton, *Phys. Rev. B*, **38**, 11296 (1988).

275. B. Abramovich and P. Indelman, *J. Phys. A: Math. Gen.*, **28**, 693 (1995).

276. G. Matheron, *Elements pour une Theorie des Milieux Poreux*, Masson, Paris, 1967.

277. P. King, *J. Phys. A: Math. Gen.*, **20**, 3935 (1987).

278. L. Landau and E. Lifhitz, *Electrodynamics of Continuous Media*, Pergamon, Oxford, 1960.
279. B. Noetinger, *Transport Porous Media*, **15**, 99 (1994).
280. T. Levy, in *Homogenization Techniques for Composite Media*, E. Sanchez-Palencia and A. Zaoui, Eds., Springer-Verlag, Berlin, 1985, p. 63.
281. J. Hearst and P. Nelson, *Well Logging for Physical Properties*, McGraw-Hill, New York, 1985.
282. J. H. Doveton, *Log Analysis of Subsurface Geology*, Wiley, New York, 1986.
283. W. Chew and P. Sen, *J. Chem. Phys.*, **77**, 4683 (1982).
284. G. Archie, *Trans. AIME*, **146**, 54 (1942).
285. P. Sen, C. Scala, and M. Cohen, *Geophysics*, **46**, 781 (1981).
286. J. Schopper, in *Landolt–Börnstein: Physikalische Eigenschaften der Gesteine*, K.-H. Hellwege, Ed., Springer, Berlin, 1982, p. 276, Vol. V/1b.
287. I. Holwech and B. Nøst, *Phys. Rev. B*, **38**, 12845 (1989).
288. R. Maute, W. Lyle, and E. Sprunt, *J. Pet. Technol.*, 103 (January 1992).
289. D. McLachlan, M. Button, S. Adams, V. Gorringe, J. Kneen, J. Muoe, and E. Wedepohl, *Geophysics*, **52**, 194 (1987).
290. G. Keller and P. Licastro, *U.S. Geol. Surv. Bull.*, **1052**-H, 257 (1959).
291. J. Poley, J. Noteboom, and P. de Waal, *The Log Analyst*, **19**, 8 (1978).
292. W. Kenyon, *J. Appl. Phys.*, **55**, 3153 (1984).
293. D. Stroud, G. Milton, and B. De, *Phys. Rev. B*, **34**, 5145 (1986).
294. R. Knight and A. Nur, *Geophysics*, **52**, 644 (1987).
295. M. Taherian, W. Kenyon, and K. Safinya, *Geophysics*, **55**, 1530 (1990).
296. D. Stroud and D. Bergman, *Phys. Rev. B*, **25**, 2061 (1982).
297. C. Yoon and S. Lee, *Phys. Rev. B*, **42**, 4594 (1990).
298. C. Böttcher and P. Bordewijk, *Theory of Electric Polarization*, Vol. II, Elsevier Scientific Publishing, Amsterdam, 1978.
299. R. Fuchs, *Phys. Lett.*, **48A**, 353 (1974).
300. R. Fuchs, *Phys. Rev. B*, **11**, 1732 (1975).
301. D. Bergman, *Physica A*, **207**, 1 (1994).
302. P. Lysne, *Geophysics*, **48**, 775 (1983).
303. J. Korringa, *Geophysics*, **49**, 1760 (1984).
304. B. U. Felderhof and R. Jones, *Z. Physik B*, **62**, 43 (1986).
305. K. Ghosh and R. Fuchs, *Phys. Rev. B*, **38**, 5222 (1988).
306. K. Ghosh and R. Fuchs, *Phys. Rev. B*, **44**, 7730 (1991).
307. D. Bergman, *Ann. Phys.*, **138**, 78 (1982).
308. R. Fuchs and K. Ghosh, *Physica A*, **207**, 185 (1994).
309. M. Thorpe, B. Djordjevic, and J. Hetherington, *Physica A*, **207**, 65 (1994).
310. C. Böttcher, *Theory of Electric Polarization*, Vol. I, Elsevier Scientific, Amsterdam, 1973.
311. G. Milton, *Commun. Math. Phys.*, **99**, 463 (1985).
312. L. Schwartz, *Physica A*, **207**, 131 (1994).

313. P. Sheng, *Geophysics*, **56**, 1236 (1991).

314. K. Mendelson and M. Cohen, *Geophysics*, **47**, 257 (1982).

315. P. Sen, *Geophysics*, **46**, 1714 (1981).

316. P. Sheng, *Phys. Rev. B*, **41**, 4507 (1990).

317. P. Kogut and J. Straley, *J. Phys. C*, **12**, 2151 (1979).

318. A. Katz and A. Thompson, *Phys. Rev. B*, **34**, 8179 (1986).

319. D. Johnson, J. Koplik, and L. Schwartz, *Phys. Rev. Lett.*, **57**, 2564 (1986).

320. J. Banavar and D. Johnson, *Phys. Rev. B*, **35**, 7283 (1987).

321. J. Auriault, *Transport Porous Media*, **2**, 45 (1987).

322. S. Whitaker, *Transport Porous Media*, **1**, 105 (1986).

323. J. Auriault and C. Boutin, *Transport Porous Media*, **7**, 63 (1992).

324. J. Auriault and C. Boutin, *Transport Porous Media*, **10**, 153 (1993).

325. J. Auriault, O. Lebaigue, and G. Bonnet, *Transport Porous Media*, **4**, 105 (1989).

326. A. Paterson, *A First Course in Fluid Dynamics*, Cambridge University Press, Cambridge, MA, 1983.

327. J. Walsh and W. Brace, *J. Geophys. Res.*, **89**, 9425 (1984).

328. L. Schwartz, P. Sen, and D. Johnson, *Phys. Rev. B*, **40**, 2450 (1989).

329. L. Rapaport, *Trans. AIME*, **204**, 143 (1955).

330. J. Geertsma, G. Croes, and N. Schwarz, *Trans. AIME*, **207**, 118 (1956).

331. F. Perkins and R. Collins, *Petroleum Trans. AIME*, **219**, 383 (1960).

332. R. Bentsen, *J. Can. Pet. Technol.*, **25** (October–December, 1978).

333. R. Larson, H. Davis, and L. Scriven, *Chem. Eng. Sci.*, **36**, 75 (1981).

334. M. Shook, D. Li, and L. Lake, *In Situ*, **16**, 311 (1992).

335. E. Peters, N. Afzal, and R. Gharbi, *J. Petrol. Sci. Eng.*, **9**, 183 (1993).

336. D. Zhou and E. Stenby, *Transport Porous Media*, **11**, 1 (1993).

337. M. Aleman, T. Ramamohan, and J. Slattery, *Transport Porous Media*, **4**, 449 (1989).

338. N. Wardlaw and M. McKellar, *Can. J. Chem. Eng.*, **63**, 525 (1985).

339. G. Willhite, *Waterflooding*, Vol. 3 of *SPE Textbook Series*. Society of Petroleum Engineers, 1986.

340. R. Lenormand and C. Zarcone, *SPE Proceedings*, Vol. SPE Conference, Tulsa, 1986, p. 23.

341. I. Chatzis, M. Kuntamukkula, and N. Morrow, *SPE Reservoir Engineering*, August 1988, p. 902.

342. M. Aleman and J. Slattery, *Transport Porous Media*, **3**, 455 (1988).

343. V. Frette, J. Feder, T. Jøssang, and P. Meakin, *Phys. Rev. Lett.*, **68**, 3164 (1992).

344. P. deGennes, *Rev. Mod. Phys.*, **57**, 827 (1985).

345. M. Zhou and P. Sheng, *Phys. Rev. Lett.*, **64**, 882 (1990).

346. P. Sheng and M. Zhou, *Phys. Rev. A*, **45**, 5694 (1992).

347. G. Jerauld and S. Salter, *Transport Porous Media*, **5**, 103 (1990).

348. H. Princen, *Colloids Surfaces*, **65**, 221 (1992).

249. C. Marle, *Multiphase Flow in Porous Media*, Editions Technip, Institut Francais du Petrole, Paris, 1981.

350. F. Kalaydjian, *Transport Porous Media*, **2**, 537 (1987).

351. J. Trangenstein, in *Multiphase Flow in Porous Media*, M. Allen, G. Behie, and J. Trangenstein, Eds., Springer-Verlag, Berlin, 1988, p. 87.

352. M. Allen, in *Muliphase Flow in Porous Media*, M. Allen, G. Behie, and J. Trangenstein, Eds., Springer-Verlag, Berlin, 1988, p. 1.

353. F. Kalaydijan, *Transport Porous Media*, **5**, 215 (1990).

354. B. Bourbiaux and F. Kalaydjian, *SPE Reservoir Engineering*, August 1990, p. 361.

355. T. Mannseth, *Transport Porous Media*, **6**, 469 (1991).

356. W. Rose, *Transport Porous Media*, **3**, 163 (1988).

357. R. Ehrlich, *Transport Porous Media*, **11**, 201 (1993).

358. N. Morrow, *J. Petrol. Technol.*, 1476 (December, 1990).

359. J. Heaviside, in *Interfacial Phenomena in Petroleum Recovery*, N. Morrow, Ed., Vol. 36 of *Surfactant Science Series*, Marcel-Dekker, New York, 1991, p. 377.

360. H. Langtangen, A. Tveito, and R. Winther, *Transport Porous Media*, **9**, 165 (1992).

361. N. Morrow, in *Interfacial Phenomena in Petroleum Recovery*, N. Morrow, Ed., Vol. 36 of *Surfactant Science Series*, Marcel-Dekker, New York, 1991, p. 1.

362. E. Haslund, *Geophysics*, in press, 1996.

SELF-REPRODUCTION OF MICELLES AND VESICLES: MODELS FOR THE MECHANISMS OF LIFE FROM THE PERSPECTIVE OF COMPARTMENTED CHEMISTRY

PIER LUIGI LUISI

Institut für Polymere, ETH-Zürich, Switzerland

CONTENTS

I. INTRODUCTION

Within the frame of molecular evolution, RNA is regarded as the most important molecular species for inducing transition to life [1–3]. The threshold of life, in the RNA world, would probably be recognized in a molecular species that is able to self-replicate and mutate (evolve) in the process—and although a few fundamental pieces of the RNA world scenario are still missing (most notably the polymerization process leading to RNA under prebiotic conditions; and the self-replication of this polymer without the help of enzymes), the scientific community seems to trust that a satisfactory general picture will be achieved shortly.

The RNA-world paradigm cannot, however, escape the question of the membrane: Life on earth is presently only cellular. It has been cellular for at least the last 3 billion years, therefore at some point in the process of

Advances in Chemical Physics, Volume XCII, Edited by I. Prigogine and Stuart A. Rice.
ISBN 0-471-14320-0 © 1996 John Wiley & Sons, Inc.

transition to life the self-replicating and evolving molecular species must become compartmented.

There are good reasons why a spherical, semipermeable boundary is important: it may maintain the necessary local concentration of reagents as well as the necessary pH and other physicochemical parameters, avoiding dilution and diffusion (this is particularly important if we are dealing with a network of spatially close-linked reactions); the semipermeable boundary may permit the traffic of selected nutrient molecules and block that of inhibitors; it may provide a microenvironment in which a reaction may occur that does not necessarily occur in bulk water; it provides a better transport of the entire internalized chemical set over relatively long distances; and so on.

Whereas everybody seems to agree on the importance of all of this, we are still uncertain about the stage at which the boundary becomes important: that is, whether RNA and the molecular evolution of this polymer unto a self-replicating molecular species came first, or whether the membrane came first.

At first sight, this sounds like the old question of the chicken and the egg, therefore is not very proficuous. Actually, it is not the temporal sequence per se that is important but a basic chemical question of whether it is the membrane-like boundary that has induced or facilitated the polymerization processes leading to RNA and eventually to its self-replication, namely whether the existence of an organized supramolecular complex was the necessary requirement for the onset of the chemistry of life.

There is no way to answer this question now, but we believe it important to investigate the chemistry of life starting from the edge of membrane chemistry. Over the last few years, our group has studied the self-reproduction of bounded structures, such as micelles and vesicles, as part of the more general investigation of the relationship between the chemistry of self-organized supramolecular structures and key molecular processes of life, such as self-reproduction. Two chemically important notions have been recognized from these studies: (1) That micelles and vesicles, due to their particular kind of compartmentation, permit reactions that would not be possible in bulk solution, reactions that in turn may lead to self-reproductive autopoietic [4] processes; and (2) that both the monolayer of micelles and the bilayer of vesicles are endowed with catalytic activity—at least towards the hydrolysis of ester and anhydride groups.

This chapter reviews this work and presents only one example for each of the three classes of supramolecular surfactant aggregates we have investigated: reverse micelles, aqueous micelles, and vesicles. All experiments have been carried out with the same class of surfactant molecules, that is simple linear carboxylates, such as caprylate or oleate. We have chosen such simple molecules because work that is germane to the origin of life must be carried out with the simplest possible structures and tools.

II. THE EXAMPLE OF REVERSE MICELLES

Of the various self-reproduction reactions carried out with reverse micelles [5–8], let us consider one based on the permanganate oxidation [6]. In this case, we have reverse micelles formed by 50 mM octanoate in isooctane, with an overall water content of 1.5 M (w_0 = 30, the molar ratio water/surfactant; the permanganate ions were localized in the water pool at an overall concentration of 97.2 mM, at 25°C.

The system also contains excess octanol, which is localized in the bulk organic solvent, and is partly partitioned with the micelle monolayer. The hydrophylic head groups of the alcohol molecules bound to the micelles are partly in contact with the water pool, so that an oxidation of the alcohol to the acid can take place, see Fig 1 for the reaction scheme and the time course of the reaction.

It is a batch reaction, which proceeds till the limiting substrate (permanganate in this case) goes to zero. The above mentioned conditions brings about a production of surfactant that yields a tenfold increase of the number of micelles (as determined by time-resolved fluorescence quenching [6, 8]). In principle, one could start the reaction over again by adding new permanganate. Note also that in this process the newly built micelles are going to be smaller than the original ones, since the initial water content must be shared by a larger number of micelles [6]. This too can in principle be corrected by adding water externally, or by coupling the oxidation with a reaction that produces water by the same rate; however, all this has not been implemented as yet. Note also that during reaction MnO_2 precipitates out of the micelles, which does not interfere negatively with the reaction. Actually, it helps in shifting the equilibrium towards the product.

The increase of the number of micelles (>10-fold) obtained in this case is remarkable. Note again that this is due to the peculiarity of the micellar system: Two per se immiscible reagents (n-octantol and per-manganate) are able to react with each other due to the interfacial properties of the reverse micelle.

Little is known about the mechanism of micellar self-reproduction. Two features of reverse micelles are certainly important. One is the propensity to be monodisperse and the other is their high dynamical character, with continuous fusion and exchange processes during their

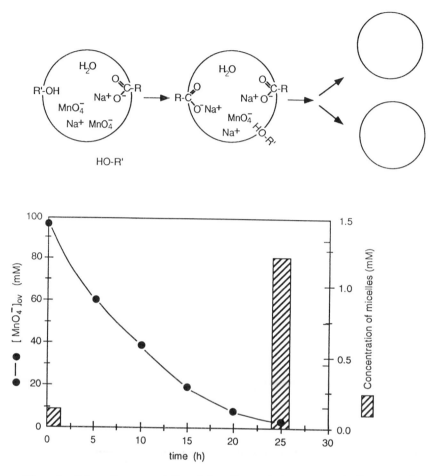

Figure 1. Schematic representation of the self-reproducing reverse micellar system that is based on the oxidation of *n*-octanol by permanganate, together with the actual time course of the decrease of permanganate concentration and increase of the number of reverse micelles. For more details see [6].

frequent encounters. Thus, presumably, the mechanism of multiplication of surfactant aggregates does not go via a larger stable micelle (or vesicle) that eventually divides up in two, but rather via intermicellar collisions and exchange of surfactant molecules. This is of course very different from what happens during a biological cell self-reproduction.

The same type of oxidation leading to micelle self-reproduction can take place with aqueous micelles of the same surfactant, with permanganate now in the bulk aqueous phase and octanol in the interior of the micelle [6]. This is again the case of segregation of one reagent that, however, due to the peculiar properties of the supramolecular aggregate, does not prevent accessibility of a reaction partner.

This type of compartmentation and corresponding reactivity is illustrated in Fig. 2, which is valid for reverse and aqueous micelles as well, and also for the work with vesicles (in this case a double layer of S must be envisaged instead of the monolayer of Fig. 2). Again, this cartoon should not literally be taken to signify that the intermediate micellar (or vesicular) species divide up in two—the process of micellar multiplication is most likely due to several intermolecular collisions involving several surfactant aggregates within a short-time interval.

III. THE EXAMPLE OF AQUEOUS MICELLES

It was already mentioned that micellar self-reproduction has been realized in water based on the same permanganate reaction illustrated above for the case of reverse micelles. Now let us illustrate in more detail another experiment in water solution, an experiment. We particularly like this experiment because it is a "water and soap" experiment (i.e., one that utilizes the simplest possible reagents) and it best emphasizes the power and beauty of self-assembly in supramolecular chemistry.

The experiment again involves caprylate (octanoic acid) as surfactant, but in this case there is a significant difference with respect to the previous example: One does not start with already preformed micelles. The question is actually: Can one start from a simple reaction that produces the microenvironment that creates micelles that then reproduce?

Now we have a biphasic system, with a water insoluble ethyl caprylate overlayed to an alkaline water solution [7]. That's all. As illustrated in Fig. 3, nothing much happens for the first several hours, just the very slow hydrolysis of the ester at the macroscopic interphase. The hydrolysis reaction produced, however, the octanoate surfactant, and there is an explosion of reactivity as soon as the critical micelle concentration is reached. The micelles formed in water are like oil droplets, which rapidly and efficiently solubilize the water insoluble ester. (Recall at this point the extremely large interfacial area that attends formation of micelles: for example, if 1 mL of ethylcaprylate added to 1 L of water would be converted into micelles, the total micellar interfacial area would be around 9900 m^2!).

Thus, the water-insoluble ester is rapidly uptaken and solubilized in the water phase with the help of the newly formed micelles, and its hydrolysis is accelerated by a very large factor. Consequently, new

THE ACTUAL WORKING PRINCIPLE

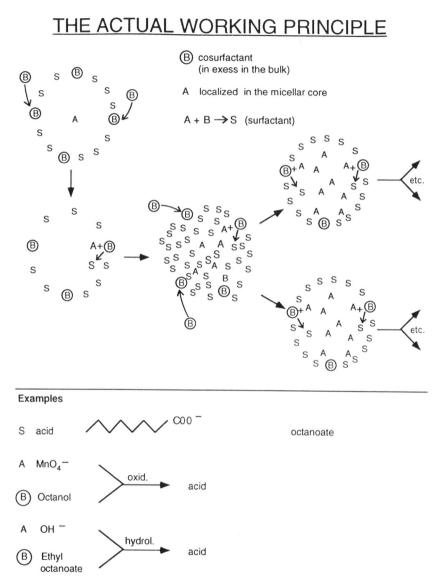

Figure 2. The actual working principle for micellar self-reproduction. One of the two reagents (B) is bound to the membrane of the surfactant aggregate (it acts as a cosurfactant), whereas the other reagent (A) is localized in the core. Reaction leads to the surfactant S, and the reagent B (in large excess in the bulk medium) is replaced on the membrane over and over again. As discussed in the test, micellar multiplication is not due to a division of a single larger micelle, but to collision processes and exchange among many micelles.

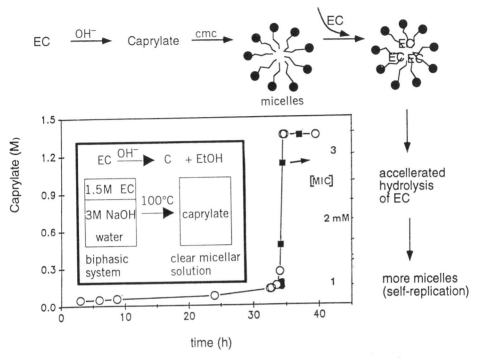

Figure 3. The biphasic system with the autocatalytic self-reproduction of aqueous caprylate micelles [7]. EC stands for ethylcaprylate.

surfactant is created in a burstlike reaction, and this immediately builds more micelles; and the more micelles that are formed, the more surfactant is produced—a truly autocatalytic process.

Whether or not, in addition to the enormous increase of interfacial area, the hydrolysis is helped also by micellar catalysis—the hydroxyl ion being a more effective nucleophile in the micelle than in bulk water— remains to be seen. In principle, one could even expect a rate inhibition (the bilayer is negatively charged and might repel OH$^-$), but apparently this effect does not play an important role.

The main point of this chapter is to pinpoint the peculiarities of the compartmented supramolecular chemistry for reactions that are relevant to the chemistry of life. In this last example these peculiarities are indeed very clear: Starting from water and an ester, a self-assembling macro-molecular system is spontaneously formed, that is able to self-reproduce. In the original paper [7], we make the point that, in view of the simplicity of this reaction and of the reagents, it is possible that aqueous micelles have been the first prebiotic spherical structures that were able to self-reproduce.

IV. AN EXAMPLE WITH VESICLES

As shown first by Deamer and coworker [9], caprylic and oleic acids as surfactants not only form micelles, but also vesicles. It is mostly the pH that determines which structure is more stable: vesicle form at a pH that equals the pK of the carboxylate, since under those conditions V-shaped dimers are formed which are suitable for vesicle formation. Figure 4 shows and electromicrograph of these vesicles.

One can now repeat the experiment as in Fig. 3, but at a pH that favors the formation of vesicles. Results are shown in Fig. 5. The initial very slow phase is now several days long [10], and again when the critical aggregate concentration (cac) of the vesicles is reached, there is an autocatalytic burst of carboxylate release attended by the formation of more and more vesicles.

The lag phase can be remarkably shortened by increasing temperature [10]—a phenomenon that is not yet understood (this is not simply due to a change of the cac with temperature). The slow initial phase can be shortened also in another way: by starting with a small amount of already performed vesicles in the water solution.

In Fig. 5 you also see the size distribution of the vesicles, as determined by painstaking analysis of many electromicrographs, at the initial point and after reaching the plateau. There is a large increase in the number of vesicles of 100-nm diameter. In this type of experiment one also has to check whether the lipidic mass balance is in order.

Another important control is to determine whether we are really dealing with a catalytic process. This control is rather simple [10]. The rate of oleic anhydride hydrolysis is compared in two parallel experiments: one is carried out in the presence of vesicles, the other without, under otherwise the same conditions. No significant reaction takes place without vesicles.

Thus, going back to our question of the importance of having a compartment: In this case, not only does the bilayer vesicle permits the reaction between a hydrophylic agent (OH^-) and a lipophylic one (the anhydride), but it also displays catalysis.

There is an interesting message to be learned from this: A lipidic membrane, in addition to the well-known general features already mentioned in the Introduction, may display catalysis. Perhaps the hydrolysis of an ester or an anhydride is not all that lipidic bilayers can do.

bar = 400 nm
⊢————⊣

Octanoic acid/octanoate vesicles
250 mM, pH 7

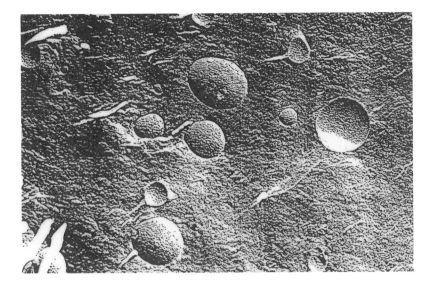

Figure 4.   Electron micrograph of caprylate vesicles.

V.   THE APPROACH TO A MINIMAL CELL

Seen from the optics of the behavior of a biological cell, the examples of self-reproduction seen thus far are simple cases for shell reproduction. If one thinks of making models of biological cell reproduction this is not enough, as in the cell both core (including the genetic material content) and shell reproduce. In fact, one can increase the complexity of our

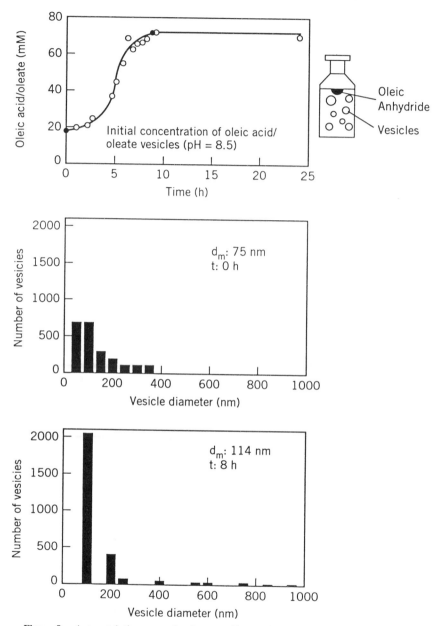

Figure 5. Autocatalytic process leading to self-reproduction of oleate vesicles, with the size distribution before and after reaction. Conditions are detailed in [10].

self-reproducing micelles or vesicles by having a "core-and-shell self-reproduction" system: A system in which simultaneously with the growth and self-reproduction of the shell, the nucleic acid interior undergoes a process of growth and multiplication.

This is part of what we call approach to a minimal cell. The relevance of this approach can be appreciated from a different perspective: Starting from the recognition that the simplest modern cell contains a minimum of a few hundred different proteins and a few hundred different nucleic acid components (think of the simple *Escherechia coli* cell!), one can ask whether it would be possible to have a cell (synthetic or not) having only an essential number of proteins and nucleic acids. In other words, how much can one reduce the complexity of a cell?

The first attempt to carry out self-replication of nucleotides inside the compartment of surfactant aggregates was carried out in our group by Böhler [11], who is now in Orgel's group at the Salk Institute. He was able to incorporate inside reverse micelles a nucleotide system similar to that originally used by von Kiedrowski [12]: In this case, the coupling of two trinucleotides is facilitated by the binding to an hexanucleotide template, as shown in the Fig. 6.

It was not easy to make the chemistry of nucleotide coupling compatible with the chemistry of micelles, and it proved impossible to have a simultaneous reproduction of micelles (the main difficulty being that the available self-reproducing micellar systems are based on reactive carboxylic groups).

This, however, opened the way to the work with vesicles, work that was carried out by Thomas Oberholzer in Zürich in collaboration with Christof Biebricher at the Max Plank Institute at Göttingen, Walde, and Monnard in our group. This work was done with oleate vesicles, the system we have seen before, and by utilizing an entrapped enzyme that produces RNA.

In one case, the self-reproduction of oleic acid vesicles was carried out simultaneously with the polymerization of ADP into poly(A) catalyzed by polynucleotide phosphorylase [13]. Poly(A) is a kind of RNA, so that production of RNA would proceed simultaneously with the multiplication of the number of oleate vesicles.

In another example, the enzyme Qβ-replicase and an RNA template was inserted in oleate vesicles together with all the necessary triphosphate substrates. In this case the enzyme operated a true RNA replication during the self-reproduction of the vesicles [14]. A successful, although weak, polymerase chain reaction (PCR) was also observed in liposomes (Oberholzer and Luisi, PNAS, in press).

In these examples, the two processes of RNA synthesis or replication

436 PIER LUIGI LUISI

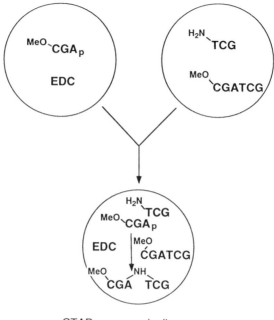

CTAB reverse micelles
in 9:1 isooctane / pentanol

Figure 6. The self-replication of nucleotides in reverse micelles. For details see [11].

and growth of the vesicles number proceed simultaneously, however, they are not chemically linked to each other: RNA does not do anything to help the process of shell reproduction, and vice versa. The next step would be to link the two processes with one another, so that, for example, the shell replication depends and is determined by the growth of RNA. Attempts in this direction are in progress. Even more challenging would be the case in which the existence of a shell would make the synthesis of RNA possible.

VI. CONCLUDING REMARKS

The above illustrated examples bring evidence of the fact that compartmented chemistry with micelles and vesicles permit reactions that would not be possible in bulk homogeneous milieu. A pictorial overview of the possible advantages of the bilayer compartmentation is given in Fig. 7. These reactions include now self-reproduction reactions and bilayer

THE IMPORTANCE OF HAVING A BOUNDARY...

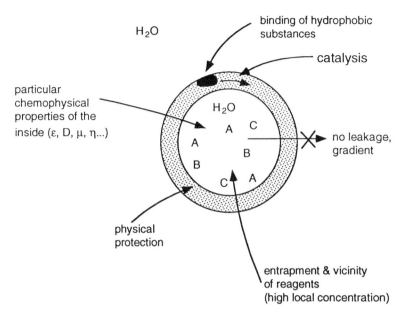

Figure 7. The importance of having a boundary.

catalysis. It has not been established whether all this is relevant to the emergence of life and whether this implies that membrane came first. More work is needed to further shed light—or at least elements of discussion—on to this point. In particular, one should establish whether micelles or vesicles may be necessary or at least helpful in inducing the synthesis of RNA and polypeptides, which is still the missing link in the narrative that goes from the occurrence of biomolecules on earth till the emergence of life. In fact, we plan to carry out experiments aimed at illustrating the possible relevance of compartmented chemistry for the polymerization of amino acids and nucleotides.

What is interesting in the experiments reviewed here with micelles and vesicles, is the extreme simplicity of the reaction conditions—water, sodium hydroxide and an ester (or anhydride) are, for example, enough to build up a self-assembling and self-reproducing supramolecular system provided with a double layer.

We believe that the approach to the origin of self-reproducing macromolecular systems should be carried out along this line, that is choosing simple "water-and-soap" conditions. The use of sophisticated chemically activated molecules defeats the very premise of the search for prebiotic conditions.

In fact, our group has two research directions that we keep rather distinct from one another. One is the search for conditions for the transition to life (and for that we like to use the simplest possible molecular components, as those used in the experiments of Figs. 3 and 5. The other is the construction of a minimal biological cell: In this case, the use of a sophisticated enzyme is allowed, as the question is not the origin of life, but how simple can the simplest possible living cell be. This is a question that maintains its validity also in "postprebiotic" times, when enzymes and RNA were already around.

REFERENCES

1. L. Orgel, *Sci. Am.*, **271**, 53 (1994).

2. T. R. Cech, *Science*, **236**, 1532 (1987).

3. G. F. Joyce, *Curr. Opin. Struct. Biol.*, **4**, 331 (1994); K. B. Chapman and J. W. Szostack, *Curr. Op. Struct. Biol.*, **4**, 618 (1994).

4. F. J. Varela, H. R. Maturana, and R. Uribe, *Biosystems*, **5**, 187 (1974).

5. P. A. Bachmann, P. Walde, P. L. Luisi, and J. Lang, *J. Am. Chem. Soc.*, **112**, 8200 (1990).

6. P. A. Bachmann, P. Walde, P. L. Luisi, and J. Lang, *J. Am. Chem. Soc.*, **113**, 8204 (1991).

7. P. A. Bachmann, P. L. Luisi, and J. Lang, *Nature*, **357**, 57 (1992).

8. A. Maliaris, J. Lang, and R. Zana, *J. Chem. Soc. Faraday Trans.*, **82**, 109 (1986).

9. W. R. Hargreaves and D. W. Deamer, *Biochemistry*, **17**, 3759 (1978).

10. P. Walde, R. Wick, M. Fresta, A. Mangone, and P. L. Luisi, *J. Am. Chem. Soc.*, **116**, 11649 (1994).

11. C. Böhler, W. Bannwarth, and P. L. Luisi, *Helv. Chim. Acta*, **76**, 1014 (1993); **73**, 1341 (1993).

12. G. von Kiedrowski, *Angew. Chem.*, **98**, 932 (1986).

13. P. Walde, A. Goto, P. A. Monnard, M. Wessicken, and P. L. Luisi, *J. Am. Chem. Soc.*, **116**, 7541 (1994).

14. T. Oberholzer, R. Wick, P. L. Luisi, and C. K. Biebricher, *Bioch. Biophys. Res. Commun.*, **207**, 250 (1995).

AUTHOR INDEX

Numbers in parentheses are reference numbers and indicate that the author's work is referred to although his name is not mentioned in the text. Numbers in *italic* show the pages on which the complete references are listed.

SUBJECT INDEX